Designing Optics Using Zemax OpticStudio®

Donald C. O'Shea
Julie L. Bentley

SPIE PRESS
Bellingham, Washington USA

Library of Congress Control Number: 2023946503

Published by
SPIE
P.O. Box 10
Bellingham, Washington 98227-0010 USA
Phone: +1 360.676.3290
Fax: +1 360.647.1445
Email: books@spie.org
Web: http://spie.org

Copyright © 2024 Society of Photo-Optical Instrumentation Engineers (SPIE)

All rights reserved. No part of this publication may be reproduced or distributed in any form or by any means without written permission of the publisher.

The content of this book reflects the work and thought of the authors. Every effort has been made to publish reliable and accurate information herein, but the publisher is not responsible for the validity of the information or for any outcomes resulting from reliance thereon.

Printed in the United States of America.
First Printing 2024.
For updates to this book, visit http://spie.org and type "PM367" in the search field.

To all our optics students, especially those who asked questions.
Donald O'Shea and Julie Bentley

Contents

Preface   xiii
Introduction   xv

1   The Basics   **1**
    1.1   Ray Calculations   2
       1.1.1   Law of refraction (Snell's law)   4
       1.1.2   Law of reflection   6
       1.1.3   The transfer equation   7
    1.2   Lenses   9
    1.3   Imaging   13
    1.4   Types of Images   15
    Additional Exercises   17
    Exercise Answers   18

2   Rays and Ray Sketching   **19**
    2.1   Collimation   19
    2.2   Thin Lenses   21
    2.3   Ray Sketching   21
       2.3.1   Finite object distance   23
       2.3.2   Object at infinity   24
    2.4   Treating Virtual Images   25
    2.5   Mirrors   26
    2.6   Planar Optics   29
    2.7   Multiple Elements   31
    2.8   Beyond Two-Lens Systems   33
    Exercise Answers   34

First Hiatus: Ledgers to Laptops   **35**
    H1.1   Simulations   35
    H1.2   Tracing Rays   36

3   How to Put a Lens in a Computer   **39**
    3.1   System Data   40
    3.2   Prescription Data   42
    3.3   Entering a Single Lens   44
    3.4   Checking the Lens   46
       3.4.1   First-order properties   50

		3.4.2	Virtual images	52
	3.5	Angle Solves		53
	3.6	Entering Mirrors		56
	3.7	Design Forms		57
	Exercises			58
	Exercise Answers			61

4 To First Order… — **65**

	4.1	Principal Surfaces and Planes		65
	4.2	What Does This Get You?		69
	4.3	Cardinal Points of Lenses and Mirrors		71
	4.4	Immersed Systems		73
		4.4.1	Nodal points for immersed systems	75
		4.4.2	The human eye	77
	4.5	A Concluding Remark		81
	Exercise Answers			81

5 Stops and Pupils and Windows, Oh My! — **83**

	5.1	Fields		83
	5.2	Special Rays		87
		5.2.1	Meridional or tangential rays	87
		5.2.2	Sagittal rays	88
		5.2.3	Skew rays	89
		5.2.4	Axial rays	89
		5.2.5	Rays for objects at infinity	89
	5.3	The Aperture Stop and Marginal Rays		90
	5.4	Chief Rays and Pupils of a Lens		97
	5.5	The Field Stop and Its Windows		101
	5.6	Tracing General Rays		102
	5.7	Field and Pupil Specifications		103
		5.7.1	Field of view	103
		5.7.2	F-number and numerical aperture	105
	5.8	Vignetting		109
	5.9	A Final Comment		113
	Exercise Answers			113

Second Hiatus: Rays and Waves — **115**

	H2.1	Rayleigh Criterion	115
	H2.2	The Pinhole Camera	117

6 Spherical Aberration — **119**

	6.1	Propagating Real Rays	119
	6.2	Third-Order Aberrations	120
	6.3	On-Axis Ray Errors for a Singlet Lens	122
	6.4	Displaying Spherical Aberration	123

	6.5	Transverse Ray Plots	125
	6.6	Third-Order Aberration Coefficients	129
	6.7	Lens Bending	131
	6.8	Going Off-Axis	136
	Exercises		137
	Exercise Answers		138
7	**Coma and Astigmatism**		**141**
	7.1	Coma	141
		7.1.1 Aberration contributions	143
		7.1.2 Coma and lens bending	146
	7.2	Aplanatic Lenses	147
	7.3	Astigmatism	149
	Additional Exercises		152
	Exercise Answers		152
8	**Aberrations of the Image Surface**		**157**
	8.1	Field Curves	157
	8.2	Petzval Curvature	159
	8.3	Field Curvature and Third-Order Coefficients	161
	8.4	An Anastigmatic Lens	164
	8.5	Distortion	166
	Exercise Answers		172
9	**Chromatic Aberration**		**175**
	9.1	Refraction and Dispersion	175
	9.2	Longitudinal Chromatic Aberration	179
	9.3	Correcting Longitudinal Chromatic Aberration	187
	9.4	An Example	189
	9.5	Secondary Color and Spherochromatism	192
	9.6	Lateral Color	194
	Exercise Answers		200
10	**Reducing Aberrations**		**203**
	10.1	The Merit Function	203
	10.2	Defocus	206
	10.3	Reducing Spherical Aberration	208
	10.4	Reducing Coma	212
		10.4.1 Stop shifting	213
		10.4.2 Flipping the lens	215
	10.5	Reducing Distortion	217
	10.6	Reducing Field Curvature	224
		10.6.1 Correcting astigmatism	224
		10.6.2 Correcting Petzval curvature	227

	10.6.3 Finishing the Job	232
	10.6.4 Final Comment	234
Explorations		235
Exercise Answers		236

11 Analyzing the Performance of a Lens — 239

11.1 Sensors	239
11.2 Spot Diagrams	242
11.3 Point Spread Function	244
11.4 Measuring Resolution	248
11.5 Modulation Transfer Function	250
Explorations	254

12 Designing a Lens — 257

12.1 Defining the Problem	257
12.2 Specifying the System	258
12.3 Step 0: The Initial Assessment	259
12.4 Step 1: Bend the Lens	263
12.5 Step 2: Shift the Stop	267
12.6 Step 3: Turn a Singlet into a Doublet	270
12.7 Step 4: Add a Field Flattener	274
12.8 Step 5: Open Up the Lens	282
12.9 Step 6: Glass Substitution	285
12.10 Wrap-up	290
Explorations	290

Third Hiatus: Building a Lens — 291

H3.1 Fabricating a Lens Element	291
H3.2 Mounting the Lens	293
H3.3 Testing the Lens	294

13 Tolerancing — 295

13.1 Statistical Tolerancing		295
13.2 The Tolerance Data Editor		297
13.3 Element Fabrication Errors		299
	13.3.1 Radius of curvature: TRAD and TFRN	299
	13.3.2 Thickness: TTHI	300
	13.3.3 Material: TIND and TABB	301
	13.3.4 Wedge: TSTX, TSTY (or TIRY, TIRX)	302
	13.3.5 Irregularity: TIRR	303
13.4 Lens Assembly Errors		304
	13.4.1 Element decenter: TEDX, TEDY, and TEDR	305
	13.4.2 Element tilt: TETX, TETY	307
	13.4.3 Airspace: TTHI	307
13.5 The Tolerance Wizard		307

13.6	Sensitivity Analysis	312
	13.6.1 Tolerancing Dialog Box Settings	312
	13.6.2 Uncompensated Sensitivity Analysis	316
	13.6.3 Compensated Sensitivity Analysis	319
13.7	Monte Carlo Analysis	322
	13.7.1 Simulation #1 (10 trials): RMS spot radius	322
	13.7.2 Simulation #2 (500 trials): RMS spot radius	326
	13.7.3 Simulation #3 (500 trials): MTF	328
	13.7.4 Advanced Monte Carlo Settings	331
13.8	Design Example: The OSsecureCam6	332
13.9	Some Final Comments	335
	Appendix: The Lens Drawing	336

Appendix: Macros: FIRST, THIRD   **339**

Index   *345*

Preface

The purpose of this text is to show you how to design an optical system using the optical design program Zemax OpticStudio®. The complete design process (from lens definition to tolerancing) will be developed and illustrated using the program. The text is organized so that a reader will be able to (1) reproduce each step of the process, including the plots for evaluating lens performance, and to (2) understand their significance in producing a final design.

This text is not a user's manual for Zemax OpticStudio (there are on-line reference guides for that). Rather, the text starts with a single lens to demonstrate the laws of geometrical optics and illustrate basic optical errors (aberrations) using Zemax OpticStudio. Then, through a series of examples and exercises, you can follow each step in the design process using Zemax OpticStudio to analyze and optimize the system to meet the required performance specifications. Once the nominal design meets these specifications, you can determine a set of tolerances that permits a large fraction of them to be manufactured with an acceptable as-built performance.

Although it is assumed that readers will follow the examples in the text and reproduce the results, you are encouraged to use them as jumping off points for an exploration of the designs. In addition to exercises with answers, we have added toward the end of the text what we call "Explorations"—open-ended problems with several possible directions in which to explore the design space. But this exploration needn't be confined to the final chapters. If there is a design feature or strategy that piques your curiosity and you want to find out what happens when you make a change in the design, go ahead and explore the consequences. You can't break anything. However, remember to save your lens before you begin to tinker with things.

One problem that will occur while exploring these various designs is maintaining a record of your work. Too often, a designer, trying out a new design or modifying a current design, can lose track of the performance and results of earlier attempts. Because Zemax OpticStudio does not currently contain a lab book or journal feature, it is highly recommended that during your design session a journal application be used to record your comments on progress, any data, and plots. This provides you with a record of your progress during a session. A journal application should capture anything typed or pasted into it and provide automatic backup so that users do not have to worry about saving the records. One of us (O'Shea), who runs Zemax OpticStudio on a Windows emulator, uses MacJournal as his journal app.

This text is written for a reader to continually interact with Zemax OpticStudio. Although any commercial optical design software can provide the tools to enter and modify designs, each program has its own interface, and it is not possible to demonstrate important optical principles with every program in a single text. For those who do not have immediate access to Zemax OpticStudio, there are two possible ways to use this text. If you're a student connected with a college or university, there are student licenses available for Zemax OpticStudio. For those who have access to other design programs, the operations and data entry may differ, but most of them will contain the same plotting, evaluation, and optimization functions as Zemax OpticStudio. So, with some translation, it should be possible to demonstrate the same operations as those used in this text.

We hope that this text will engage your curiosity and provide directions that will encourage you to work through all of our examples and then continue exploring optical design on your own. Designing optics is much like a game, where the rules are laid down by the laws of physics, where the pieces are surfaces, airspaces, and glass, where aberrations are obstacles to be overcome, and where the goals are set by the practical requirements of a design. Have fun playing!

Acknowledgments

We would like to express our sincere thanks to both Hsiu-An Lin and Kevin Scales from Zemax/Ansys's customer support team and to University of Rochester students Ankur Desai and Owen Lynch for reviewing the entire text and doing all of the examples. Their suggestions and feedback were invaluable in helping us provide accurate descriptions of the program and its uses.

Donald C. O'Shea   **Julie L. Bentley**
Atlanta, Georgia   Rochester, New York

December 2023

Introduction

Our approach is different from most texts on optical design because the operations and analyses in an optical design software package (Zemax OpticStudio) are described in great detail so that you can use the software to duplicate our examples. Zemax OpticStudio uses a Graphical User Interface (GUI) to input data, execute various analyses, and plot results. It is best to perform the operations as they are described in the text. From these results, you will be able to follow our discussion, familiarize yourself with program operations, and understand the information created there. "Humming along" by just reading the text won't do that.

To simplify the number of figures needed to show a particular GUI operation, we have chosen to use a shorthand notation for the GUI navigation. For example, to start a new design, File > New is used. This path format tells you to click on the File tab and select New from the available menu icons. The light blue background behind the text and the "greater than" character tells you that this is a Zemax OpticStudio operation, and this is the path you use to complete it. Additionally, many operations involve opening a new window, pushing a button, or selecting a drop-down menu option. These actions will be given a gray background, such as **Select Preset**.

Zemax OpticStudio uses an optical shop specification sheet format for its plots depicting lens performance. Most of these plots were intended for letter-size ($8.5'' \times 11''$) paper in a landscape format and have a frame containing additional information such as the date that the plot was created, the name of the lens file used for the plot, and some significant analysis values for that plot. Simply reducing these plots to fit in a 5-inch-wide area for this text can't be done without a loss in detail and legibility. Because this text relies heavily on these plots to convey information, we have reformatted them. We eliminated the outside frame, extracted the figures in vector format, enlarged them, and added legible labels for placement within the figure. For those who have been using Zemax OpticStudio for some time, the switch between its native plots and our extracted versions may be disconcerting, but we believe our work ("It ain't easy!") adds to our descriptions and explanations on designing optics. Examples of modifications to the native OpticStudio graphic output are shown below for the lens data tables, lens cross-sections, transverse ray plots, and spot diagrams.

Modifications to the Lens Data Editor (LDE)

By default, the Lens Data Editor has a large number of columns, many of which only contain data for special types of lenses not used in this text. As shown below, we have hidden the lens parameters that are not needed for our discussion (e.g., Coating, Chip zone, Conic, and TCE), resulting in a much more compact display for the LDEs shown in this text. We've also added a title bar to the bottom of the LDE that lists the title of the lens and the settings for the aperture, field, and wavelength of the optical system that are entered in the System Explorer.

	Surface	Type	Comment	Radius	Thickness	Material	Clear Semi-Dia
0	OBJECT	Standard ▼		Infinity	Infinity		0.000
1	STOP	Standard ▼		120.000	5.000	N-BK7	10.000
2		Standard ▼		-50.000	0.000		9.898
3	IMAGE	Standard ▼		Infinity	-		9.748

Title (OSlens); EPD (20 mm); Field (0°); Wavelength (d-line)

Modifications to Lens Cross-sections (Layouts)

The default lens cross-section for a double-Gauss lens is shown below.

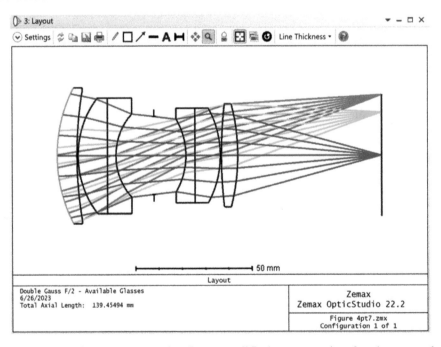

For comparison, an example of our modified cross-section for that same lens is shown on the next page. Some obvious differences can be seen when comparing the two figures. First, the outside frame has been suppressed to save space.

Second, to provide an immediately recognizable reference within the plot, the optical axis ray has been replaced by a black dashed line, designating the optical axis of the lens. Third, the default colors of the rays (blue, red, and green) for each field have been replaced by different shades of blue (the on-axis field is the darkest).

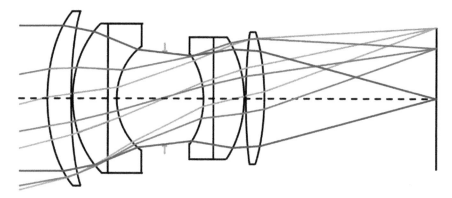

The other more subtle differences that you may notice when comparing the original figure to the modified figure is that in the default cross-section you can't see the rays coming into the lens (this occurs when the object is at infinity) and each lens is only drawn as big as the largest ray height at the lens. Many users find it useful to increase the viewed diameter of the lens elements and add an offset surface to observe the rays in object space when looking at these cross-section lens plots. Chapter 3 describes how to make these changes to your lens cross-sections. Note: The majority of the lens cross-sections in this text will be created with these two modifications in place even though we may not explicitly show the offset surface in the LDE and/or state that a specific aperture increment has been added to the lens diameters to create the figures.

Modifications to Other Graphic Output Plots

Other graphic output plots have been modified in a similar manner (primarily, to save space in the text). For example, the default plot of an array of ray fan diagrams for a lens with three field angles is shown on the next page. Note that there is a great deal of white space, and the labels for each field and the Maximum Scale are barely readable. The ray fans are shown in a landscape format with the on-axis field in the upper left corner and the full field at the bottom.

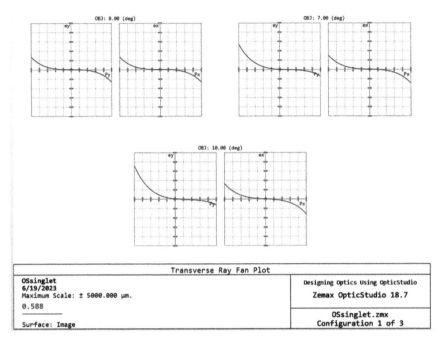

In this text, these plots are reformatted with the ray fan diagrams stacked vertically, with the on-axis field at the top. They are often placed next to the spot diagrams to save space, as shown below. This is Fig. 7.2 in the text.

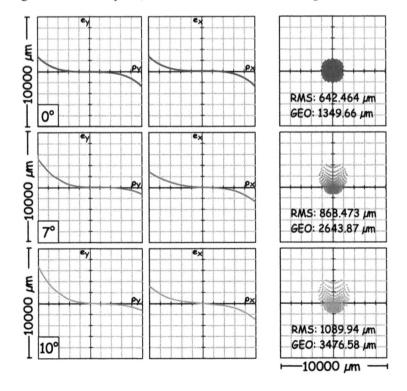

The ray fan plots have been squared up, and the e_x and p_x, e_y and p_y axis labels that Zemax OpticStudio uses are retained. The vertical flat-ended arrow displays the range of error (units are included). In a similar manner, plots for field curvature, distortion, color aberration, and other performance plots have been extracted from their Zemax OpticStudio versions and replotted to increase detail and readable values.

A Note on Ray Aiming

Zemax OpticStudio uses an algorithm, called ray aiming, which determines the rays at the object that correctly fill the stop surface for each field for a given stop size. Generally, ray aiming is only required when the image of the stop as seen from object space is considerably aberrated, shifted, or tilted. Ray aiming can be turned on in the Ray Aiming tab of the system explorer SysEx > Ray Aiming > Paraxial by selecting Paraxial in the drop-down menu. This is discussed in detail in Chapter 5. Paraxial ray aiming is used for all examples and demonstrations from Chapter 5 to the end of the text. Since the default setting for ray aiming is ray aiming "off" when you first start Zemax OpticStudio, if you notice small differences between your results and ours, check your ray aiming setting—if it is off, turn it on!

Comments by Nicholas Herringer (Ansys, formerly Zemax)

Ansys and our optics-focused teams are excited to support Don O'Shea and Julie Bentley in the publication of this text. Through this book, it is our great pleasure to contribute to the education of both the next generation of optical designers as well as seasoned veterans. Over Zemax's thirty-year history, the field of optics has become an exciting scientific domain, fueled by innovation and simulation. We look forward to continuing that journey with you in the decades to come.

Note that this text was created with Zemax OpticStudio versions 22.2 and 23.1. One author (Bentley) ran it on a PC; the other author (O'Shea) ran it on a MacBook Pro in emulation with Parallels 16.5 to run first on Windows 7 and later on Windows 10. In the spirit of continual improvement, Ansys is regularly enhancing and updating OpticStudio; over time, some features and functionality may change, including compatibility with certain hardware configurations. If you are using a more recent version of Zemax OpticStudio than was used in this text and notice a difference in output from the examples, please contact Zemax.Support@ansys.com. Our support team will be happy to discuss with you any questions or concerns you may have.

Chapter 1
The Basics

Until recently, there was only one application for ray tracing, the modeling and analysis of an optical system. However, today's students have a different application in mind. They think that they are going to be taught how to use powerful computers to generate realistic scenes like those that animation studios use to create movies (see Fig. 1.1).

Both applications, optical system modeling and realistic scene generation, simulate rays traveling through space to create images. Some of these images may be very simple (such as a point or a line), while those in computer-generated images (CGI) are extremely complex.

In the latter case, thousands of rays are traced to build the image of a scene. To reduce the time to compute the scene, only those rays that will reach the eye are traced. The easiest way to do this is to trace the rays in reverse. Starting at an eye or a camera sensor, a ray is traced from a point on the sensor, through the lens, and out to the scene, where its intersections with the surfaces defined by the computer model of the scene reflect, refract, and scatter the light in the scene.

Figure 1.1 "Bosque Morning," by John Fowler. Reprinted from Wikimedia under the Creative Commons Attribution 20. Generic license.

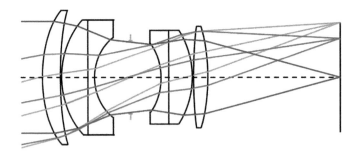

Figure 1.2 Ray trace of a double Gauss lens using an optical design program.

The CGI procedure is designed to use as few rays as possible so that the images can be rapidly generated. Similarly, system analysis ray tracing tries to use the fewest rays possible to determine how well an optical system, such as that shown in Fig. 1.2, would perform if it were built. In this case, the scene doesn't change. Instead, the same rays are traced through many different variations of an optical system whose dimensions and other variables are changed to find the best performance under specific conditions. This operation is called optimization, and a substantial part of this text is devoted to describing and demonstrating how optimization works.

Although computers provide us with enormous power to solve complex and sophisticated problems, they represent our biggest concern in teaching optical design. Once you sit down in front of a computer, the rest of the world can disappear, and time can pass quickly. Useful ideas and concepts fade when you are typing and looking at numbers that are supposed to tell you how well things are going. The physics can get lost in front of a keyboard; basic insights can disappear inside the workings of the design program.

To resist this tendency, we will introduce certain basic concepts to help you visualize what is going on in a design or that can provide insight and guidance toward improving its performance. Some of these concepts are simple, but they are just as useful as the ability to run a series of advanced analyses on a system.

1.1 Ray Calculations

We use computers that perform complex calculations and, for the most part, do not wonder what the calculations are or which ones are being used. Until a ridiculous result or a bug presents itself, we're fine with the way things are going.

In the case of ray tracing, the equations are so simple and their approximations are sometimes so useful that it would be a shame to ignore them. First, we need to set up a coordinate system that will be used from here on out— even when we resort to computing. The coordinates are established by defining the z axis as the axis of symmetry of a lens (Fig. 1.3) or another optical component, such as a spherical mirror. Light from an object is directed into the optical system in a positive direction.

The Basics 3

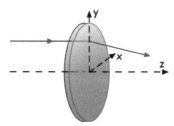

Figure 1.3 Right-handed coordinate system used in ray tracing.

Using the convention that light initially travels from left to right (see the text box, "The Ways of Rays"), the positive z axis points to the right. In the right-handed coordinate system used in optics, the x and y axes are perpendicular to the z axis. They are oriented with the positive y axis pointing up and the z axis to the right, and the x axis points away from the y-z plane. (If you curl the fingers of your right hand so that they point from the x axis to the y axis, your thumb will point along the z axis). Whew!

The index of refraction n is a characteristic of a transparent medium (gas, liquid, or solid). When light travels through a vacuum, it travels at a velocity of 3×10^8 meters per second (m/s). This velocity is denoted by a special symbol c. In any other medium, the interaction of the light with its atoms, molecules, or structures causes the light to slow down to a speed of v. To account for the velocity changes, each medium has a refractive index, which is the ratio of the speed of light divided by its speed in the medium:

$$n = c/v. \qquad (1.1)$$

Even air, the medium between lenses and mirrors in most optical systems, slows the progress of light (it has an index of 1.000273 at standard room temperature and pressure). These indices are known as *absolute* indices. Because air (with an index very close to 1.0) is a medium present in nearly every lens design, it is standard practice in optical design software to redefine the materials in terms of their *relative* index, where the speed of light used in Eq. (1.1) is the speed of light in air (not vacuum), making the refractive index of air exactly equal to one and simplifying the material entry of the optical system. All standard glass catalogs also have indices expressed relative to this unit index of air.

The Ways of Rays

There are many mathematical and graphic conventions in optics and optical engineering, but one that is unstated, yet almost universal, is that light rays in an optical design always travel from left to right, beginning at the object plane or source and entering the initial surface of the optical system. What happens after that is, of course, dependent on that particular system. Still, this convention means that you should expect to read an optical system as you would a sentence. Anyone who draws an optical system with the source on the right side of the design is, to our mind, guilty of a violation of optics grammar.

1.1.1 Law of refraction (Snell's law)

The laws of ray optics can be expressed in three equations. The first law, known as Snell's law or the law of refraction, expresses the amount by which a light ray is bent when it crosses the interface between media of different refractive indices (Fig. 1.4). In the case of a plane surface, there is no unique optical axis as there is in the case of a lens. Yet, to be able to specify the quantities that describe the rays and the optics, a local coordinate system needs to be established. In this case, we use an incident ray to locate the origin. Starting at the interface that separates two media whose refractive indices are n and n', the origin of the coordinate system is located at the point where the incident ray intersects the interface. The z axis is defined as the normal (perpendicular) to the interface at the origin, as shown in Fig. 1.4. See the text box, "Sign Conventions," for a summary of definitions for optical systems.

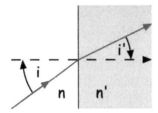

Figure 1.4 Refraction of light at an interface.

If the angle of the incident ray to the z axis is i, then the angle of refraction i' is given by the relation

$$n' \sin i' = n \sin i. \tag{1.2}$$

In Fig. 1.4, the signs of the angles i and i' are both positive. The sign convention is that the angle between a ray and a reference axis is positive if the rotation of the ray into the axis is clockwise. (For a horizontal reference axis, rays progressing upward are positive, and those going downward are negative.) Note that the equation is not solved for i' but expressed as sine–index products, which helps to emphasize the fact that this product is constant across index boundaries, and in the case of a parallel plate of material, it can simplify some of the calculations (Exercise 1.1).

Exercise 1.1 Parallel slab

A ray enters a 1-cm-thick pane of glass in air ($n = 1$) at an angle of 30° to the surface normal. Its surfaces are parallel to each other. If the refractive index of the pane is 1.666, what is the exiting angle? The answers are given at the end of the chapter. (Hint: Examine the sine–index product at each interface.)

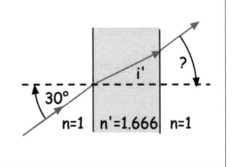

Sign Conventions

Optical axis
The optical axis, the z axis, is the axis of symmetry of the optical system. The positive direction points along the direction of the initial propagation of light (to the right).

Coordinate origin
The coordinate system for ray tracing is right-handed. Its origin is located at the intersection of a surface and the optical axis. As a ray is propagated to the next surface, a new origin is established there.

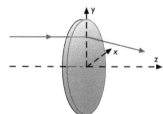

Distances
The distance to the next surface (thickness) is positive if it lies along the initial direction of light propagation (to the right) and negative if it is directed against the initial ray propagation (to the left). The radius of curvature of a surface is positive if the center of curvature is located to the right of the surface vertex and negative if it is located to the left of the surface vertex.

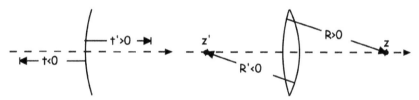

Rays
The height of a ray above the axis is positive; below the axis, it is negative. The angle of the ray with respect to a reference line is positive if the rotation of the ray into the reference is clockwise and negative if the rotation is counterclockwise. This is true for angles of incidence and refraction (i, i') measured relative to the surface normal and for ray angles (u, u') measured relative to the optical axis.

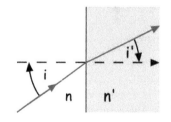

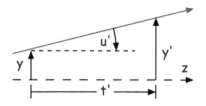
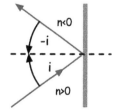

Reflections
All sign conventions remain the same except that the sign of the refractive index is changed after each reflection.

1.1.2 Law of reflection

The second law, the law of reflection, states that the angle of reflection equals the angle of incidence. But how is this expressed algebraically? We can treat it as a special case of Snell's law, provided we establish another convention. One of the problems with ray tracing is that when a ray is reflected, its general direction must be reconciled with the coordinate system that we initially established with the z axis pointing in the direction of the light propagation. And it also has to be consistent with our angle convention (positive, if the ray rotates clockwise into the axis).

In Fig. 1.5(a), the refractive index in the space in front of the mirror n' should be the same for the reflected ray as it is for the incident ray. But the angles i and i' should have opposite signs ($i' = -i$). If we insert these values into the equation for Snell's law (Eq. 1.2), we obtain

$$n' \sin(-i) = n \sin i,$$

or

$$-n' \sin(i) = n \sin i.$$

Therefore,

$$n' = -n.$$

The only time that this is true is when $i = 0$. It would appear that we couldn't use Snell's law universally and would have to set up a special case for reflections, which would make ray tracing more difficult because each time the ray encountered a reflecting surface, some additional branching would have to be added. One way that these computations can be completely consistent is if the refractive index were to change sign after each reflection. So, $n' = -n$. When this is inserted into the equation for Snell's law, the law is satisfied [Fig. 1.5(b)]. In nature, the refractive index of optical materials is always positive (exceptions only occur in very exotic situations), but in ray tracing, the signed refractive index is used as a geometrical bookkeeping trick that reflects (pun intended!) the physics of light.

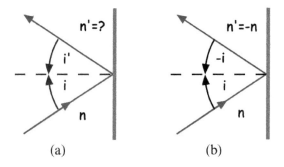

Figure 1.5 Reflection of light off a mirror.

The Basics

1.1.3 The transfer equation

The third and simplest of the laws is that in a medium of constant refractive index, light travels in a straight line. If a ray is a distance y above the optical axis and at an angle u' to the optical axis (Fig. 1.6), after traveling a distance t' it will be a distance y' above the axis, where

$$y' = y + t' \tan u'. \tag{1.3}$$

Equation (1.3) is called the transfer equation, and in conjunction with Snell's law, it is all we need to trace rays through a complex optical system.

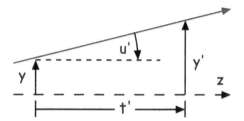

Figure 1.6 Transfer of a ray through a distance t'.

Exercise 1.2 Transfer equation

A ray is traveling at a height of 10 mm above the optical axis with a slope angle of +0.2 radians (rad). If it travels 50 mm farther along the optical axis, what will be its new height? When its height is 75 mm, how far will it have traveled from the point where it started at a height of 10 mm?

As an example of how is this done, let's look at a ray at an angle u to the optical axis that strikes a spherical surface with a radius of curvature R, as shown in Fig. 1.7. The ray is incident on the spherical surface at a height y and an angle of incidence i with respect to the surface normal, which is an extension of the radius of the spherical surface. The ray angle after refraction i' can be calculated using Snell's law.

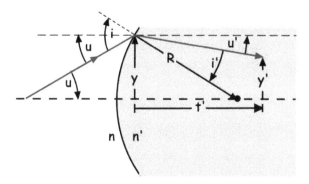

Figure 1.7 Ray trace at a spherical surface.

Although it is possible to compute the direction of the ray after refraction using trigonometry, the analysis becomes much simpler for small ray angles in the paraxial region (close to the axis) where the sine and tangent of an angle are approximately equal to the angle in radians. As you can see in Fig. 1.8, the three functions (plotted as a function of the angle u) are coincident for a small range of angles about the origin such that we can replace sin i with its angle i.

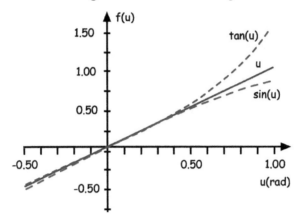

Figure 1.8 Demonstration of the small-angle approximation of trigonometric functions close to the origin.

Snell's law [Eq. (1.2)] is then reduced to

$$n'i' = ni, \qquad (1.4)$$

and the transfer equation [Eq. (1.3)] is expressed as the paraxial transfer equation:

$$y' = y + t'u'. \qquad (1.5)$$

To continue the ray trace, the angle that the ray makes with the optical axis after refraction u' must be determined. Because the refraction angles depend on where the ray hits the surface, Eq. (1.4) can be rewritten in terms of the ray angles with respect to the optical axis u and u' as

$$n'u' = nu - y\phi. \qquad (1.6)$$

This is called the paraxial refraction equation. The term ϕ is the optical power of the surface and is related to the difference in the refractive indices across the surface and its radius of curvature R or its curvature c (the reciprocal of R) as

$$\phi = \frac{n' - n}{R} = (n' - n)c. \qquad (1.7)$$

Note that for a flat surface ($R = \infty$) the optical power is zero and the paraxial refraction equation reduces to Snell's law because $u = i$ and $u' = i'$. Given the initial ray angle u, ray height y, and the power of the surface ϕ, one can calculate the ray angle on the other side of the surface u' after refraction at the surface.

This information is then fed into the paraxial transfer equation to determine the ray height at a surface a distance t' away.

If there were more than one surface in the lens system, given the ray height y' and ray angle at the second surface u', the paraxial refraction equation can be used again to calculate the ray angle after refraction at the second surface followed by a transfer to the next surface, and so on until the ray is traced through all of the surfaces in the optical system.

If that's all there is to ray tracing, what's the big deal? For one thing, not all ray angles are small, and not all surfaces are spherical. Therefore, the computation needed to *rapidly* trace a ray through multiple surfaces with a high degree of accuracy requires well-written software running on a high-performance computer. Furthermore, a ray tracing program allows us to improve a design by modifying its curvatures, thicknesses, and glass types.

> **Exercise 1.3 Small-angle approximation**
> At what angles, expressed in both radians and degrees, are the values of the $\sin(u)$ and $\tan(u)$ functions within 1% of their angles?

1.2 Lenses

Although optical systems may be composed of numerous lenses, mirrors, filters, and other components, a single lens can be used to define and illustrate many concepts that will be applied to more elaborate systems. We begin by using the terms defined in the text box, "Anatomy of a Simple Lens," to describe the passage of light rays though the lens. For example, the optical axis, the axis of symmetry through the center of a lens, provides a line of reference for our simple optical system, a single lens.

Light from a faraway source in front of the lens, made up of rays parallel to the optical axis, will be focused to a point behind the lens on the optical axis (the focal point F′) by a positive lens [Fig. 1.9(a)], or will appear to diverge from a point in front of the lens on the optical axis from a focal point F′ by a negative lens [Fig. 1.9(b)]. (For a discussion of what constitutes a faraway source and how it is treated in optical design, see the box, "Far Away," at the end of Section 2.1). This is the focal point of the lens. The distance between the lens and its focal point is the focal length f of the lens.

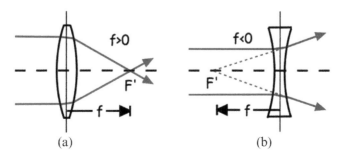

Figure 1.9 Parallel rays (a) focused by a positive lens and (b) diverged by a negative lens.

Anatomy of a Simple Lens

Most everyone knows what a lens is. Some of us know how it functions. But few people can describe one in any detail. To be able to understand how it works and to be able to modify its construction, a single lens should be described so that its performance and ailments (aberrations) can be diagnosed. That being the case, we provide, herewith, the anatomy of a single lens. It provides a glossary of terms that will be used in the rest of the text.

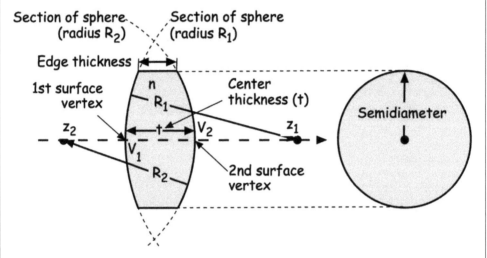

A simple lens consists of a piece of glass shaped by two opposing surfaces, each being a section of a sphere. The shape of the first surface is defined by its **radius of curvature** R_1, which is the radius of the sphere whose **center of curvature** is located at z_1. The second surface has a radius of curvature R_2 with a center of curvature at z_2.

A line through the two centers of curvature defines the **optical axis** of the lens, and the separation between the two surfaces along the optical axis is the **center thickness** t of the lens. The point where the optical axis intersects a surface is the **vertex** for that surface. The two vertices are labeled V_1 and V_2 in the above figure.

The size of the lens is specified by its **semidiameter**, shown on the right side of the figure. To avoid any confusion with the radius of curvature, we will use the term "semidiameter" for the lens rather than the radius. Once the semidiameter is given, the **edge thickness** of the lens is determined, as can be seen on the left side of the figure.

Not all lens surfaces are spherical. Some are parts of conic surfaces or more elaborate geometrical functions. However, the radius of curvature measured at the vertex of the surface provides a base radius of curvature.

The Basics

In practice, the focal length f of the lens is not as well defined as is shown in Fig. 1.9. Although the focal point would seem easy to locate (not so, as we'll see later), the plane that represents the lens cannot be defined without further analysis. But for an initial design or for a preliminary layout, it is easiest to use the **thin lens approximation**. In this case, the lens is treated as a thin optical component located on a plane in the middle of the actual lens. One definition of a thin lens is a lens whose thickness is smaller (say, one-tenth) than its focal length.

In Section 1.1, we discussed the optical power of a surface [Eq. (1.7)]. We expect each of the surfaces in a lens to contribute to the overall power of the lens. Because we are ignoring its thickness, the optical power of the thin lens ϕ is simply the sum of the surface powers [Eq. (1.7)], or

$$\phi = \phi_1 + \phi_2 = \frac{(n-1)}{R_1} + \frac{(1-n)}{R_2} = (n-1)\left(\frac{1}{R_1} - \frac{1}{R_2}\right). \quad (1.8)$$

The result is that the power of a thin lens is dependent only on the radii of curvature of its two surfaces (R_1 and R_2) and its refractive index n. These variables are shown in Fig. 1.10. The subscripts of the radii of curvature are assigned to the variables in order from left to right, the direction of the light entering the system or, in this case, the lens. Each of these is a directed distance and is depicted as an arrow in Fig. 1.10. The radius of curvature R_1 is a **positive** quantity because its center of curvature z_1 is to the **right** of the first lens surface. R_2 is **negative** because its center of curvature z_2 is to the **left** of the second surface.

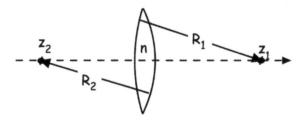

Figure 1.10 Variables that determine the focal length of a thin lens.

The optical power of a lens is the reciprocal of the lens focal length: $\phi = 1/f$. (This definition is valid unless the object or image space medium is not air. Special cases, such as a lens immersed in a liquid or a reflective optical element, will be treated later.) The unit of power is the diopter, which is the reciprocal of the focal length when it is expressed in meters. Therefore, a +100 mm (or 0.1 m) focal-length lens has a power of 10 diopters. The result is that the focal length of a thin lens can also be calculated from the three lens variables using the **lensmaker's formula**:

$$\frac{1}{f} = (n-1)\left(\frac{1}{R_1} - \frac{1}{R_2}\right). \quad (1.9)$$

By substituting curvatures c for the reciprocal of the radii of curvature, the formula can be written compactly as

$$\phi = (n-1)(c_1 - c_2) = (n-1)\beta, \qquad (1.10)$$

where β, the bending factor, is the difference of the curvatures $c_1 - c_2$. This form of the lensmaker's formula will be useful when we discuss the correction of color error (chromatic aberration) in a lens in Chapter 9.

For example, we can use the lensmaker's formula to calculate the focal length of the thin lens shown in Fig. 1.10, given that the refractive index of the lens is 1.5, the radius of curvature of the first surface is $+150$ mm, and the radius of curvature of the second surface is -100 mm. By inserting $n = 1.5$, $R_1 = 150$, and $R_2 = -100$ into the lensmaker's formula, we get

$$\frac{1}{f} = 0.5\left(\frac{1}{150} - \frac{1}{-100}\right) = \frac{1}{2}[0.0066 - (-0.01)] = \frac{0.0166}{2} = \frac{1}{120}.$$

According to the lensmaker's formula, the focal length of the lens is 120 mm.

In a second example of a plano-concave lens (Fig. 1.11), the first surface is flat. The radius of curvature of a flat surface is infinite, and the term $1/R_1$ in the lensmaker's formula equals zero. The center of curvature of the concave surface is to the right of the surface, so the radius of curvature is positive. If the refractive index of the lens is also 1.5, and the radius of curvature of the concave surface is $+100$ mm, then the focal length of this lens is

$$\frac{1}{f} = 0.5\left(0 - \frac{1}{+100}\right) = \frac{1}{2}(-0.01) = \frac{-0.01}{2} = \frac{1}{-200}.$$

This is a negative lens with a focal length of -200 mm. Note that the thickness of a negative lens is smaller at its center than at its edges compared to the positive lens in Fig. 1.10, where the opposite is true.

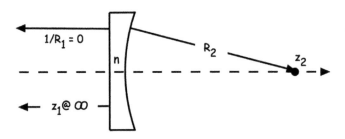

Figure 1.11 Variables for a plano-concave lens.

1.3 Imaging

In addition to focusing parallel light to a point, a lens can collect light from many points on an object and focus these points on a plane to create an image of that object (Fig. 1.12).

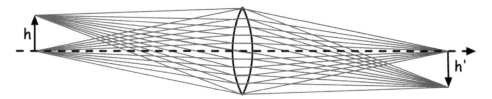

Figure 1.12 Imaging of an object with ray bundles.

In terms of imaging, the thin lens equation is the corresponding equation to the lensmaker's formula:

$$\frac{1}{t'} = \frac{1}{f} + \frac{1}{t}. \tag{1.11}$$

It can be used to find the location of the image a distance t' from the lens, for a lens of focal length f and an object located a distance t from the lens. For example, consider an object located 150 mm in front of a 100-mm-focal-length lens. Using Eq. (1.11), we insert the focal length and object distance:

$$\frac{1}{t'} = \frac{1}{100} + \frac{1}{(-150)} = \frac{1}{100} - \frac{1}{150} = \frac{3}{300} - \frac{2}{300} = \frac{1}{300},$$

and the result is that the object will be imaged at a distance 300 mm beyond the lens. Note that the object distance was entered as a negative quantity (−150) because of our sign convention (the object is to the left of the lens). This may be different from the relation that you learned in sophomore physics. There, the equation was given as

$$\frac{1}{o} + \frac{1}{i} = \frac{1}{f}, \tag{1.12}$$

where o was the object distance, and i was the image distance. The discrepancy between these two equations is that the object distance o in the second equation is considered a positive quantity, whereas the object distance t is a negative quantity because the origin of the coordinate system is located at the lens, as shown in Fig. 1.13.

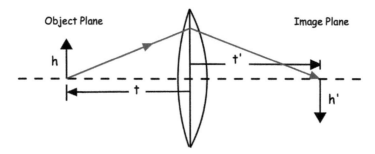

Figure 1.13 Imaging of an object by a lens of focal length f.

Now that we know how to find the location of the image, it would be nice to be able to determine its size. To start, we examine the point in the object plane that is at the top of the arrow and represents an object point at a distance h from the optical axis. The bundle of rays emitted from that point, traced in Fig. 1.12, is focused by the lens to a point on the image plane at a point h' from the optical axis. We can select a specific ray from this bundle, one through the center of the lens that is undeviated by the lens (called a **center ray**), as shown in Fig. 1.14, and trace it to the image plane.

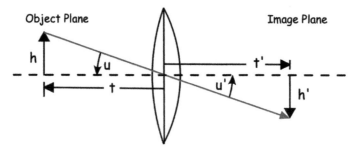

Figure 1.14 Magnification of a lens.

The two triangles in the figure have equal interior angles $u = u'$, so the tangents of these angles are equal:

$$\tan u = \frac{h}{t} = \tan u' = \frac{h'}{t'}.$$

The magnification m of the optical system is the ratio of the image height h' to the object height h. Solving for the ratio h'/h produces the law of magnification:

$$m = \frac{h'}{h} = \frac{t'}{t}. \qquad (1.13)$$

For our example, the magnification $m = 200/-150$, or $-4/3$X magnification. The negative sign indicates that the image is inverted in relation to the object. Although Eq. (1.13) is a simple equation, it establishes significant limits in the design of an optical system. The simplest application of the law shows that it is impossible to locate a magnified real image close to an imaging lens.

The definition of a center ray whose direction is undeviated by the lens is based on the assumption that the lens, however thick it may be drawn in a figure, is considered a thin lens whose front and back surfaces nearly touch. Because the points where the surfaces will nearly touch are on the optical axis, a ray aimed at that point will necessarily pass through the center of the lens. In that region, the tangents to the two surfaces are parallel to each other, so the center of the lens appears to be a parallel slab, and a ray entering the slab at some angle will emerge on the other side at the same angle, i.e., undeviated (see Exercise 1.1).

> **Exercise 1.4 Thin lens equation and magnification**
> In Section 1.3, the object was located 150 mm in front of the 100-mm-focal-length thin lens. Where will the image be located if the object is relocated to a point 200 mm in front of the lens? What will be the new magnification?

1.4 Types of Images

When an object is located outside the focal point of a positive lens, as illustrated in Fig. 1.15(a), the rays from each object point *converge* to its corresponding image point, resulting in a **real image** of the object. If a white surface is placed behind the lens in the image plane, a real image can be seen there, just as we see an image projected on a screen in a movie theater. There are other optical arrangements where the lens doesn't focus light rays from an object. Instead, the rays from an object point diverge as they emerge from the lens. For example, when an object is located inside the front focal point of a positive lens, the rays diverge, as shown in Fig. 1.15(b). The image plane is located by tracing the diverging rays backwards to where they meet at a point. In this case, a screen located behind the lens will not show any image. You need to use another optical system, such as the eye, to see the image. This type of image is called a **virtual image**. In Fig. 1.15(b), the lens acts as a magnifier, displaying a larger image h' than the original object h. In comparing the lens systems in Fig. 1.15, we see that for a positive lens the real image space is on the image side of the lens (positive t'), whereas the virtual image space is on the object side of the lens (negative t'), which means that the sign of the image distance indicates whether the image produced by the lens is real (positive t') or virtual (negative t').

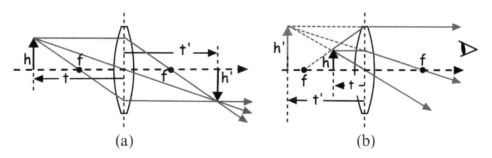

Figure 1.15 Types of images: (a) real and (b) virtual (t' is a negative quantity).

All of the quantities that have been defined and discussed are signed quantities. In the case shown in Fig. 1.13, the image height is inverted relative to the object, and the sign of the ratio of the heights (h'/h) is negative, as is m. The ratio of the distances gives the correct sign of the magnification because t' is positive and t is negative, as we noted earlier [Eq. (1.13)].

This introduction to ray tracing has established some of the basic concepts needed to understand and analyze optical systems. But before showing you how to enter a system into a ray tracing program, in the next chapter we describe a graphic technique we call ray sketching that can give you a feel for a system before any keys are pressed and routines are run.

Additional Exercises

Exercise 1.5 Equiconvex lens
$n = 1.5$; $t = -400$ units; $R_1 = +100$ units and $R_2 = -100$ units. Determine its focal length f, the location t' of the image, and the magnification m of the system, and indicate if the resulting image is R (real) or V (virtual). Sketch the lens shape.

Exercise 1.6 Biconvex lens
$n = 1.4$; $t = -200$ units; $R_1 = +200$ units and $R_2 = -100$ units. Determine its focal length f, the location t' of the image, and the magnification m of the system, and indicate if the resulting image is R (real) or V (virtual). Sketch the lens shape.

Exercise 1.7 Positive meniscus lens
$f = 266.66$ units; $t = -200$ units; $R_1 = +100$ units and $R_2 = +200$ units. Determine the refractive index of the lens n, the location t' of the image, and the magnification m of the system, and indicate if the resulting image is R (real) or V (virtual). Sketch the shape of the lens.

Exercise 1.8 Negative meniscus lens
$n = 1.5$; $t = -200$ mm; $R_1 = +200$ mm and $R_2 = +100$ mm. Determine its optical power ϕ in diopters, the location t' of the image, and the magnification m of the system, and indicate if the resulting image is R (real) or V (virtual). Sketch the shape of the lens.

Exercise 1.9 Biconcave lens
$n = 1.5$; $t = -400$ units; $R_1 = -75$ units and $R_2 = +300$ units. Determine its focal length f, the location t' of the image, and the magnification m of the system, and indicate if the resulting image is R (real) or V (virtual). Sketch the lens shape.

Exercise 1.10 Equiconcave lens
$\phi = -13.33$ diopters; $t = -400$ mm; $R_1 = -100$ mm and $R_2 = +100$ mm. Determine the refractive index of the lens n, the location t' of the image, and the magnification m of the system, and indicate if the resulting image is R (real) or V (virtual). Sketch the shape of the lens.

Exercise 1.11 A rule for lenses
Based on an observation of the **shapes** of lenses in the Exercises 1.5–1.10, how could you determine whether a lens on a laboratory table is positive or negative by just picking it up, even before you look through the lens?

Exercise 1.12 Transfer equation II
A ray is traveling along the optical axis at a height of 10 mm above the coordinate origin with a slope angle of $+0.2$ rad. Where does the ray cross the optical axis?

Exercise Answers

Ex. 1.1

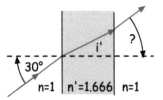

The index–sine product is the same for all three media, so the exit angle equals the entrance angle, 30°.

Ex. 1.2
$y' = y + t' \tan u' = 10 + 50 \tan(0.2)$
$\quad = 10 + 10.14 = 20.14;$
$y' = y + t' \tan u',$
$75 = 10 + t' \tan(0.2),$
$65 = t' \cdot 0.203,$
solving for $t' = 65/0.203 = 320.66$ mm.

Ex. 1.3
Depending on round-off, the value of the angle or the sine is 0.24 rad or 14°. For the tangent, $i = 0.17$ rad or 10°.

Ex. 1.4
The image is a real image located 200 mm behind the lens. Its magnification $m = t'/t = -200/200 = -1$. The image is inverted and the same size as the object.

Ex. 1.5

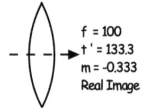

f = 100
t' = 133.3
m = -0.333
Real Image

Ex. 1.6

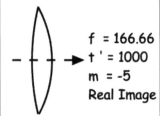

f = 166.66
t' = 1000
m = -5
Real Image

Ex. 1.7

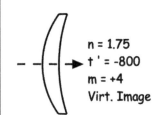

n = 1.75
t' = -800
m = +4
Virt. Image

Ex. 1.8

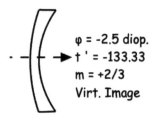

φ = -2.5 diop.
t' = -133.33
m = +2/3
Virt. Image

Ex. 1.9

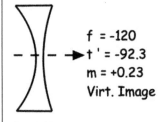

f = -120
t' = -92.3
m = +0.23
Virt. Image

Ex. 1.10

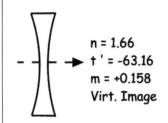

n = 1.66
t' = -63.16
m = +0.158
Virt. Image

Ex. 1.11
The center thickness of a positive lens is larger than its edge thickness, whereas for a negative lens the reverse is true.

Ex. 1.12
Solving $0 = 10 + t \tan(0.2)$, the ray is traveling upward and had already crossed the optical axis at −49.33 mm.

Chapter 2
Rays and Ray Sketching

The first thing you do when you start a new lens in an optical design program is to enter the parameters that describe its design. But how can you tell if you entered the design correctly? How do you determine if the results are correct? And how can you gain some sense of how well your lens performs? These questions express the daily concerns of all optical designers. Some of the answers come from an understanding of how a system *should* operate, some from experience, and some from common sense. This chapter is intended to give you some tools to help answer these questions. When it comes to experience, that is something you will have to gain on your own.

2.1 Collimation

A lens (composed of a single element or multiple elements) separates the design space into **object space** and **image space**. Object space is also referred to as the **front** of the lens and image space as the **back** of the lens (Fig. 2.1). One of the reasons for doing so is that a lens has two focal points, and it is useful to be able to distinguish between them as the front and back focal points.

Both of the figures in Fig. 2.1 illustrate **collimated light**, a bundle of light whose rays are parallel to each other. Collimated light is sometimes referred to as parallel light. This concept is useful because it provides a simple way to represent light from a distant source, which is usually described as a source at infinity. (A laser beam that can travel long distances without much broadening (divergence) may also be modeled as a collimated beam but is usually simulated using physical optics concepts rather than geometrical optics.) In reality, collimation is only an approximation, but it simplifies the problem, and we will take advantage of that.

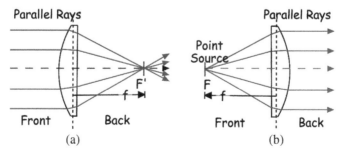

Figure 2.1 Collimation: (a) Collimated light focuses at the back focal point F′. (b) A point source at the front focal point F produces collimated light.

In Fig. 2.1(a), light from an on-axis source at infinity produces parallel rays, which are directed by the lens to a point on the axis. This point is called the **back focal point** F′ of the lens and is located a focal length f beyond the lens. If a **point source** is located on-axis a distance f in front of the lens at the **front focal point** F, then the lens will collimate the rays emitted from the point source, as shown in Fig. 2.1(b). A point source, by the way, is an infinitesimally small source of light that radiates in all directions. It is an unrealistic, but convenient, model for a small light source. As to what constitutes a distant source and how it is treated in optical design, see the text box, "Far Away."

Far Away

We have a problem with very large or very small numbers. This is true when it comes to infinity. For some, infinity is the number after three (1, 2, 3, infinity!). Others, mainly mathematicians, make up names for large numbers. One large number, for example, is the googol, which equals 10^{100} or 1 followed by 100 zeroes. (The name of the search engine, Google, is a pun on the term.)

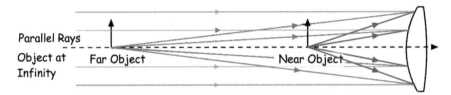

In the above figure, we show fans of axial rays from three objects. Two of the objects are shown in the figure; the third one, located at infinity, could not be shown, given the size of the page. Note that as the object is located farther from the lens, the angles that its axial rays make with respect to the optical axis get smaller. In the limit of an object at infinity, the rays are parallel to the axis.

For the optical designer, an object that is very far away is considered to be at infinity. Infinity, for the purposes of lens design, is determined by the object distance in relation to the focal length of the lens. We see that in the thin lens equation (Eq. 1.11), as the object distance t is increased, the $1/t$ term goes to zero, and the image distance t' equals the focal length f. If the object distance is much larger than the focal length of the lens (say, 100X), the image can be considered to be located in the back focal plane. In OpticStudio®, the software application we use to illustrate our examples, an "infinite" object distance is set to 10^{10} lens units. Although not quite infinity, this is still a very large distance—well over six million miles (if the lens units are millimeters). With a number like that, you are assured that the object is located far away and the image will be located in the focal plane.

This may seem like overkill, but there is a very practical aspect to this definition. Most objects that we image with an optical system will be located at a variety of large distances compared to the focal length f of the imaging system. In all cases, the image will be located in the focal plane. Instead of needing to specify the exact object distance to any lens, one very large number serves the same purpose for all distant objects.

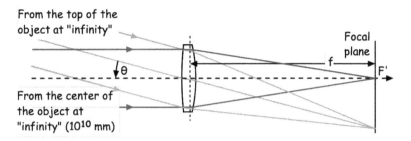

Also, the height of a faraway object is often difficult to measure or discover, but the angle that the object makes with the optical axis of the system can be determined or estimated. (For example, when you're looking toward the sky at the moon, it is hard to measure its exact size, but the angle the moon subtends at your eye can be easily estimated.) As a result, the location and size of faraway objects in optical designs are usually specified by the initial thickness of 10^{10} and the angle θ between the optical axis and a ray from the top of the object to the center of the lens.

2.2 Thin Lenses

Many lens systems contain multiple surfaces and materials, the details of which may impede simple calculations needed for a preliminary evaluation of a system. In Chapter 1, we defined a thin lens as a lens whose center thickness is much smaller (say, one tenth) than the focal length. From an operational standpoint, a thin lens is a lens whose center thickness is sufficiently small compared to its radii of curvature and to the object and image distances used in the thin lens formula [Eq. (1.11)].

A thin lens allows the simplification of the graphic depiction of a lens during the initial design and also when sketching rays to assess the performance of that design (ray sketching). This is done by drawing the lens as a single line with open arrows to indicate a positive or negative lens, as shown in Fig. 2.2. When a rigorous analysis is performed, this thin lens approximation is supplanted with the exact ray traces, and each lens component is drawn with the correct center thickness. Note in Fig. 2.2 that the primed focal points F′ determine where incident parallel light will be directed, and the unprimed focal points F determine where light should be directed to emerge as parallel light.

2.3 Ray Sketching

Ray sketching is a simple method used to describe the basic operation of an optical system. In contrast to ray tracing, which uses highly accurate computations to precisely model a system, ray sketching provides an approximate graphical model with only a few rays. The point of ray sketching is to provide a quick look at the components and ray locations before approaching the computer keyboard. A sheet of graph paper, a restaurant napkin, or the back of an envelope is the usual medium for ray sketching.

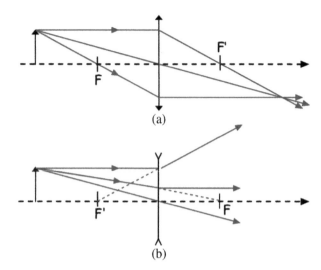

Figure 2.2 Graphic symbols for thin lenses: (a) positive and (b) negative.

In Fig. 2.2(a), an object, represented by the vertical arrow, is imaged by a positive lens. The image plane is located by tracing rays through the lens from a point on the object and looking for the intersection of the rays in image space. Although the lens focuses all of the rays from each point on the object to a corresponding point on the image, there are a few special rays that can be used to easily determine the location and size of the image given the lens focal length. After we define these rays, we will use them to analyze some *really* simple optical systems (a single lens or mirror).

The first of these special rays is the **center ray**, a ray from the top of the object that travels undeviated through the center of the lens. The center of the lens resembles a thin slab of glass, and so, as in Exercise 1.1 (Parallel slab), the ray exits the center of the lens at the same angle that it entered. The second ray is a **parallel ray** from the top of the object that travels parallel to the axis and, upon encountering a lens, crosses the axis at the back focal point F′. Another ray from the top of the object crosses the axis at the front focal point F and emerges from the lens parallel to the optical axis. Although it becomes a parallel ray after passing the lens, this third ray will be called a **focal ray** to distinguish it from the previous ray. This way, the rays are defined by their behavior and location before (or, in the case of the center ray, at) the lens. These special rays are illustrated individually in Fig. 2.3 for the positive lens.

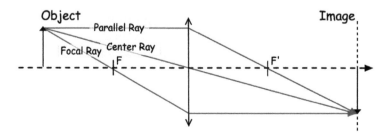

Figure 2.3 Special rays used to locate an image in ray sketching.

The parallel ray from the top of the object crosses the back focal point and continues into image space. The focal ray from the top of the object is directed to the front focal point and exits the lens parallel to the axis. Their intersection determines the top of the image (finding both the image location and size). As a check (and, in some cases, as a useful alternative) to either of the first two rays, the center ray also intersects the top of the image, as it should, because it was emitted from the same object point as the other two rays. These rays cross beyond the positive lens and converge to the real image, as we described in Section 1.4.

What about the on-axis image point? If we assume that all of the points in the image are located in a vertical plane (indicated by the dashed line in Fig. 2.3), then the on-axis image point is located where the image plane crosses the axis. The drawing of the rays to the on-axis image consists of nothing more than launching a ray at an angle to the axis and then, at the point where it hits the lens, drawing the ray to the on-axis image point. This leads to the definition of a more general ray used in graphical analysis of a system. An **axial ray** is any ray that begins at the point on the optical axis where the object is located and travels at an angle u to the optical axis through the system (Fig. 2.4). Upon emerging from the system, it will cross the axis at the image plane.

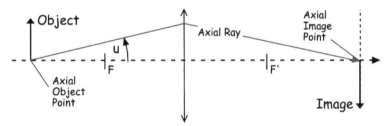

Figure 2.4 Axial rays cross the optical axis at the image plane.

In ray sketching, an axial ray cannot be used to determine the image location because there is no specific construction for it as there was for the special rays from an off-axis point. (You need to know how it bends at the lens, and for that you need to do a ray trace.) However, once an image plane is established with the other three rays, it is certain that an axial ray will cross the axis there, and including an axial ray helps to complete the ray sketch.

Considering how little work is needed, ray sketching gives you lots of information about a lens so long as you know or can approximate its focal length. Not only does it give you some idea of the image distance relative to the object distance, it also demonstrates the magnification or demagnification of the image. For example, in Fig. 2.3 the image is inverted and magnified relative to the object and is located a bit farther from the lens than the object.

2.3.1 Finite object distance

Next, we will look at some different arrangements and then let you test your understanding of ray sketching. Let's begin with a camera lens imaging a scene close to the lens. In this case, the object is only a few focal lengths from the lens (Fig. 2.5).

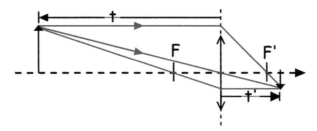

Figure 2.5 Ray sketch of a camera lens system.

Note that the center and focal rays intersect the lens we have drawn, but the parallel ray does not. For the purposes of ray sketching, this doesn't matter. We could have scaled down the object height or scaled up the lens, and all of the rays would make it through the lens, but this wouldn't change the results. However, drawing the lens at the appropriate size can help with visualizing clearances and relative component sizes. In Fig. 2.5, the object distance is three times the image distance, the image is real, and its magnification $m = -1/3$. Take the time and do the following exercise using graph paper before moving on.

Exercise 2.1 Projector
$f = +3$ units; $t = -4$ units. The diameter of lens is 6 units. Sketch the rays and determine the location of the image and magnification of the system. Is this a real or virtual image?

2.3.2 Object at infinity

If the object is far from the lens (at infinity), then the special rays need to be treated a little differently, as was noted in the "Far Away" text box earlier. Because the object is so far away, all of the rays emitted from a single point on the object that intersect the lens are parallel to one another. The construction for the on-axis object point is shown in Fig. 2.1(a): rays parallel to the axis of the lens focus at the back focal point.

A point at the top of an object at infinity is not described by an object height h. Rather, it is given as the angle θ that the object subtends at the lens. In Fig. 2.6, the ray sketch construction for the off-axis point uses the center ray as the "off-axis" axis plus the focal ray that is parallel to the center ray "axis" to locate the image plane. An extra ray parallel to the center ray is added to show the path of a bundle of rays through the lens. This extra off-axis ray will also cross the center ray at the focal plane. In fact, the parallel ray bundle for each point on an object at infinity will focus at a point in the **back focal plane** of the lens.

When the off-axis object point is combined with the rays parallel to the lens axis, the real image of the object at infinity is found in the back focal plane (Fig. 2.7). The height h' of the image can also be determined from the object angle θ and the focal distance f:

$$h' = f \tan \theta. \tag{2.1}$$

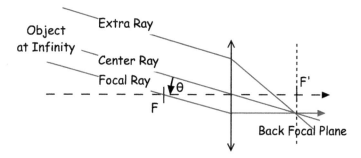

Figure 2.6 Construction of an off-axis ray from an object at infinity.

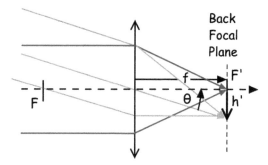

Figure 2.7 Determining the image height for an object at infinity.

2.4 Treating Virtual Images

Virtual images were first introduced in Chapter 1 [Fig. 1.15(b)] for a positive lens. Virtual images can also be produced by a negative lens. When compared to a positive lens, the locations of the focal points for a negative lens are switched: F is to the right of the lens and F' is to the left of the lens, as shown in Fig. 2.8.

So, to ray sketch a negative lens, the center ray still goes through the center of the lens undeviated, but the focal ray is directed at the focal point F to the right of the lens and results in a ray parallel to the axis. The parallel ray is directed parallel to the axis as before, but it is refracted by the lens so that it appears to come from the focal point F' to the left of the lens. The image location is the intersection of the three rays emerging from the lens. Because these rays are diverging from each other, the image is virtual. It is located at a distance t, which is a negative distance according to the sign convention because the origin of the axes is at the lens.

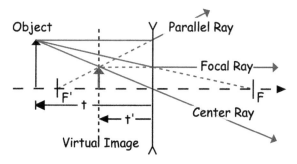

Figure 2.8 Ray sketch for a negative lens.

Here is another exercise involving virtual images.

> **Exercise 2.2 Simple magnifier**
> A positive lens with focal length $f = +6$ units is located 4 units to the right of an object ($t = -4$), and its semidiameter is 3 units. The symmetrical object is 4 units in diameter ($h = \pm 2$ units). Using graph paper, sketch rays from the top, center, and bottom of the object and determine the location of the image. (Note that it is not possible to sketch a focal ray, but if you look at the two other rays that can be sketched and at Fig. 2.1(b), you can figure out where any other ray goes.)

2.5 Mirrors

Compared to a lens whose properties are determined by two surfaces, a central thickness, and the refractive index of its glass, a spherical mirror is a simpler optical element. It consists of a single reflective surface defined only by its radius of curvature R. What is not so simple with mirrors, as we shall see, is getting the signs right.

The optical power of a mirror surface can be derived from the same equation [Eq. (1.7)] from Chapter 1 that we used to calculate the optical power of a refracting surface. If the mirror is located in air ($n = 1$), according to our sign convention, the index of refraction after reflection is a negative quantity $n' = -1$, and the result is that

$$\phi = \frac{n' - n}{R} = \frac{-1 - 1}{R} = \frac{-2}{R}. \tag{2.2}$$

Therefore, a concave mirror with a negative radius of curvature has positive power and focuses light like a positive lens, whereas a convex mirror with a positive radius of curvature has negative power and diverges the light like a negative lens (see Fig. 2.9).

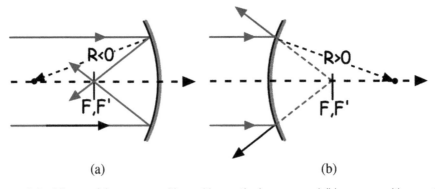

Figure 2.9 Mirrors: (a) concave with positive optical power and (b) convex with negative optical power.

The focal length f of a mirror is also directly related to its radius of curvature R, where

$$\phi = \frac{-n}{f} = \frac{-1}{f} = \frac{-2}{R}; f = \frac{R}{2}. \qquad (2.3)$$

In this case, we used the relationship between power and focal length for immersed systems (see text box, "Immersed Systems") since the index of refraction is no longer equal to 1 in both object and image space for a mirror. We can also find the distance f' from the mirror to the rear focal point F′, where

$$\phi = \frac{n'}{f'} = \frac{-1}{f'} = \frac{-2}{R}; f' = \frac{R}{2}.$$

The final result is that the focal length of the mirror is half its radius of curvature R, and its front and rear focal points F and F′, respectively, are located on top of each other, as shown in Fig. 2.9.

Immersed Systems

In Chapter 1, we assumed that all of the lens systems were located in air. There are, however, instances when this is not so, and the optical system is immersed in a medium other than air, such as water. In some cases, only one side of the lens system is immersed. For example, in an oil-immersion microscope, one end of the objective lens and the specimen of interest are immersed in a transparent oil of high refractive index. This technique increases the resolution of the microscope.

You can also consider a reflective optical system to be an immersed medium where the assignment of a negative sign to the refractive index of a space after the reflection of the ray is part of a set of rules used to calculate its trajectory as it travels through an optical system. The laws of physics do not govern this rule; rather, it is a convention worked out by optical designers over a number of years.

If the optical system is immersed in a medium other than air, the indices of refraction of the object- and image-space media need to be taken into account in our imaging formulae. In particular, the definition for optical power of an immersed lens is written as

$$\phi = \frac{-n}{f} = \frac{n'}{f'}, \qquad (2.4)$$

where f is the distance to the front focal point F, f' is the distance to the rear focal point F′, n is the object-space index of refraction, and n' is the image-space index of refraction. Under such conditions, the imaging equation becomes

$$\frac{n'}{t'} = \frac{n'}{f'} + \frac{n}{t}.$$

The equation for magnification also needs to be modified as follows:

$$m = \frac{h'}{h} = \frac{nt'}{n't}. \qquad (2.5)$$

Both focal points of a positive-powered mirror with a negative radius of curvature are located to the left of the mirror. Because the focal length is a directed distance, whereas the optical power of a component is a constant, a mirror with positive power has a negative focal length. At first this sounds odd and perhaps confusing (remember that a positive-powered lens has a positive focal length), but it is simply a result of our choice of sign convention. To further complicate the situation, lens systems and mirrors are often referred to by their effective focal length (EFL), which is defined to be

$$\phi \equiv \frac{1}{\text{EFL}} \qquad (2.6)$$

and is discussed in more detail in Chapter 4. With this definition, both a positive-powered lens and a positive-powered mirror have a positive EFL, and a "mirrormaker's" formula may be written as

$$\text{EFL} = -R/2. \qquad (2.7)$$

As with thin lenses, the graphic depiction of a mirror in the initial design and the sketching of paraxial rays to assess the performance of that design can be simplified by drawing the mirror as a single line with bent tips to indicate a positive or negative mirror, as shown in Fig. 2.10. Later in the design process, this thin mirror depiction is replaced with the exact ray traces, and its surface is drawn as a curved surface.

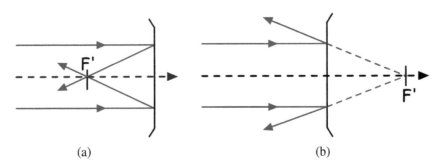

Figure 2.10 Graphic symbols for thin mirrors: (a) positive and (b) negative.

Ray sketching mirrors is similar to ray sketching lenses. The intersection point of a parallel ray and a focal ray from the top of an object locates the top of the image (Fig. 2.11). However, for a single reflection, the real image space is located in the same volume as object space. [Remember that in the case of lenses, the virtual space was on the object side of the lens, and the real space was on the image side (Fig. 2.2).] Although these two spaces overlap each other, optically they are different spaces. For one thing, light runs in opposite directions in object space and real image space. The negative-index convention not only makes the calculation easy but also serves to keep track of the space in which the ray is traveling.

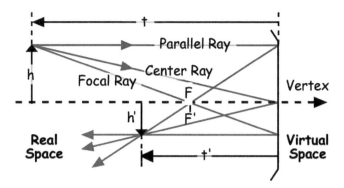

Figure 2.11 Ray sketching a concave thin mirror (h and h' are conjugate pairs).

To complete the sketch, the center ray for the system is directed at the **vertex** of the mirror, the point where the optical axis intersects the mirror. So this ray might also be called a **vertex ray**. Instead of the center ray going through the lens undeviated, in this case, the center ray is reflected off the vertex point with the optical axis serving as the normal to the surface (Fig. 1.5). In the example shown in Fig. 2.11, where the object is located outside the focal point of the mirror at a distance t from the mirror, the rays from a point on the object of height h come to a focus at an image distance t' and height h' as a demagnified, inverted ($m = -t'/t$, where the negative sign is a result of using the equations for immersed systems; see box, "Immersed Systems") real image.

Notice in Fig. 2.11 that if the arrow below the optical axis were relabeled as the object h and the one above the axis as h', the locations of the object and image would be switched. This situation is true for all pairs of objects and their corresponding images. Each pair of locations is called a **conjugate** pair, and the two locations are referred to as **conjugates** of one another. The magnifications of the pair are the reciprocals of one another. (Do you see why?)

Exercise 2.3 Simple mirror

A concave mirror with a 6-unit focal length is located 12 units in front of an object ($t = -12$). The mirror diameter is 6 units. In this case, the object is 3 units high. Sketch rays from the top of the object. Determine the location of the image, its orientation, and the magnification of the system.

2.6 Planar Optics

Although most optical components magnify or demagnify an object, some components such as plane mirrors, windows, and prisms redirect the light with no change in magnification. We have already touched on one example of planar optics with the window in Exercise 1.1 (Parallel slab). Another example is a bathroom mirror (Fig. 2.12). Ray sketching a planar optical element usually involves no special rays. A ray can be launched from any object point and be reflected off the plane surface at the same angle to which it was incident to the surface normal at that point according to the law of reflection.

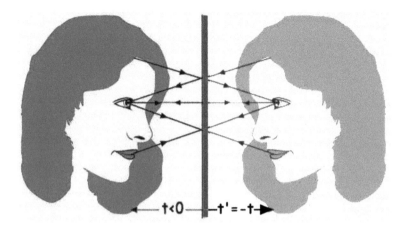

Figure 2.12 Plane mirror (bathroom mirror).

When an observer is in front of the surface, the light rays scattered from the observer's face are reflected by the mirror, but only a small set of rays will enter her eyes. These are rays that determine what the observer sees, as depicted in Fig. 2.12. The image of the person, located in front of the mirror at a distance t, is seen by the eye as a virtual image that is erect and located a distance t' behind the mirror where t' equals $-t$.

When a plane mirror is inserted into an optical system, the optical axis is folded just as a ray would be when reflecting off the mirror. Consider the system shown in Fig. 2.13(a), which magnifies the object by a factor of four. If a mirror is inserted behind the lens at an angle of 45° to the optical axis of the lens [Fig. 2.13(b)], the rays from the top of the object that exit the lens are traced to the mirror surface, where they and the optical axis are reflected according to the law of reflection. The reflected rays continue to their intersection point, where the top of the image is located [Fig. 2.13(b)]. The result of the reflection of the rays and the optical axis can be seen as the original paths being folded along the surface of the mirror. In many optical designs, mirrors and prisms can be used to create more compact systems.

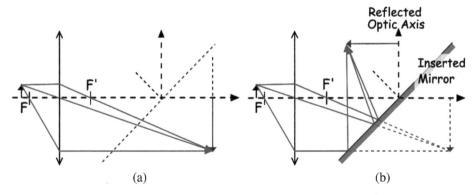

Figure 2.13 Reflection of a real image by a plane mirror: (a) Location of the mirror along the optical axis. (b) Traced reflective rays determine the location of the reflected real image.

2.7 Multiple Elements

Most optical systems contain more than a single element, and the rules of ray sketching can often be extended to cover designs with multiple elements. Two good examples are the Keplerian and Galilean telescopes. The Keplerian telescope [Fig. 2.14(a)] consists of two positive lenses separated by a distance equal to the sum of their focal lengths. The focal length of the first lens, the objective lens, is larger than the focal length of second lens, the eyepiece lens. A bundle of rays from an on-axis object point at infinity will consist of rays parallel to the optical axis. The two rays entering at the top and bottom of the lens focus at the back focal point by the first lens and then diverge. Because the front focal point of the second lens is located at the same point, the second lens will collimate the rays. Thus, the image is located at infinity.

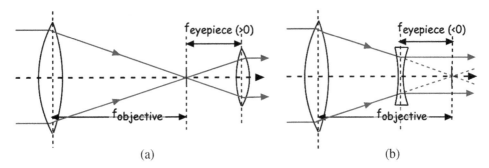

Figure 2.14 Types of refractive telescopes: (a) Keplerian and (b) Galilean.

In contrast to the other optical systems that produce a finite image, the object and image distances are both located at infinity. That is, light from an object at infinity will not be imaged at a finite distance from the final element. Because there is no focal point or system focal length, these instruments are referred to as **afocal** systems.

In the case of the Galilean telescope [Fig. 2.14(b)], the objective lens is also a positive lens, but the eyepiece lens is a negative lens. The negative lens is located so that its front focal point is coincident with the back focal point of the positive lens. The rays from the objective lens that would focus at the back focal point are intercepted by the negative lens and collimated, as were those for the Keplerian telescope. For either telescope to produce an image at a finite distance, additional optics are required to focus these collimated rays to an image. This may be a set of lenses called, appropriately, an imager, or it may be the lens of your eye placed just behind the eyepiece.

Because both the object and image of a telescope are located at infinity, the magnification of a telescope cannot be expressed as the ratios of image height to object height. Instead, a telescope magnifies the angles that an object and image subtend (make) with respect to the optical axis. Rays entering the objective lens at an angle θ will exit the eyepiece at an angle θ'. An object perceived to subtend an angle of θ without the telescope will appear to subtend an angle θ' when looking through the eyepiece.

The fact that the entering rays enter at a downward angle in the Keplerian telescope and the exiting rays have an upward slope means that the image will appear upside down, as is depicted in Fig. 2.15. There is no image inversion in the Galilean telescope. The **angular magnification** for both telescopes is the ratio of the exit angle to the entrance angle. It can be shown (see Exercise 2.4) that angular magnification also equals the negative of the ratio of the focal lengths:

$$M_\theta = \frac{\theta'}{\theta} = -\frac{f_{\text{objective}}}{f_{\text{eyepiece}}}. \tag{2.8}$$

For visual instruments (lens systems used with the eye), the angular magnification is also known as the magnifying power (MP). The MP determines how much bigger (or smaller) something will look to the user when they look through the instrument.

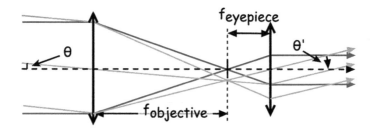

Figure 2.15 Angular magnification of a Keplerian telescope.

Exercise 2.4 Angular magnification
In Fig. 2.15, the intermediate image is located between the two telescope lenses. If h_i is the height of the intermediate image from the optical axis to the intersection of the rays at the image, show that the angular magnification is as it is defined in Eq. (2.8). (You will need to use the paraxial approximation $\tan\theta \approx \theta$ to arrive at the correct expression.)

The sign of the angular magnification of a system is often omitted in the name of a telescope (e.g., 15X rifle scope). However, the sign of the angular magnification indicates the orientation of the image relative to the original object and should not be ignored in the general description of an optical system.

Exercise 2.5 Telescopes
Suppose you had a +100-mm-focal-length objective lens. (a) If you wanted to create a Keplerian telescope with a 4X angular magnification, what is the focal length of the eyepiece lens you would need? (b) What would be the overall length of the telescope? (c) If you wanted to use the same objective lens for a Galilean telescope with a 4X angular magnification, what focal length eyepiece lens would you select? (d) What would be the length of this telescope?

2.8 Beyond Two-Lens Systems

Most optical systems are more elaborate than what we have described here. In some cases, many lenses are combined into one lens system and treated as such. This will be explained in considerable detail in Chapter 4 regarding cardinal points and first-order quantities. Other times, the set of imaging rays will be intercepted by another component with optical power before they can focus, and the simple rules given here cannot be applied. However, reasonable guesses as to the focal lengths of the lenses and their separations will let you sketch a rudimentary optical system that can help you decide on the practicality of the design for a given situation.

Exercise Answers

Ex. 2.1

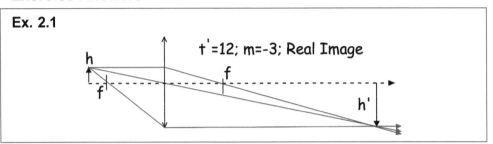

Ex. 2.2

Ex. 2.3

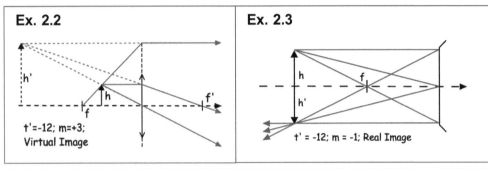

Ex. 2.4
In Fig. 2.15, the central rays entering the objective and leaving the eyepiece make the same angles with respect to the optical axis between the lenses as they do outside. These central rays cross at the top of the intermediate image at a height of h_i. You must keep track of signs. Note that the image is inverted, so h_i is negative, as is the entering angle θ. The focal lengths and the exiting angle θ' are positive. The height can be written as

$$h_i = f_{\text{objective}} \tan\theta \text{ and } h_i = -f_{\text{eyepiece}} \tan\theta'$$

using Eq. (2.1). By equating these two expressions and using the paraxial approximation, we get

$$f_{\text{objective}}\theta \text{ and } h_i = -f_{\text{eyepiece}}\theta'.$$

Solving for $M_\theta = \theta'/\theta$ gives Eq. (2.3).

Ex. 2.5
Because $M_\theta = -4$ (the Keplerian inverts the image), inserting -4 into Eq. (2.8) for the angular magnification and $+100$ for $f_{\text{objective}}$, the focal length of the eyepiece is $+25$ mm. (b) Its length is the sum of the focal lengths, or 125 mm. (c) In the case of the Galilean scope, $M_\theta = +4$, and a -25-mm eyepiece should be used. (d) The telescope length would again be the sum of the focal lengths of the lens, or 75 mm.

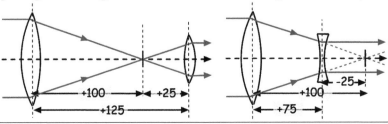

First Hiatus
Ledgers to Laptops

A hiatus is an interruption. The word is used to describe a break in the continuity of a work, a series, or some action. When a TV network says that your favorite TV series is "on hiatus," there is the prospect that at some future time the series will resume, and its fans will be pleased (which, of course, hardly ever happens).

In optics, the hiatus (Fig. H1.1) is the distance between two lens planes, the principal planes. (These will be discussed in Chapter 4.) When an axial ray enters a lens and hits the front principal plane P_F, it jumps across the hiatus at the same height to the back principal plane P_B and exits the lens.

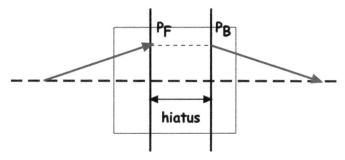

Figure H1.1 Illustration of "hiatus" in optics. Hiatus is the distance between the two principal planes of the lens, P_F and P_B.

So, this section of the text is an interruption. It is not directly related to what has just been discussed and what follows it. However, for some readers, the information provided here will give you a better sense of the development of optical design. It will also provide the tools and ideas needed to design and perfect an optical design. And it will give an appreciation of the work that others have done in this field.

H1.1 Simulations

The calculation of results began once humans learned how to quantify things. And once it was possible to describe some physical process using an equation, it was possible to model the process so that certain rules of thumb gained through much trial and error could be replaced by mathematical calculations. Thus began the practice of simulating these processes to be able to predict an outcome without performing the act itself.

Almost certainly, the first place that simulation was used was in a military application [Fig. H1.2(a)]. Once the range of a cannon could be calculated using the equation for a parabolic trajectory and the distance to the target by triangulation, the results of artillery fire would be greatly improved. With time, corrections for wind and air resistance were added to the basic simulation.

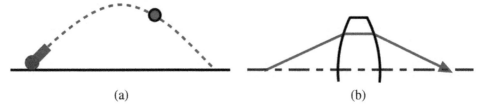

Figure H1.2 Modeling physics (a) mechanics and (b) optics.

Another field where simulations could be applied was in ray tracing. The equations required to model the path of a light ray traveling through an optical system are quite simple: (1) straight line trajectory of light in a constant medium and (2) the law of refraction. These two equations plus the correct description of the optical surfaces and the correct refractive indices will permit the designer to trace a ray through the optical system [Fig. H1.2(b)]. In order to analyze an optical system, a series of rays from several points on an object are traced through the system and the results in image space are examined to determine if the system performs properly or if it needs to be improved. With the increase in computational power, it is possible not only to analyze the performance of an optical system, but also to optimize the design for best performance.

H1.2 Tracing Rays

In the days before there were any mechanical or electronic aids to computation, ray traces were computed by hand using the equations for light propagation. Because a number of ray traces were needed to determine the lens performance, a standard procedure was developed that used financial ledgers (spreadsheets) to record the calculations and summarize the results. These calculations were aided first by logarithmic tables of trigonometric functions (log-trig tables) that reduced multiplication of terms in the ray trace equations to addition. Later, the tables were supplemented by adding machines so that a person could do a large set of ray traces that would permit analysis of the lens. To speed up the work and reduce the error that would occur with long calculations, short cuts were devised that gave valuable information using simplified equations.

One technique was paraxial ray tracing. When rays were traced close to the optical axis, the equations could be written in terms of linear functions of the ray angles and ray heights. This is much easier to compute. Further, the values computed for these paraxial traces can be used to compute the aberrations, or ray errors, of the lens, extracting the maximum amount of information about a lens with minimum amount of work.

Still, these ray traces demanded meticulous attention to detail. And because perfecting the performance of a lens required many, many rays to be traced for each revision of the design, a large number of people were needed to make the calculations. The work of systematic ray tracing was done with rooms of people, generally women, who were called "computers." The preference for women may have been because of a belief, or an observation, that they performed these tasks meticulously and, perhaps, because they would work for less. Later, when mainframes were used for the calculations, there were references in the optics literature to the fact that the calculations were done on "electronic computers" so as to distinguish them from the human computers that had previously been employed at the task.

As computing technology advanced, the platforms on which ray traces could be performed became faster and smaller—able to handle computations in seconds that had taken days for the human computers. In the early 1970s, the chief lens designer at Kodak, Rudolph Kingslake, taught a summer course, Fundamentals of Lens Design, at the Institute of Optics in Rochester. It was clear once the class began that he enjoyed his work in a fascinating field and shared it with anyone who was interested. That summer, Texas Instruments and Hewlett-Packard began their battle of programming calculators. Kingslake could not contain himself as he bounced across the front of the auditorium, narrating with great glee his calculations for ray tracing through a series of lenses using one of these new marvels. After all, for those who had punched their cards and brought their offerings to the high priests of the IBM computers of that era, the invention of a handheld programmable calculator was like the removal of shackles. From there, the input terminals to the mainframes were replaced by desktop workstations. These were followed by personal computers. Now, as is the case for this text, many of the calculations are done on a laptop computer.

Chapter 3
How to Put a Lens in a Computer

To trace rays with a computer, we start by entering the imaging system into the design program. The basic layout of OpticStudio's® graphic user interface (GUI) is shown in Fig. 3.1. The menu banner at the top of the main window has a series of menu tabs (e.g., File, Setup, Analyze, Optimize, and Tolerancing) used to enter, analyze, refine, and tolerance a design. Each menu tab has a row of icons to indicate the available functions and actions for that tab. Hovering your cursor over one of the icons opens a window showing the icon name and a short explanation of its function or action. In the next few chapters, we will focus on the File and Setup tabs. Analyze will be discussed in Chapters 6–9, Optimize in Chapters 10–12, and Tolerancing in Chapter 13. The last tab, the Help tab, provides links to OpticStudio resources.

The area below the menu banner is divided into the System Explorer on the left and a display space on the right for the Lens Data Editor (and any other editors or analysis windows). At the end of the Setup tab is a series of icons that help you control how these windows are displayed in the workspace (e.g., tiled, floating, or docked). At the very bottom of the program window is a status bar containing four-letter operands (discussed in Chapters 10–13) whose values update as the design is changed.

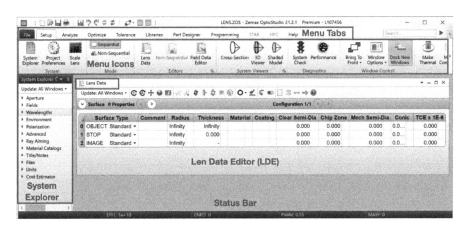

Figure 3.1 The basic layout of OpticStudio's graphic user interface.

There are two primary sets of information needed to describe any new optical design: the system data and the lens prescription. The system data describes how the lens will be used (e.g., object size, wavelength, aperture size), whereas the prescription consists of the shape, location, and material of the lens elements. In this chapter we will demonstrate how to enter the system data into the System Explorer and how to enter the prescription data into the Lens Data Editor, and then run some simple checks on the design to make sure everything has been entered properly.

Navigating OpticStudio

There are many tasks that you must perform to enter, analyze, and optimize a lens design. This text is intended to introduce you to OpticStudio and teach you how to use it. To do this, these operations must be described with sufficient detail that you can correctly perform the required actions. Otherwise, we have failed.

But many of these actions require you to open a series of tabs, windows, and input boxes that are used again and again in the text. So, to reduce repetition and to provide easy recognition of the actions, we invented a number of text formats that provide the information you need to navigate OpticStudio. For example, to start a new design, File > New, is used. This path format tells you to click on the File tab and select New from the available menu icons. The light-blue background behind the text and the "greater than" character tells you that this is an OpticStudio operation and this is the path you use. While we could have also shown this action with a number of figures, our formatted text can do it more compactly.

Additionally, many operations involve the opening of a new window, the push of a button, or the selection of a drop-down menu option. These actions will be given a gray background, such as Select Preset. Finally, in Chapter 12, there are a sufficient number of operations that the various actions are collected and presented in a boxed format.

3.1 System Data

Open up OpticStudio and start a new lens by selecting the File tab in the menu banner and clicking on the first icon, New, in the ribbon (File > New). Before any rays can be traced, three pieces of system data must be specified:

1. the size of the light bundles that enter the optical system,
2. the location(s) of points in the object plane to serve as an object, and
3. the wavelengths at which this lens is intended to operate.

In OpticStudio these data are specified using the first three items (Aperture, Fields, and Wavelengths) of the System Explorer, as described below.

Let's start by looking at the settings for the aperture and the field of the lens. The default Aperture Type is Entrance Pupil Diameter (EPD), which defines the size of the light bundle accepted by the lens. The default Aperture Value is zero (no light). As shown in Fig. 3.2, you change this value from 0 to 20 by entering "20" in the Aperture Value box (System Explorer > Aperture >

Aperture Value). The default field Type is Angle with only one field point (Field 1) defined with a value of zero. This specification represents a point object on the optical axis with an angle of 0°. For now, we will only consider the on-axis lens behavior of objects at infinity, so this single axial field can be left as it is. Note: other types of apertures (e.g., Image Space F/#) and fields (e.g., Object Height) are available in their respective drop-down menus and will be discussed in Chapter 5.

The third required system setting is wavelength. The default wavelength is set to 0.550 µm. This wavelength is used because it is in the middle of the visible light spectrum (0.4–0.7 µm). Unlike aperture and field, it would make no sense to have a default wavelength of zero. The wavelength value can be easily changed by entering a new value directly in the Wavelength box or by selecting one of the many presets available for different light sources. For the design examples in this chapter, we're going to change it to 0.5876 µm. This is known as the d-line, a standard spectral line emitted by an excited helium atom (see Chapter 9). Under System Explorer > Wavelengths > Settings > Preset, select **d (0.587)** from the drop-down menu under Preset and then click **Select Preset** (see Fig. 3.2).

Finally, the default unit of length for a lens system is a millimeter. This can also be checked (and changed, if needed) in the System Explorer (System Explorer > Units > Lens Units).

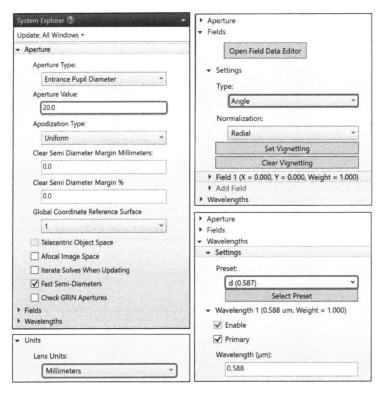

Figure 3.2 System Explorer items used to set the aperture, fields, wavelengths, and units for an example design.

3.2 Prescription Data

Now that the system data has been entered in the System Explorer, we can turn our attention to the spreadsheet titled Lens Data. There are a number of such spreadsheets in OpticStudio that are referred to as *Editors*. To distinguish the current spreadsheet from the others, we will refer to it as the Lens Data Editor or LDE for short. The initial LDE for a new lens consists of three rows and many columns. Each row represents a surface in the optical design with the data for that surface in the individual columns. For the purposes of this text, only six of the columns (Surface Type, Comment, Radius, Thickness, Material, and Clear Semi-Dia) will be shown in all LDE figures. The rest of the columns contain lens parameters for special types of lenses (e.g., aspheres), coatings, and mechanics that are not needed for our discussion, resulting in a much more compact display.

Figure 3.3 shows the current LDE with the three initial surfaces of any new lens: **OBJECT**, **STOP**, and **IMAGE**. The first row lists the values for the object surface. It is a planar surface (its radius of curvature is infinite) that is located at infinity (its thickness is infinity). The Material cell is blank, indicating that no glass has been assigned to object space. The default medium is air, and the refractive index is set to unity. The next term in the surface number column, **STOP**, may not be familiar. The designation of this surface as a stop will be the subject for considerable discussion later (Section 5.3). For now, consider it to be an important surface needed for a complete analysis of the lens. The last row in the LDE is always the image surface. In summary, there are three required surfaces that each design must have: object, stop, and image.

Typically, we need at least one more surface in a design (e.g., for a lens element we need a front and back surface). To insert a surface after the first (stop) surface, select the stop surface (S1) row and right-click on it. This opens a drop-down menu, where you can select **Insert Surface After**. When this action is taken, a new row with a "2" on the left side of the LDE will appear between the stop and the image surface, as shown in Fig. 3.4. New surfaces can also be added (or deleted) quickly by clicking on an existing surface row and using the insert and delete keys on your keyboard. The Insert key enters a surface **before** the current surface, while Ctrl-Insert inserts a surface **after** the current surface.

	Surface Type	Comment	Radius	Thickness	Material	Clear Semi-Dia
0	OBJECT Standard ▼		Infinity	Infinity		0.000
1	STOP Standard ▼		Infinity	0.000		10.000
2	IMAGE Standard ▼		Infinity	-		10.000

Figure 3.3 The lens "spreadsheet" (LDE) for a new lens with an EPD = 20 mm.

How to Put a Lens in a Computer   43

	Surface Type		Comment	Radius	Thickness	Material	Clear Semi-Dia
0	OBJECT	Standard ▼		Infinity	Infinity		0.000
1	STOP	Standard ▼		Infinity	0.000		10.000
2		Standard ▼		Infinity	0.000		10.000
3	IMAGE	Standard ▼		Infinity	-		10.000

Figure 3.4 Second surface inserted in the LDE.

This initial framework is used to build a complex design by adding surfaces and then entering the quantities that describe each of the surfaces and the spaces that separate them. Before entering a complete lens, we will first describe the inputs for each of the columns in the LDE, beginning with Surface Type. This column describes the form of the surface. The simplest surface, Standard, is based on a conic section whose vertex radius is given in the next column and by default has a zero conic constant (resulting in a simple spherical surface). For some surfaces, the shape is more elaborate, such as a complex asphere described by a polynomial. Expect to see Standard as the surface type for all designs in this text.

We can now specify the quantities that describe the makeup of the lens elements (radii of curvatures, thicknesses, and glasses) in the order that they are encountered by light from an object. The radius and thickness values are signed and determined by the coordinate system whose positive z axis is in the direction of initial light propagation (left to right, see Section 1.1). One odd thing about this coordinate system is that it is really a set of many coordinate systems with origins along the z axis. Each defined surface has an accompanying coordinate system whose origin is at the intersection of the surface and the optical axis.

The lens values, labeled with a surface number k, are thickness Tk, radius of curvature Rk, and glass Gk, starting with the object plane and continuing to the image plane. The values are entered into their corresponding cells in the LDE. References to the surfaces of the lens in the text will be labeled as Sk. In OpticStudio, the quantities related to the object and image planes have letter labels (O and I, respectively). The surfaces between them have numerical subscripts starting with 1, as shown in Fig. 3.5.

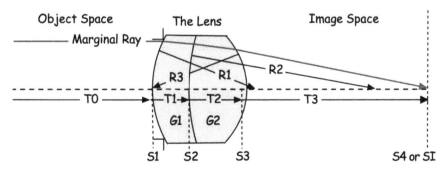

Figure 3.5 Describing a lens.

The R1 cell in the LDE contains the value of the first radius of curvature, R_1. Once the shape of the surface is defined, the distance to the next surface is entered. As was noted in Chapter 1, the sign of a thickness t to the left of a lens surface is negative and that to the right is positive. Unless a reflective surface is inserted, the lens thickness and the separating spaces are positive numbers entered in the Thickness column. So, typing 5 into T1 enters a thickness of 5 mm between the first and second surfaces.

Beyond the last lens surface is the image surface at a distance T3 or TI, where the image produced by the lens is located and analyzed. One other feature of this lens drawing is the blue ray entering at the edge of the lens. This is a **marginal ray** that describes the size of a bundle of light rays entering a lens. A full description and a discussion of its importance is given in Section 5.3.

To cause any ray bending in a lens, there must be changes in the refractive index across the lens surfaces. While the value of the refractive index could be entered, we would need to determine what glass is being used and then find the index of refraction at each design wavelength by looking it up or calculating it using a formula. Although it is possible to do this, in most lens design programs, a glass name is usually entered in the Material column, and the program calculates the refractive index for the wavelengths that have been chosen in the specification phase. For example, to assign a common optical glass, N-BK7, to the lens, its name, N-BK7, is entered in the G1 cell. An extensive discussion of glasses and wavelengths is given in Chapter 9 on chromatic aberration.

The last column, Clear Semi-Dia, displays the semi-diameters for each surface. (The term "semi-diameters" is used instead of "radius" to remove any confusion between the radius of a lens opening and the radius of curvature of a surface.) The semi-diameters are values that can be assigned by a user or computed by the program. For the present, all semi-diameters will be automatically calculated. User-entered apertures can be defined by changing the solve type (click on the small box next to the semi-diameter value to get to a drop-down menu) from automatic to fixed. A more detailed discussion of the handling of apertures and their effects on ray tracing is given in Chapter 5.

3.3 Entering a Single Lens

Start the entry of a single lens (often called a singlet by designers) by opening a new lens (File > New). We will call this lens "OSlens." Enter this name as its title in the System Explorer (System Explorer > Title/Notes > Title:). The "OS" label, standing for OpticStudio, is used to designate designs created, modified, and discussed throughout the text. As described in Section 3.1 System Data, you will give the lens a 20-mm-diameter entrance aperture and change the operating wavelength to the d-line. Insert a surface (see Section 3.2) after the stop surface.

Our singlet is to be made of N-BK7. The first surface is convex with a 120-mm radius of curvature. It is separated by 5 mm from the second surface, also convex, with a –50-mm radius of curvature. We have listed each of the parameters in Fig. 3.6 with the values to be entered to construct the singlet. Figure 3.7 shows the resulting LDE along with a compact notation for system specification data shown in the area below the LDE grid that will be used throughout the remainder of the text.

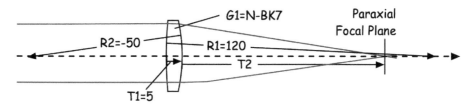

Figure 3.6 Graphic depiction of the values for a singlet lens to be entered in a new lens LDE.

	Surface Type		Comment	Radius	Thickness	Material	Clear Semi-Dia
0	OBJECT	Standard ▼		Infinity	Infinity		0.000
1	STOP	Standard ▼		120.000	5.000	N-BK7	10.000
2		Standard ▼		-50.000	0.000		9.898
3	IMAGE	Standard ▼		Infinity	-		9.748

Title (OSlens); EPD (20 mm); Field (0°); Wavelength (d-line)

Figure 3.7 The LDE after the values shown in Fig. 3.6 are entered with the specifications listed below the LDE.

In most designs, the paraxial image location (the distance from the last lens surface to the paraxial image plane) can be computed by the program. For example, the image of an on-axis object point can be found by calculating where a paraxial ray from the base of the object crosses the optical axis beyond the last surface. This is done by setting up a "solve" that calculates the distance between S2 and the image plane SI. This action is a called a **thickness solve** because it calculates the thickness T2 between the second surface and the where the ray crosses the optical axis.

In OpticStudio, this is achieved by clicking on the small box next to the T2 value (Fig. 3.8), which opens a new dialog box, titled **Thickness solve on surface 2**. The button next to Solve Type is labeled **Fixed**. Clicking on the drop-down arrow next to it reveals a menu of different solves. Select **Marginal Ray Height** (leaving the values for Height and Pupil Zone set to 0 as shown in Fig. 3.8), and an "M" will appear in the solve box. It is important to note that this solve traces a *paraxial* marginal ray when the Pupil Zone is set to zero and finds the image distance where that ray crosses the optical axis (Height = 0). This is the paraxial image distance. Note also that if the pupil zone is set to a non-zero value, then real (aberrated) marginal rays are traced (see Chapters 5 and 6 for more information).

	Surface Type	Commer	Radius	Thickness	Material	Clear Semi-Dia
0	OBJECT Standard ▼		Infinity	Infinity		0.000
1	STOP Standard ▼		120.000	5.000	N-BK7	10.000
2	Standard ▼		-50.000	68.006 M		9.898
3	IMAGE Standard ▼		Infinity			

Thickness solve on surface 2
Solve Type: Marginal Ray Height
Height: 0
Pupil Zone: 0

Figure 3.8 The LDE when a paraxial image thickness solve is applied to S2.

In the LDE (Fig. 3.8), the cell for the thickness of surface 2 now shows a value of 68.006, with "M" indicating that the thickness T2 was determined with a marginal ray height thickness solve. We will use this type of solve many times in this text and refer to it as a **paraxial image thickness solve**. Save this lens (File > Save As) with the file name **OSlens**. Later on, we will be using this lens again. To assist you with these lens files, the file names will be boldfaced whenever a lens is either Saved or Opened again.

3.4 Checking the Lens

Once you've entered the lens parameters, the easiest way to get an initial check on the lens to see if you entered it properly is to plot the lens. This is done by using the System Viewers on the Setup tab. The simplest plot is a y-z cross-section of the lens (Setup > Cross-Section). This type of plot can also be generated with the shortcut key Ctrl-L. What you will notice when you plot the lens (Fig. 3.9) is that the rays at the top and bottom of the lens, called marginal rays, do not focus at the image plane. This is because the image plane is located along the axis at the paraxial focal point, and the (real) marginal rays, produced in the Cross-Section, cross in front of the paraxial focal plane because of spherical aberration, which is discussed in detail in Chapter 6.

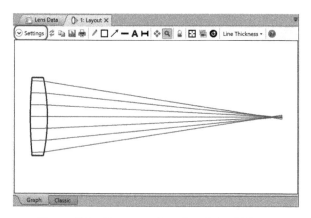

Figure 3.9 Cross-section plot of the OSlens.

In the upper left-hand corner of the cross-section window (see Fig. 3.9) is a Settings drop-down arrow. Clicking on this arrow opens a dialog box that allows you to change a number of settings for the cross-section, including the first surface, the last surface, and the number of rays to be plotted. Try it out for yourself. For instance, change the number of rays in the fan by either entering the number of rays directly or increasing or decreasing the counter value with the arrows. Next to the settings drop-down arrow is a ribbon of icons that can also be used to change the plot. As with the main menu icons, hovering the mouse over any icon will open up a window with more information on what each icon does.

The other thing you may notice about Fig. 3.9 is that you can't see the rays coming into the lens (this occurs when the object is at infinity) and the lens is only drawn as big as the largest ray height. Many users find it useful to increase the viewed diameter of the lens and add an offset surface to observe the rays in object space when looking at these cross-section lens plots. See the box "Making Your Lens Cross-Section Look Good" for details on how to do this. Note: When new surfaces (like an offset surface) are added to the LDE, the surface range (First and Last) of any existing layout plot will need to be updated to properly display the added surfaces.

Making Your Lens Cross-Section Look Good

Figure 3.9 shows a basic cross-section of the OSlens (Setup > Cross-Section). Most users of OpticStudio find it useful to add an initial offset surface to the LDE to observe the entering ray bundles when the object is at infinity. This is done by inserting a surface after the Object surface and giving it a reasonable thickness as shown below. You can label it as "Offset" in the Comment column for S1.

		Surface Type	Comment	Radius	Thickness	Material	Clear Semi-Dia
0	OBJECT	Standard ▼		Infinity	Infinity		0.000
1		Standard ▼	Offset	Infinity	10.000		11.000
2	STOP	Standard ▼		120.000	5.000	N-BK7	10.000
3		Standard ▼		-50.000	68.006 M		10.887
4	IMAGE	Standard ▼		Infinity	-		0.547

You can then change how the offset surface is displayed in the lens layout by modifying its visibility with the Surface Properties drop-down dialog box at the top left of the LDE window. Use the < > arrows next to Properties to move the surface counter to Surface 1 (if not already at 1). Then open the Surface 1 Properties dialog box by clicking on the drop-down arrow to the left. This will show all of the surface properties associated with the offset surface, S1. Choose the **Draw** tab and check the **Do Not Draw This Surface** (as shown below) to hide this surface in the lens layout.

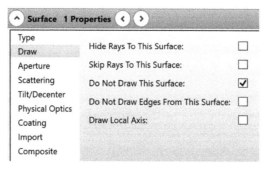

By default, the lens surfaces in the cross-section are only drawn out to the maximum aperture needed to pass all rays (See Chapter 5 for more detailed information on the calculation of apertures). However, in practice, lenses are typically fabricated at diameters larger than their default clear apertures. With this in mind, the lens cross-section can also be significantly improved by adding a small increment (margin) to the drawn lens diameters by changing the settings in the Aperture tab of the System Explorer.

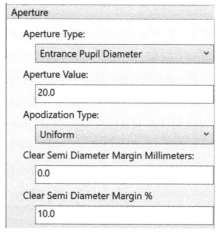

After adding an offset surface and a 10% Clear Semi Diameter % on the lens apertures (System Explorer > Aperture > Clear Semi Diameter Margin %), the improved cross-section for the OSlens is shown as below.

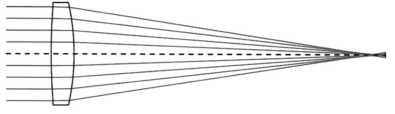

Note: The majority of lens cross-sections in this text are created with these two modifications in place even though we will *not* always explicitly show the offset surface in the LDE and/or state that a specific aperture increment has been added to the lens diameters to create the figures.

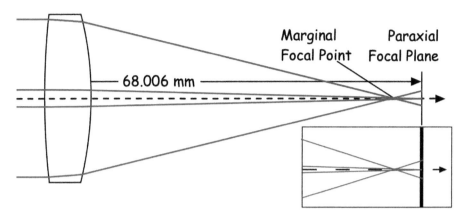

Figure 3.10 Plot of the lens with the image plane at paraxial focus, showing marginal and paraxial ray traces.

In Fig. 3.10 we have plotted two marginal rays (changing the Number of Rays in Settings from seven to three) along with two paraxial rays that cross at the paraxial focal plane (located at the far right of the trace). The paraxial rays were created with another Cross-Section plot where the upper and lower pupils were set at 10% of the total aperture in the Settings (values of 0.1 and -0.1, respectively). The two plots were then superimposed on each other. The image distance (68.006 mm) is the distance between the second lens surface S2 and the paraxial focal plane.

Since we located the object at infinity, the paraxial image distance also equals the distance to the rear focal point or the **back focal length** (BFL) of the lens. This value can be compared with either a focal length value given with the lens prescription or a quick thin-lens estimate of the focal length for another check on the lens entry. For example, remember that the focal length of a thin lens is the distance from the lens to its focal point, the place where incoming collimated rays cross the optical axis. We can use the lensmaker's equation [Eq. (1.9)] from the first chapter,

$$\frac{1}{f} = (n-1)\left(\frac{1}{R_1} - \frac{1}{R_2}\right), \tag{1.9}$$

to find the thin-lens focal length of the OSlens ($R_1 = 120$ mm, $R_2 = -50$ mm, and the refractive index of the N-BK7 glass at the design wavelength = 1.5168), and we get

$$\frac{1}{f} = (1.5168 - 1)\left(\frac{1}{120} - \frac{1}{-50}\right) = 0.5168(0.008333 + 0.02) = 0.0146427.$$

The reciprocal of this value gives a thin-lens focal length equal to 68.294 mm, which is close to the image distance of 68.006 mm.

> **OpticStudio Macros**
>
> OpticStudio is a ray trace program that can simulate the propagation of light through an optical system and analyze the system using a range of plots and data tables to assess design performance. There are, as you will see, a large number of tools that can be used to determine how well your system performs. However, there are instances where the available tools may not be appropriate to evaluate the performance of your system. For example, the standard output may not contain enough data (or have too much!) to be able to evaluate your design.
>
> To remedy this, OpticStudio provides a Zemax Programming Language (ZPL), resembling Basic, to take advantage of the computational power of OpticStudio. ZPL is used to create macros to get the specific data that you want. In this text, several macros will be used to collect and present data. They are accessed through the Programming tab and will be described as they are introduced in the text. The macro programs are run from the Programming tab. When the leftmost icon, Macro List, is selected, a drop-down list of the available macros appears. The list of macros shown in the figure below has been shortened to save space. The first macro we will use, FIRST.ZPL, has been boxed. The full path to access this macro is Programming > Macro List > FIRST.ZPL. Since we will be using this macro many times in the text, the path will be shortened to a simple **FIRST**. [Note: If your macros list does not include the **FIRST** macro, you can create one using the listing and instructions in the Appendix.]
>
>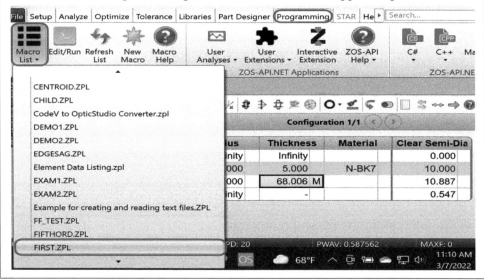

3.4.1 First-order properties

Open the **OSlens** saved at the end of Section 3.2 (File > Open). We can also check our lens entry using the macro **FIRST** (See box "OpticStudio Macros"). This macro creates a list of the basic first-order properties of the lens (Table 3.1). While this information is also available in a full prescription listing for the lens (Analyze > Reports > Prescription Data), the **FIRST** macro creates a much more compact (and easily digestible) listing of just the first-order data.

Table 3.1 First-order properties of the OSlens from the FIRST macro.

```
Infinite Conjugates
   Effective Focal Length      68.9849
   Back Focal Length           68.0055
   Front Focal Length         -66.6344
   F/#                          3.4492
   Image Distance              68.0055
   Lens Length                  5.0000
   Paraxial Image
      Height                    0.0000
      Angle                     0.0000
   Entrance Pupil
      Diameter                 20.0000
      Location                  0.0000
   Exit Pupil
      Diameter                 20.7055
      Thickness                -3.4127
```

FIRST lists the Image Distance and the Back Focal Length as well as two other focal lengths (Effective Focal Length and Front Focal Length). The Effective Focal Length (EFL) of the lens will be discussed in detail at the beginning of the next chapter. For now, EFL represents a focal length that is independent of which way the lens is facing the light. In this case, it is 68.9849 mm. The Lens Length is defined as the distance from the first surface vertex to the last surface vertex (before the image plane). In this case, it is simply the thickness of the singlet, e.g., 5 mm. The other quantities listed will be described in the next two chapters.

If you leave the Text Viewer window open after you run the **FIRST** macro, you can view any changes in the first-order properties of your design by clicking on the double-arrow update icon in the upper left-hand corner to refresh the window.

In the above example, the object is located at infinity, so the back focal length and the image distance are the same. But the object can be placed anywhere between infinity and the front face of the lens. When we change the object thickness to 125 mm in the LDE, the image will not be located in the back focal plane.

The output from the **FIRST** macro (Table 3.2) shows that the focal lengths, listed under Infinite Conjugates, remain the same, but the image distance, listed under At Used Conjugates, is now located at 149.5418 mm (Table 3.2), and in this case the object and image are nearly equidistant from the lens (Fig. 3.11). The first-order listing for finite object distances also includes the Total Length, the distance from object plane to image plane. Total Length is the sum of the three values above it in Table 3.2. It tells the designer whether or not the entire optical system (object to image) will fit within a certain length for the device.

Table 3.2 First-order properties of the OSlens with an object distance of 125 mm.

```
Infinite Conjugates
    Effective Focal Length           68.9849
    Back Focal Length                68.0055
    Front Focal Length              -66.6344
    F/#                               3.4492
At Used Conjugates
    Magnification                    -1.1819
    Paraxial Image Height             0.0000
    Paraxial Working F/#              7.3928
    Object Distance                 125.0000
    Lens Length                       5.0000
    Image Distance                  149.5418
    Total Length                    279.5418
```

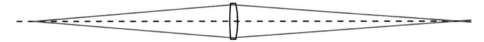

Figure 3.11 Ray trace of the OSlens for a finite object distance.

The EFL in the **FIRST** listing is 68.9849. This differs from the thin lens value of 68.2936 given earlier. However, if we give the lens a zero thickness, **FIRST** gives 68.2936 for the EFL, as we would expect. Don't forget to change the thickness of the lens back to 5 mm before moving to the next section.

3.4.2 Virtual images

Finally, let's move the object inside the front focal point by setting the object surface thickness to 35 mm in the LDE. According to the first-order values listed in Table 3.3, the image distance T2 is now –82.4290 mm. The negative sign indicates that the image is to the left of S2 and is therefore a virtual image.

Table 3.3 First-order properties of the OSlens with an object distance of 35 mm.

```
Infinite Conjugates
    Effective Focal Length           68.9849
    Back Focal Length                68.0055
    Front Focal Length              -66.6344
    F/#                               3.4492
At Used Conjugates
    Magnification                     2.1807
    Paraxial Image Height             0.0000
    Paraxial Working F/#              3.8271
    Object Distance                  35.0000
    Lens Length                       5.0000
    Image Distance                  -82.4290
    Total Length                    -42.4290
```

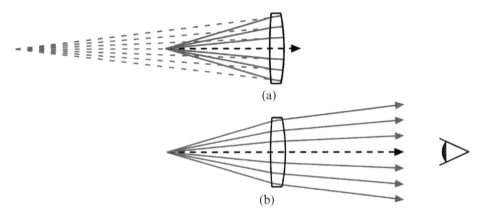

Figure 3.12 A combination of two ray traces of the OSlens for an object inside the front focal point. (a) Plot with a paraxial image thickness solve and (b) plot with a fixed "image" thickness 40 mm after the lens.

The ray trace for this lens is shown in Fig. 3.12. To understand the ray trace, we have generated two separate traces in OpticStudio using a fan of seven rays in each trace. Ray trace (a) uses the paraxial image thickness solve and traces rays back to the virtual image plane; whereas, in ray trace (b), the thickness solve is deleted and 40 mm is entered for the thickness of S2 to provide some distance to the "image" plane S3. The ray trace shows the rays diverging from the lens. An observer sees the rays exiting the lens [Fig. 3.12(b)], and it appears as though it is at the virtual image point [Fig. 3.12(a)].

3.5 Angle Solves

In the design of optical systems, it is not unusual to modify a lens to meet some particular design requirement. For example, the new system may need a lens to have a specific EFL. As we noted in the previous section, the EFL is one of several different focal lengths of a lens. A complete discussion of this and other characteristics of lenses is given in Chapter 4. For the present, the EFL of a lens is related to its EPD by the slope of the marginal ray (see Section 5.3) or angle u' (see Fig. 3.13), where

$$u' = -\frac{EPD/2}{EFL}. \tag{3.1}$$

The negative sign is required because the ratio of the right side of Eq. (3.1) is positive, whereas the ray angle u' is negative.

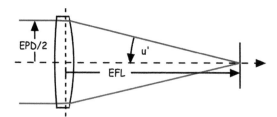

Figure 3.13 EFL of a lens.

The modification of the EFL of a lens can be done using another solve. An **angle solve** can be used to set the EFL of a lens once the EPD of the lens has been specified. To generate a specific EFL, the ray angle emerging from the last surface must have a slope u' that equals the entrance pupil radius (EPD/2) divided by the EFL, as expressed in Eq. (3.1). If the refractive indices on both sides of the final lens surface have been specified, the only quantity that can be varied is the radius of curvature of the last surface. This is determined by solving the paraxial refraction equation [Eq. (1.6)] with the expression for the power of the surface [Eq. (1.7)] inserted:

$$n'u' = nu - y\left(\frac{n' - n}{R}\right). \qquad (3.2)$$

In effect, the radius of curvature of the final surface needed to deliver the ray at the required angle u' can be determined by Eq. (3.2), a calculation that is easily carried out by an angle solve within the program.

As an example, let's determine the radius of curvature R2 needed to change the EFL of the OSlens from 68.98 mm to 50 mm (Fig. 3.14). First, open the **OSlens**. Using Eq. (3.2), the angle u' that we need is $-10/50$, or -0.2. Like the image thickness solve, this solve is done by clicking on the solve box next to R2, which opens a dialog box titled, **Curvature solve on surface 2**. Choose **Marginal Ray Angle** from the drop-down menu and enter "-0.2" for the Angle (Fig. 13.15).

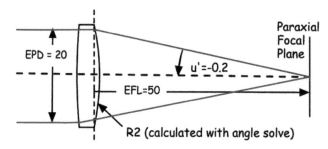

Figure 3.14 Angle solve to set the EFL to 50 mm.

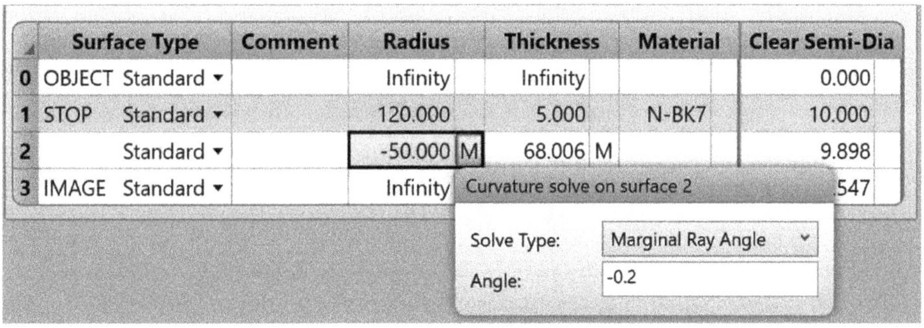

How to Put a Lens in a Computer 55

	Surface Type		Comment	Radius	Thickness	Material	Clear Semi-Dia
0	OBJECT	Standard ▼		Infinity	Infinity		0.000
1	STOP	Standard ▼		120.000	5.000	N-BK7	10.000
2		Standard ▼		-50.000 M	68.006 M		9.898
3	IMAGE	Standard ▼		Infinity			.547

Curvature solve on surface 2
Solve Type: Marginal Ray Angle
Angle: -0.2

Figure 3.15 Applying an angle solve for a 50-mm EFL.

The angle solve changes the radius of curvature of the second surface in the LDE to –32.464 mm. And when we run the FIRST macro again, the results in Table 3.4 show that the EFL is indeed 50 mm and the BFL is now 49.3 mm. Both the R2 and T2 cells in the LDE also show an "M" in the square cell next to them, indicating that each of these quantities was calculated using a solve.

Table 3.4 First-order properties of the OSsinglet (EFL = 50 mm).

```
Infinite Conjugates
    Effective Focal Length      50.0000
    Back Focal Length           49.2902
    Front Focal Length         -47.3762
    F/#                          2.5000
    Image Distance              49.2902
    Lens Length                  5.0000
```

Now that we have a 50-mm EFL lens, we can delete the angle solve by clicking on the solve box next to R2, which reopens the **Curvature solve on surface 2** dialog box. Choose **Fixed** from the drop-down menu and the "M" disappears, indicating that the lens parameters are fixed for a 50-mm lens (Fig. 3.16). Because the angle solve was deleted, we can now change the EPD of the lens without needing to calculate the marginal ray slope angle to maintain the EFL of the lens at 50 mm. To distinguish this from the OSlens, we will call it the OSsinglet.

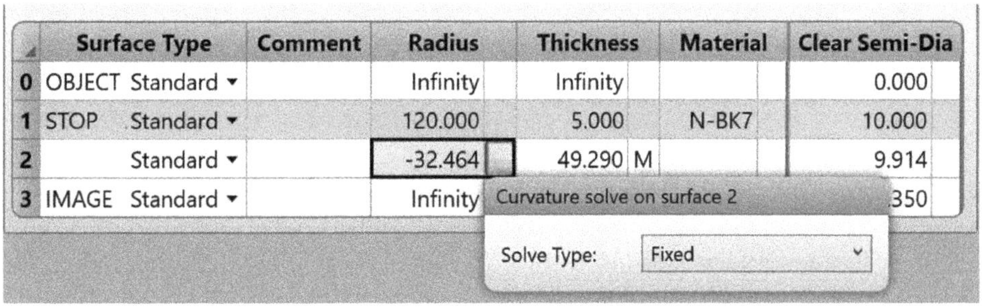

	Surface Type		Comment	Radius	Thickness	Material	Clear Semi-Dia
0	OBJECT	Standard ▼		Infinity	Infinity		0.000
1	STOP	Standard ▼		120.000	5.000	N-BK7	10.000
2		Standard ▼		-32.464	49.290 M		9.914
3	IMAGE	Standard ▼		Infinity			.350

Curvature solve on surface 2
Solve Type: Fixed

Figure 3.16 LDE of the OSsinglet after the angle solve for a 50-mm EFL was deleted.

Change the title of our new lens by entering "OSsinglet" in the Title in the System Explorer (System Explorer > Title/Notes > Title) and save your lens as **OSsinglet**. This lens will be used for a number of examples and demonstrations in the next seven chapters. Additional lenses will be introduced, used, and then saved for further use. A discussion of these saved lens files can be found in the box, "OS Lens File Formats."

OS Lens File Formats

Throughout the text, a small number of lenses will be used to demonstrate various aspects of ray tracing and lens design. To provide a consistent format and eliminate repetitive instruction when one of these lenses is saved, a few rules should be followed:

1. The lens should not have any solves except a paraxial image thickness solve on the final thickness. This provides a lens whose EPD, $f/\#$, and other attributes can be changed while the EFL remains the same.
2. The names of the demonstration lenses will have an "OS" prefix followed by a descriptive label (e.g., OSsinglet). Once you have saved a copy, it should never be saved again. If you want to keep a copy of a lens that you have modified, it should be saved under another descriptive name.

3.6 Entering Mirrors

Not all optical systems rely solely on refraction to bend and focus light. To solve some optical problems, mirror surfaces provide superior answers. For example, it would have been impractical to fabricate and lift into space a refractive lens system that would provide the resolution and light-collecting capability of the Hubble Space Telescope. When entering the parameters for a mirror in OpticStudio, we use the same sign conventions described earlier with a single change in entering the data. In contrast to the case of the single lens, we will begin by entering a spherical mirror.

We begin the mirror entry in exactly the same manner as we did at the beginning of the singlet entry in Section 3.3. Open a new lens (File > New) and establish a 20 mm entrance aperture and 5.876 nm wavelength (*d*-line). The initial screen should look like Fig. 3.3. The initial difference between the singlet lens and the mirror is that because there is only a single surface, no additional surfaces are needed for the mirror as was the case for the lens.

Because this is to be a 50-mm-EFL mirror, the radius of curvature of the mirror should be twice the focal length, as shown in Section 2.5 on mirrors and stated in Eq. (2.5). However, for light traveling in the positive *z* direction, the center of curvature of a positive mirror is located to the left of the mirror surface, so the radius of curvature is negative. The value entered in the Radius cell (R1) is –100 and MIRROR is typed in the G1 cell. The image distance is established using a paraxial image thickness solve on T1. If you forgot how to do this, see Fig. 3.8. It is found to be –50 mm. The negative sign indicates that the paraxial image plane is located to the left of the mirror surface S1.

	Surface Type	Comment	Radius	Thickness	Material	Clear Semi-Dia
0	OBJECT Standard ▼		Infinity	Infinity		0.000
1	Standard ▼		Infinity	60.000		10.000
2	STOP Standard ▼		-100.000	-50.000 M	MIRROR	10.000
3	IMAGE Standard ▼		Infinity	-		0.051

Title (OSmirror); EPD (20 mm); Field (0°); Wavelength (d-line)

Figure 3.17 LDE for a +50-mm-EFL spherical mirror.

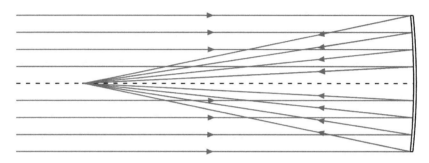

Figure 3.18 Ray trace of a spherical mirror.

The LDE for this mirror is shown in Fig. 3.17, where a 60-mm offset surface was added to show the incoming rays in any cross-sections (see the box "Making Your Lens Cross-Section Look Good" at the beginning of Section 3.4 if you need help remembering how to do this). The ray trace (Setup > Cross-Section) of this simple mirror system is shown in Fig. 3.18.

3.7 Design Forms

Thus far, we have demonstrated how to enter a single lens and a single mirror into OpticStudio. If those were the only two elements needed to solve any optical problem, this would be a very short text. As it is, there are a myriad of **design forms** that are available in a designer's toolbox. Some of them are constructed by adding additional lenses. The simplest are labeled as singlets, doublets, and triplets, for example. But, as we progress through the text, making greater and greater demands on the optics, a change of design forms may be required to achieve the needed performance.

In the following exercises, a number of design forms are introduced to provide practice entering them. Three of them—the rapid rectilinear, the Protar, and the Schwarzschild mirror system—were designed for specific applications. The OStriplet, OSschwarzschild, OSprotar, and OSdoubleGauss will be used later in the text so these lenses should be saved after entry.

Exercises

Because this chapter describes the entry of a lens into the design program, there were no exercises proposed until now. Instead, all of the exercises are given here with the answers following them. You may want to save the lenses in these exercises, but reserve the names of the files, OSlens and OSsinglet, for the design as they were saved in Sections 3.3 and 3.5, respectively. They will be used in future chapters of the text.

Exercise 3.1 Plano-convex lens
Restore the **OSlens** and replace its second surface with a flat surface. Determine the BFL and the paraxial image distance for this lens. If the object distance is changed to 500 mm, where is the paraxial image plane located? Did the BFL change?

Exercise 3.2 Meniscus lens
Replace the second surface of the **OSlens** with a concave surface of 200-mm radius of curvature. Determine the BFL and the paraxial image distance for this lens. If the object distance is changed to 1 m, where is the paraxial image plane located? What is the BFL?

Exercise 3.3 75-mm singlet
Restore the **OSlens** and use an angle solve on the R2 radius to change it to a 75-mm-EFL lens.

Exercise 3.4 Changing the singlet to a doublet
Restore the **OSsinglet** and make the following changes to the design:
(a) Insert a surface between S1 and S2 of the singlet.
(b) Give the new surface, S2, a radius of +200 mm and a thickness of 5 mm with SF1 glass.
(c) Determine the BFL and EFL of the new lens.

The following four exercises provide training in entering a system's prescription. The values for the radii, element thicknesses, airspaces, and materials are given in the figures. Assume that the object is at infinity, and use a thickness solve to find the image location. All lenses have EFLs at or near 50 mm at 587.56 nm (this text's default wavelength). For those lenses with multiple glasses, the glasses are given by the color-coded legend. When you have the correct configuration, save the file for future use, using the names given with each exercise.

In all of the previous exercises, the stop was located at S1. This will not be the case for these next four exercises. The stop will be elsewhere. After the necessary number of surfaces have been entered into the LDE, click on the surface where the stop should be. Then click on the drop-down arrow above

How to Put a Lens in a Computer   59

Surface Type to open the surface properties dialog box for that surface. Check the box in the upper right-hand corner of the **Type** tab, **Make Surface Stop**. The stop should now move from S1 to the correct surface.

Exercise 3.5 Triplet
The EPD is 10 mm, and the stop is located at the first surface of the second element.

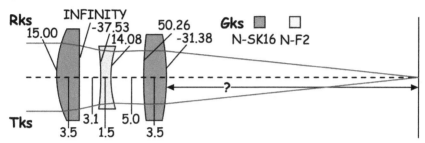

Use a thickness solve to find the image distance. What is the EFL? BFL? Save this lens as **OStriplet**.

Exercise 3.6 Rapid rectilinear
The EPD is 10 mm. The second doublet is the mirror image of the first.

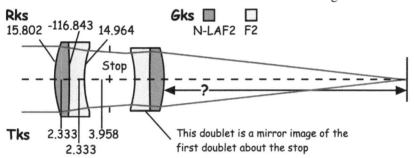

Use a thickness solve to find the image distance. What is the EFL? BFL? Save this lens as **OSrapidrect**.

Exercise 3.7 Protar
The EPD is 5 mm. The stop is between the two doublets, 2.126 mm from the first.

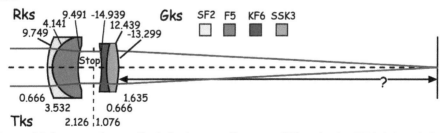

Use a thickness solve to find the image distance. What is the EFL? BFL? Save this lens as **OSprotar**.

Exercise 3.8 Schwarzschild mirrors

The EPD is 10 mm, and the stop is located at the first mirror surface (this is the smaller of the two mirrors and has a radius of 61.803 mm). To get a plot like the one shown with entering rays that begin before the large secondary mirror, a dummy surface (a surface with air on both sides) must be inserted before the first mirror with a thickness greater than the 100-mm mirror separation.

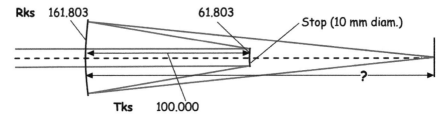

Use a thickness solve to find the image distance. What is the EFL? BFL? Save this lens as **OSschwarzschild**.

Exercise 3.9 Double Gauss

The EPD is 50 mm. The stop is in the middle of the lens.

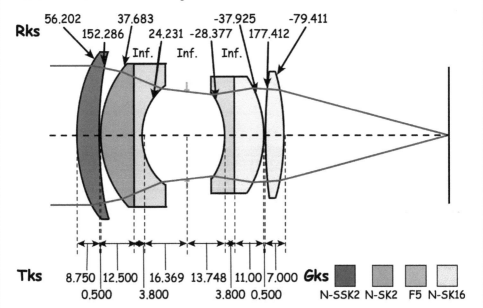

Use a thickness solve to find the image distance. What is the EFL? BFL? Save this lens as **OSdoubleGauss**.

Exercise Answers

Ex. 3.1
Open the **OSlens** and change the R2 radius to flat by entering either "I" or "1E11" in the R2 cell of the LDE. Both the paraxial image distance and the BFL are now 228.9017 mm. If the object distance is changed to 500 mm, the paraxial image distance becomes 430.2295 mm. The BFL is unchanged.

Ex. 3.2
Open the **OSlens** and enter +200 in the cell for R2. Both the paraxial image distance and the BFL are 560.3223 mm. When the object distance is changed to 1 m (1000 mm), the paraxial image distance becomes 1317.3371 mm. The BFL is unchanged.

Ex. 3.3
Open the **OSlens** and put a curvature solve on the R2 radius. Set the Marginal Ray Angle to $-10/75 = -0.133333$.

	Surface Type	Comment	Radius	Thickness	Material	Clear Semi-Dia
0	OBJECT Standard ▾		Infinity	Infinity		0.000
1	STOP Standard ▾		120.000	5.000	N-BK7	10.000
2	Standard ▾		-56.440 M	73.935 M		9.894
3	IMAGE Standard ▾		Infinity	-		0.433

Ex. 3.4
(a) Open the **OSsinglet**, click on the Stop surface, and Ctrl-Insert to insert a surface between S1 and S2 of the singlet.

(b) Enter a radius of +200 mm in R2, 5 mm in T2, and SF1 in G2. The LDE should look like this:

	Surface Type	Comment	Radius	Thickness	Material	Clear Semi-Dia
0	OBJECT Standard ▾		Infinity	Infinity		0.000
1	STOP Standard ▾		120.000	5.000	N-BK7	10.000
2	Standard ▾		200.000	5.000	SF1	9.863
3	Standard ▾		-32.464	36.292 M		9.762
4	IMAGE Standard ▾		Infinity	-		1.812

(c) What are the BFL and EFL of the new lens? FIRST gives an EFL = 37.40 mm and a BFL = 36.29 mm.

Ex. 3.5
OStriplet
    EFL   50.0107
    BFL   38.5152

	Surface Type	Comment	Radius	Thickness	Material	Clear Semi-
0	OBJEC Standard ▼		Infinity	Infinity		0.000
1	Standard ▼		15.000	3.500	N-SK16	5.000
2	Standard ▼		Infinity	3.100		4.647
3 STOP	Standard ▼		-37.530	1.500	N-F2	4.015
4	Standard ▼		14.080	5.000		3.812
5	Standard ▼		50.260	3.500	N-SK16	3.936
6	Standard ▼		-31.380	38.515 M		3.894
7 IMAGE	Standard ▼		Infinity	-		6.381E-03

Ex. 3.6
OSrapidrect
    EFL   50.3304
    BFL   41.5683

	Surface Type	Comment	Radius	Thickness	Material	Clear Semi-Dia
0	OBJEC Standard ▼		Infinity	Infinity		0.000
1	Standard ▼		15.802	2.333	N-LAF2	5.000
2	Standard ▼		-116.843	2.333	F2	4.800
3	Standard ▼		14.964	3.958		4.325
4 STOP	Standard ▼		Infinity	3.958		4.107
5	Standard ▼		-14.964	2.333	F2	3.881
6	Standard ▼		116.843	2.333	N-LAF2	4.062
7	Standard ▼		-15.802	41.568 M		4.157
8 IMAGE	Standard ▼		Infinity	-		0.392

Ex. 3.7
OSprotar
    EFL   50.0056
    BFL   43.8037

	Surface Type	Comment	Radius	Thickness	Material	Clear Semi-Dia
0	OBJEC Standard ▼		Infinity	Infinity		0.000
1	Standard ▼		9.749	0.666	SF2	2.500
2	Standard ▼		4.141	3.532	F5	2.387
3	Standard ▼		9.491	2.126		2.122
4 STOP	Standard ▼		Infinity	1.076		2.116
5	Standard ▼		-14.939	0.666	KF6	2.113
6	Standard ▼		12.439	1.635	SSK3	2.160
7	Standard ▼		-13.299	43.804 M		2.202
8 IMAGE	Standard ▼		Infinity	-		9.383E-04

Ex. 3.8
OSschwarzschild
    EFL   49.9996
    BFL   211.8026

		Surface Type	Comment	Radius	Thickness	Material	Clear Semi-Dia
0	OBJEC	Standard		Infinity	Infinity		0.000
1		Standard		Infinity	150.000		5.000
2	STOP	Standard		61.803	-100.000	MIRROR	5.000
3		Standard		161.803	211.803 M	MIRROR	21.148
4	IMAGE	Standard		Infinity	-		1.190E-05

Note: To get OpticStudio to plot rays before the convex mirror, a dummy surface is inserted before the mirror (similar to an offset surface). This is why the first surface has no material after it.

Ex. 3.9
OSdoubleGauss
    EFL   100.0038
    BFL   61.4875

		Surface Type	Comment	Radius	Thickness	Material	Clear Semi-Dia
0	OBJEC	Standard		Infinity	Infinity		0.000
1		Standard		56.202	8.750	N-SSK2	25.324
2		Standard		152.286	0.500		24.433
3		Standard		37.683	12.500	N-SK2	23.101
4		Standard		Infinity	3.800	F5	21.249
5		Standard		24.231	16.369		16.929
6	STOP	Standard		Infinity	13.748		15.853
7		Standard		-28.377	3.800	F5	14.764
8		Standard		Infinity	11.000	N-SK16	16.021
9		Standard		-37.925	0.500		17.111
10		Standard		177.412	7.000	N-SK16	16.815
11		Standard		-79.411	61.488 M		16.500
12	IMAGE	Standard		Infinity	-		0.088

Chapter 4
To First Order...

In Chapter 3, we did a quick check to see if the initial entry of our OSlens was correct using the **FIRST** macro. The first-order listing for the OSlens is shown again in Table 4.1. The first three numbers may be puzzling because there are three separate focal lengths (effective focal length, back focal length, and front focal length) listed, and one of them is negative. Furthermore, none of these values exactly matches the thin lens focal length value of 68.294 mm calculated in Section 3.4. So, what is the correct value for the focal length?

Table 4.1 First-order properties of the OSlens using the FIRST macro.

```
Infinite Conjugates
   Effective Focal Length    68.9849
   Back Focal Length         68.0055
   Front Focal Length       -66.6344
   F/#                        3.4492
   Image Distance            68.0055
   Lens Length                5.0000
   Paraxial Image
      Height                  0.0000
      Angle                   0.0000
   Entrance Pupil
      Diameter               20.0000
      Location                0.0000
   Exit Pupil
      Diameter               20.7055
      Thickness              -3.4127
```

4.1 Principal Surfaces and Planes

To help explain why several focal lengths are computed and used for a single lens, we begin with an exaggerated lens example. Figure 4.1 shows a thick lens in air with a very short radius of curvature on its back surface focusing a fan of parallel axial rays. The solid blue lines show the standard trace of the rays through the lens. Paraxial rays, those close to the axis, focus at the **back focal point F'**. A plane perpendicular to the axis at this point is defined as the **back focal plane (BFP)**.

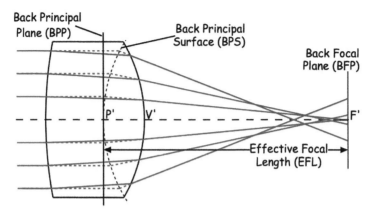

Figure 4.1 Demonstration of the definitions of the back principal surface and plane. The solid blue lines are the ray traces, and the dashed blue lines are the extensions whose intersections locate the principal surface.

If the entering parallel rays are extended and their exiting counterparts are extended backward as dashed blue lines, their intersections form a surface called the **back principal surface (BPS)**. A plane tangent to the BPS at the point P′ on the optical axis is defined as the **back principal plane (BPP)**. The distance from the BPP, P′, to the back focal point F′ is the **effective focal length (EFL)**. (This is true for lenses in air where $n = n' = 1.0$; a more general treatment will be given in Section 4.4.) It appears as if a thin lens located at P′ had focused the rays. This becomes useful, as we shall see, when modeling any lens, no matter how complicated its design. The EFL is also the first focal length listed in our first-order tables created by the **FIRST** macro.

Now let's go back to the OSlens so we can understand how the three different focal lengths in the first-order listing are defined. First, we need to restore the lens that we saved at the end of Section 3.3. The Open icon on the File tab (File > Open) gives you two ways of retrieving the lens. Clicking the icon itself opens the last folder you used to save a lens. If you click the arrow below the icon, a list of recent lenses will be displayed. Either way, find and select **OSlens**. If you can't find it (or didn't save it), you can start a new lens and type in the lens prescription and system data from Figs. 3.7 and 3.8 in Chapter 3.

Ctrl-L opens a Layout window showing the lens cross-section with a default set of seven rays. Decrease the number of rays in the fan to five by selecting the drop-down arrow next to Settings in the upper left corner of the Layout window and then changing the Number of Rays from 7 to 5. The result is shown in Fig. 4.2. Note that the marginal rays (axial rays at the edge of the lens) do not focus in the paraxial image plane due to spherical aberration (Chapter 6).

The paraxial rays focus at the back focal point F′ and determine the location of the back focal plane. The distance between the back vertex of the lens V′ and the back focal point F′ is 68.0055 mm, as shown in the T2 cell of the LDE (Fig. 3.8). This distance is the **back focal length (BFL)** of the lens and is the second focal length value listed in the **FIRST** output in Table 4.1.

To First Order... 67

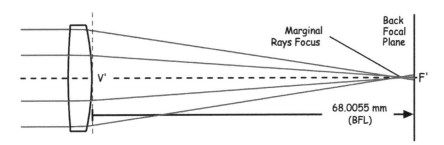

Figure 4.2 Ray fan focused by the OSlens. The distance from the lens vertex V' to the back focal point F' is the BFL.

We can use the two focal lengths (EFL and BFL) in the first-order data table (Table 4.1) to determine where the BPP of the OSlens is located. The positive directed distance P'F' (from P' to the back focal point F') equals the EFL. The location of the BPP can then be determined by calculating the difference between the BFL (68.0055 mm) and the EFL (68.9849 mm). Specifically, the distance from the vertex V' to the BPP is V'P'. This locates the BPP at –0.9794 mm from the back surface vertex (Fig. 4.3). The segment V'P' should also be treated as a directed distance rather than a length.

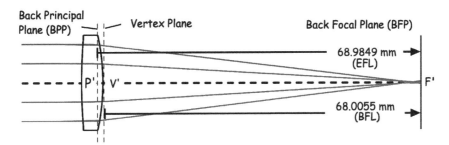

Figure 4.3 Location of the BPP at P'.

The first-order table also lists a third focal length, the **front focal length (FFL)**, with a value of –66.6344 mm. The FFL is the distance from the first lens vertex V to the **front focal point F**. The negative sign indicates that F is to the left of the lens. If a point source is placed at F, the lens will collimate the rays in image space. Note: When using the **FIRST** macro with a lens with an offset surface (to draw the incoming rays like those shown in Figs. 4.2 and 4.3), the value of the FFL may be confusing at first as it is reduced by the thickness of the offset surface since the macro calculates the FFL from the vertex of the first *surface* in the lens system and *not* the first *lens* vertex.

Exercise 4.1 Reversed lens
Open the **OSlens** and reverse the radii (R1 = 50 and R2 = –120). Find the BFL for this lens. It should equal the negative of the FFL from the original OSlens. Why?

Now that we have a value for the FFL, we can find the location of the **front principal plane (FPP)** with respect to the vertex V of the first surface of the lens by adding the EFL and FFL (the two values are added because of our sign convention). For the OSlens, this is (68.9849) + (–66.6344) mm, or 2.3505 mm (VP) to the right of the first surface.

The separation between the principal planes PP′, another directed distance, is called the **hiatus** H of the lens. The term is used to describe a physical gap or a pause between two events. Given the thickness of the lens VV′, the hiatus H for OSlens can be determined using our calculated values for VP and V′P′, where

$$H = \text{PP}' = \text{VV}' - \text{VP} + \text{V}'\text{P}' = 5.000 - 2.3505 - 0.9794 = 1.6701 \text{ mm}.$$

The principal plane locations and hiatus for the OSlens are shown in Fig. 4.4.

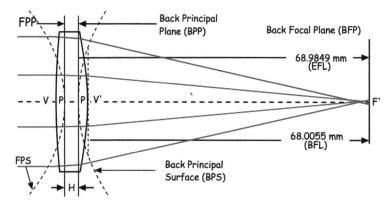

Figure 4.4 Principal planes and hiatus of the OSlens.

The separation between the principal planes PP′ can also be determined directly from the first-order data because the distance between the focal points can be calculated in two ways. One way is to sum the directed distances, FFL, BFL, and Lens Length (LL). (Lens Length is the distance from the first surface of the lens to the last surface of the lens before the image plane; for this singlet, the Lens Length is simply the lens thickness T1.) Another way is to add the hiatus to twice the EFL:

$$-\text{FFL} + \text{LL} + \text{BFL} = \text{EFL} + H + \text{EFL}. \tag{4.1}$$

To compute the hiatus, solve this equation for H and insert the values from the **FIRST** output listing. Note that the negative sign in the equation arises because the value for the FFL is negative in keeping with the sign conventions for distances measured relative to the lens surfaces. The basis for Eq. (4.1) is shown in Fig. 4.5, where the quantities on the left side of the equation are shown above the optical axis, and those on the right side of the equation are shown below the optical axis (vertical and horizontal axes have different scales).

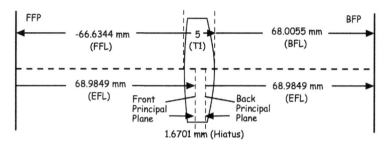

Figure 4.5 Distances and planes for the OSlens (not to scale).

4.2 What Does This Get You?

This approach reduces the individual surfaces, thicknesses, and materials of a complex lens to a set of four planes. These four planes (two principal planes plus two focal planes) can now replace and greatly simplify a complex lens consisting of many surfaces. For example, for lens systems with the same object- and image-space index of refraction (a system in air, for example), a ray from an off-axis point that intercepts the optical axis ($y = 0$) at the FPP at an angle v will be translated across the distance of the hiatus and emerge from the BPP at the same height ($y' = y = 0$) and same angle v to the optical axis, as shown in Fig. 4.6. These rays are the thick lens equivalent of the center rays used for ray sketching in Chapter 2. The planes also simplify the depiction of an axial ray through the lens. An axial ray from an object point O directed at the FPP with an angle u and height y will emerge from the BPP at the same height ($y' = y$). The emerging angle u' obeys the law of magnification: $m = u'/u$.

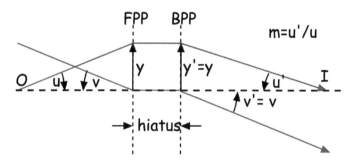

Figure 4.6 Ray tracing using principal planes.

Of course, the example we have used consists of a single glass bounded by two surfaces. A more convincing example of the approach can be demonstrated with a classic complex lens, the double Gauss. This lens has a total of six elements, a doublet and singlet on each side of the stop, with similar materials and geometries, resulting in a lens that is nearly symmetric about the aperture stop (Fig. 4.7) This is the same lens that was used to illustrate ray tracing (Fig. 1.2) using an optical design program in Chapter 1. There are several similar double Gauss lenses available in the OpticStudio® design templates (Libraries > Design Templates).

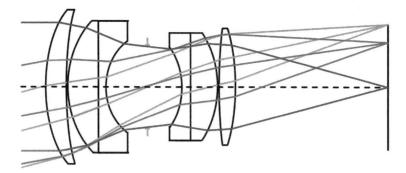

Figure 4.7 A more complex lens, the double Gauss.

Restore the **OSdoubleGauss** from Exercise 3.9. The first-order values for this lens are shown in Table 4.2. From these values we can determine the locations of the focal planes and principal planes for the lens and plot them atop the components that make up the lens, as shown in Fig. 4.8.

Table 4.2 First-order properties of the OSdoubleGauss.

```
Infinite Conjugates
    Effective Focal Length      100.0038
    Back Focal Length            61.4875
    Front Focal Length          -29.3208
    F/#                           2.0001
    Image Distance               61.4875
    Lens Length                  77.9674
```

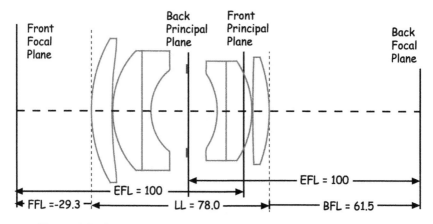

Figure 4.8 Distances and focal planes for the OSdoubleGauss lens.

Exercise 4.2 Hiatus
What is the hiatus of the **OSdoubleGauss** lens? Note that the order of the principal planes is reversed.

In Fig. 4.9, two axial rays are traced from an object plane O located a finite distance from the lens to the image plane I. A magnified view of the rays and the principal planes overlays the full ray trace. The actual ray traces through the lens elements are shown in blue, whereas the extensions of the rays traced to the principal planes are shown as dashed black lines. The individual components of the lens are shown in light gray. Once the principal planes are located, they can be used to trace rays in a lens without knowing its internal construction. Remember that a ray from an object point O directed at the FPP with an angle u will emerge at the same height y at which it entered from the BPP. In this case, the BPP is to the left of the FPP, but the procedure still works. A ray is directed at the FPP from the axial object point in the plane O. The ray height at which it hits the FPP is then translated across the hiatus parallel to the optical axis, and then it is directed at the axis point in the image plane I.

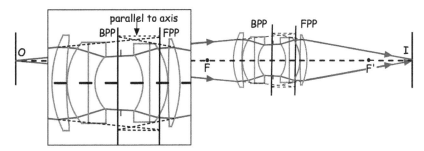

Figure 4.9 Ray traces through the OSdoubleGauss lens with a finite object distance and their simplified ray paths using the principal planes. A magnified view of the lens overlays the full ray trace.

4.3 Cardinal Points of Lenses and Mirrors

The intersections of the four planes (two focal planes and two principal planes) with the optical axis determine four points, which are four of the six **cardinal points** of a lens. The other two cardinal points are called **nodal points** (**N** and **N'**). Nodal points are the two points on the optical axis of a lens about which the lens can be rotated, and a ray passing through the nodal point will be unaffected by the lens rotation (Fig. 4.10). To be more concrete, when you look through a lens that is rotated about one of its two nodal points, the image remains steady. At any other point, the image moves as the lens rotated.

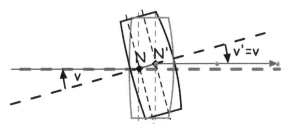

Figure 4.10 When a lens is rotated about its front nodal point, a ray entering at an angle v to the optical axis will exit at the same angle.

If the refractive indices in object space and image space are the same (for most lens systems, this is usually the case), then the nodal points, N and N′, lie on top of the principal points, P and P′, respectively. This means that a ray entering the lens at an angle v to the optical axis and intersecting the point P will emerge from the point P′ with the same angle $v' = v$ to the optical axis. The entering and exiting rays are thus parallel to each other. (These are the points where the lens acts like a parallel slab of glass; see Exercise 1.1.) Located along the optical axis of the lens, the six cardinal points represent a simplified version of an optical system that is useful to a designer (Fig. 4.11).

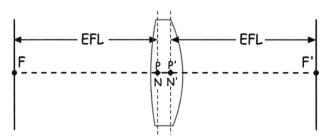

Figure 4.11 The cardinal points for the OSlens.

For lens systems in air, one might wonder, "Why bother to define nodal points if they are just going to overlap the principal points?" In the lab, you can easily measure the BFL of a lens by sending a collimated laser beam into it and measuring the distance from the back vertex to the focused spot on a screen. But what if you wanted to experimentally measure the EFL of a lens? How can you find the principal plane? On a test bench, it would be difficult to "trace" rays forward and backward and find their intersection point like we did in Fig. 4.1. However, there are instruments (e.g., a nodal slide) that can determine the location of the nodal points of a lens. The nodal slide lets you rotate a lens around an axis perpendicular to its optical axis at different points along the optical axis. When you reach the nodal point, the image on a screen does not move as the lens is rotated. Because the nodal points and principal points of the lens coincide, the distance from the rotation point to the screen is the EFL.

This geometry is useful for constructing a panorama from a set of images taken while rotating the camera lens about its nodal point. However, this is necessary only if there are nearby objects in the panorama, such as room interiors and other confined spaces. For most panoramas, such as landscapes, such care is not required. If the photographer stands facing the middle of the scene and twists at the waist while taking a series of pictures, a decent panorama can be captured and assembled from them without needing to pay attention to the nodal points of the camera lens.

> **Exercise 4.3 OSsinglet**
> Find the six cardinal points of the **OSsinglet** using the first-order data. Sketch the locations of the principal planes and focal planes relative to the front and back vertices (V and V′). Determine the hiatus of the lens.

The arrangement of cardinal points for spherical mirrors is quite different from that for lenses (see Fig. 4.12). Because the sign of the index changes upon reflection ($n' = -1$), the refractive indices of object and image space differ, causing the nodal points to shift off the principal points. The principal planes P and P' are located at the mirror because it is a single surface, and it is the only location where any ray bending can occur (the back focal length then equals the EFL). As discussed in Chapter 2, the focal length of a mirror is half its radius of curvature R, so its front and rear focal points F and F', respectively, are located on top of each other at half the distance to the center of curvature ($R/2$). A ray through the center of curvature C is always perpendicular to the mirror surface, so rotation about C will always return the ray at the same angle. Therefore, both nodal points N and N' are located at the center of curvature C.

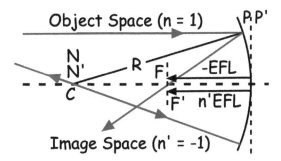

Figure 4.12 Cardinal points for a spherical mirror.

Note: Although any calculation for mirror systems must employ the sign changes due to reflection to arrive at the correct values, OpticStudio always lists the refractive indices as positive.

4.4 Immersed Systems

Although most optical systems are designed to operate in air, occasionally you may encounter an immersed system where either or both of the object and image spaces have a refractive index that does not equal one. (The earlier spherical mirror example is a special case of an immersed system.) This section is devoted to developing the equations that determine the locations of the principal and nodal points of an immersed lens. As such, it may be reserved for study at a later time, when immersed designs are being considered.

One common example of an immersed system is the human eye, where the object-space refractive index is that of air and the image-space refractive index is close to that of water ($n' \approx 1.333$). For such systems, the relationships between cardinal points are more complex, and the principal points and nodal points may no longer coincide.

We will first examine a simple immersed system, a single spherical surface at a water–air interface. In Chapter 1, the paraxial refraction equation was given as

$$\phi = \frac{n' - n}{R} = (n' - n)c = \frac{1}{\text{EFL}}, \qquad (1.7)$$

where an additional relation had been added: the reciprocal of the power ϕ is the EFL [as defined in Eq. (2.4)]. The reciprocal of the power of the surface or its EFL should be the same whether the light is traced from the water into air or air into water. We can use this fact to help us to understand the distances between the cardinal points for immersed systems.

Figure 4.13 shows the simple geometry of a single ray traced from air into water through a spherical interface whose radius of curvature is +50 mm for an EPD of 20 mm. The principal points P and P' are naturally located at the surface vertex of this simple system because this is the only location where the ray is bent (similar to the previous mirror).

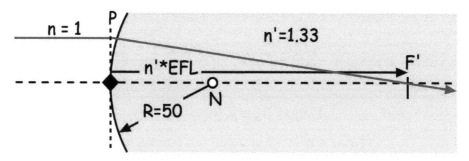

Figure 4.13 Geometry for a ray trace of an air–water interface.

Open a new lens (File > New). Give the lens a 20-mm-diameter entrance aperture and change the operating wavelength to the *d*-line. At S1, 50 mm is entered for R1. There is no "Water" in the current Schott glass catalog, but when WATER is entered in the material column for G1, a notice pops up (Fig. 4.14) telling you that. It notes that there is a miscellaneous catalog (MISC) that does contain the data for water. Clicking Yes adds the catalog, enters the material data for water, and closes the notice.

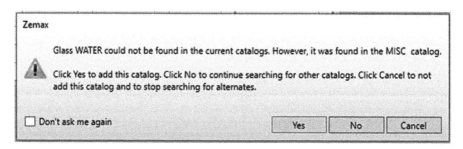

Figure 4.14 OpticStudio notice about the miscellaneous catalog (MISC).

There is no thickness value assigned to define a distance beyond S1. If you plot this "lens," the image surface will be on the first surface. To locate the actual image surface, a paraxial image thickness solve is applied to T1, as shown in Fig. 3.8. Using this solve, the rays through the air–water interface are traced to the image surface 200.130 mm from the vertex. The resulting LDE is shown in Fig. 4.15 and a ray trace in Fig. 4.16.

	Surface Type	Comment	Radius	Thickness	Material	Clear Semi-Di:
0	OBJECT Standard ▾		Infinity	Infinity		0.000
1	STOP Standard ▾		50.000	200.130 M	WATER	10.000
2	IMAGE Standard ▾		Infinity	-		0.115

Title (OSairwater); EPD (20 mm); Field (0°); Wavelength (d-line)

Figure 4.15 LDE for an air–water interface.

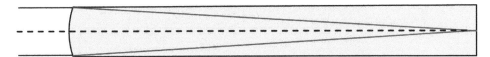

Figure 4.16 Ray trace through an air–water interface.

The first-order listing from the **FIRST** macro (Table 4.3) indicates that the EFL is 150.1303 mm. This may be surprising as it does not match the 200.13-mm image distance value (T1) for an object at infinity, but this can be explained by the fact that we have so far dealt with specific cases of lenses in air. The more general relationship that covers all lenses (including immersed lenses) is that the distance between the principal plane of a lens and its corresponding focal point equals the *product* of the refractive index of that space and the EFL of the lens (i.e., $FP = -n \cdot EFL$ and $P'F' = n' \cdot EFL$). Because most systems are "immersed" in air, the distances P'F' and PF equal the EFL and –EFL, respectively. In this air–water surface case, the refractive index at the d wavelength is 1.333044, so the rear focal point is located $1.333044 \cdot 150.1303 = 200.1303$ mm from the BPP.

Table 4.3 First-order values for the air–water interface.

```
Infinite Conjugates
    Effective Focal Length      150.1303
    Back Focal Length           200.1303
    Front Focal Length         -150.1303
    F/#                           7.5065
    Image Distance              200.1303
    Lens Length                   0.0000
```

4.4.1 Nodal points for immersed systems

If the front and back indices of an optical system are different, then the nodal points and principal points are no longer coincident, and there will be six distinct cardinal points instead of four. This was first discovered by Johann Listing, a student of Gauss, who was interested in the optical properties of the human eye.

The simplest way to illustrate this is to examine our current example. When you look at Fig. 4.13, it is obvious that rotating the system about the principal point at the vertex of the surface would not permit a ray to stay stationary during the rotation. The obvious location of the nodal point N is at the center of curvature of the surface, indicating that the principal point and nodal point are not coincident in this simple immersed system.

For a general lens system, an equation for the nodal point shifts can be derived from the concepts we have presented earlier. The basic geometry of the system is shown in Fig. 4.17 and consists of two principal points P and P′ separated by the hiatus H and two focal points F and F′ that are not equidistant from their respective principal planes. F is a distance $-n \cdot$ EFL from P, and F′ is $n' \cdot$ EFL from P′. If a ray is launched from the front focal point F to the front principal plane P, it enters the hiaitus parallel to the optical axis with a height h' from the optical axis, according to the definition of principal planes.

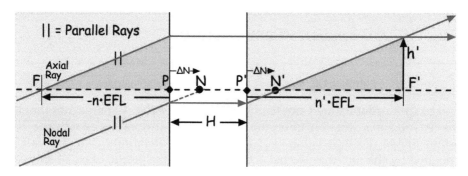

Figure 4.17 Cardinal points of a lens system immersed in two different media.

Starting at a point in the back focal plane with a height h' from the optical axis, a second ray is traced backwards in image space parallel to the first ray in object space. Because this ray crosses the initial ray at the same point in the back focal plane, its corresponding object point must also be located at infinity, which means that the continuation of the ray in object space must also be parallel to the first ray there. Because the slope of this ray is the same on both sides of the lens, it must be a nodal ray, and the intersections of these ray segments with the optical axis mark the locations of the nodal points N and N′, neither of which are coincident with the principal points. From the geometry shown in Fig. 4.17, it easy to show that the nodal point shifts (PN and P′N′) are equal. They are labeled as a signed quantity ΔN, which can be calculated from the geometry using the fact that the dark shaded areas in Fig. 4.17 are identical triangles. Therefore, both of the sides adjacent to the optical axis are equal:

$$n \cdot \text{EFL} = \text{FP} = \text{N}'\text{F}' \text{ (identical triangles)}.$$

The distance from P′ to F′ (P′F′) equals $n' \cdot$ EFL. This equals the side of the triangle N′F′ plus the nodal point shift ΔN:

$$n' \cdot \text{EFL} = \text{N}'\text{F}' + \Delta N.$$

But $N'F' = n \cdot EFL$, so $n' \cdot EFL = n \cdot EFL + \Delta N$. Solving for ΔN, we arrive at the equation for the nodal point shift, whose sign indicates the direction of the shift:

$$\Delta N = n' \cdot EFL - n \cdot EFL = (n' - n)EFL. \tag{4.2}$$

Some simple geometry will show that the ΔN values are equal, so that NN' equals the hiatus PP'. Several important conclusions can be drawn from the figure and from Eq. (4.2):

1. There is no nodal point shift if the refractive index is the same in both spaces.
2. The shift ΔN is in the direction of the higher refractive index.
3. Because both nodal points are shifted by the same amount, the nodal points are separated by the hiatus H, as are the principal points.

From Eq. (1.7), we see that for a single surface $EFL = R/(n' - n)$. If we insert this into Eq. (4.2), we find that the displacement of the nodal point from the principal point ΔN equals R for the air–water interface problem. When we add the cardinal points to the air–water interface figure (Fig. 4.13), as shown in Fig. 4.18, you can see that the nodal points N and N' are at the center of curvature of the surface, as expected.

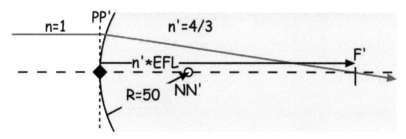

Figure 4.18 Air–water interface showing the principal, nodal, and focal points.

4.4.2 The human eye

Certainly, the most common immersed optical systems are our eyes. In Fig. 4.19, the optically important components are labeled on a model of the human eye. The outer surface of the eye (the cornea) is very thin (about 0.5 mm) and encapsulates the vitreous humor. The cornea provides the major contribution to the eye's optical power. Behind it is the crystalline lens whose power changes to focus light onto the retina as objects at different distances are observed. Behind the lens (and filling the rest of the eyeball) is the aqueous humor. Both humors are liquids with properties very close to water. Besides establishing the optical properties of your eye, the humors also maintain the rigidity of the eye.

Everyone's eyes are different, and they change as we age. There is no single optical prescription for the eye, but there are a number of models that represent a fairly realistic version of the eye. The Emsley model (Fig. 4.19) is one of the simplest models in that it ignores the thickness of the cornea. It also models liquids in the eye (humors) using water.

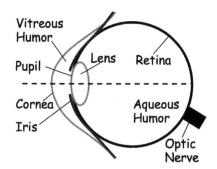

Figure 4.19 Cross-section of the human eye using the Emsley model.

Open up a new lens (File > New) and enter the prescription for the model eye from Figure 4.20. The stop is located at the iris at the front surface of the lens (S2). Change the EPD to 6 mm and the operating wavelength to the *d*-line. Because the material in the eye lens is not a catalog glass, OpticStudio permits the user to enter a model glass. For the present, the *V*-number is a property of glass that, along with its refractive index, identifies a specific glass. This will be covered in Chapter 9 on chromatic aberration. An extensive discussion of model glasses can be found in Section 12.6.1. In this simple model of the eye, the lens is assigned a refractive index of 1.415 with a *V*-number of 47. This is entered in the Material column of S2 by clicking on the box to the right of the cell and from the Solve Type menu by selecting Model, then entering 1.415 in the Index Nd space and 47.0 in the Abbe Vd space. Add a paraxial image thickness solve to the T3 thickness to locate the image surface at the paraxial image plane. The resulting LDE is shown in Fig. 4.21.

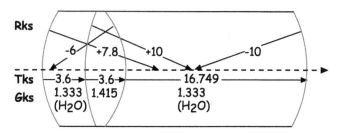

Figure 4.20 Simple model of the human eye for computer analysis.

	Surface Type	Comment	Radius	Thickness	Material	Clear Semi-Dia
0	OBJEC Standard ▾		Infinity	Infinity		0.000
1	Standard ▾	Cornea	7.800	3.600	WATER	3.150
2 STOP	Standard ▾	Lens	10.000	3.600	1.42,47.0 M	2.657
3	Standard ▾	Aqu. Humor	-6.000	16.749 M	WATER	2.466
4 IMAGE	Standard ▾		-10.000	-		0.241

Title (OSeye); EPD (6 mm); Field (0°); Wavelength (d-line)

Figure 4.21 LDE for a simple model of the human eye.

A cross-section of the eye model is shown in Fig. 4.22. In this cross-section, a 5-mm-thick (hidden) offset surface and a 10% clear aperture margin were added to the lens to make the cross-section "look" better. If you've forgotten how to do this, see the box in Chapter 3 on making your lens cross-sections look good. To draw the curved retinal image surface, a fixed 3.3-mm semi-aperture was added to the image surface in the LDE by clicking on the solve box next to the Clear Semi-Dia and changing it from **Automatic** to **Fixed** and entering a value of 3.3. The modifications to this design are shown in its LDE in Fig. 4.23.

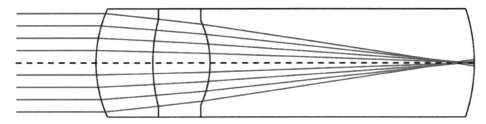

Figure 4.22 Lens cross-section for a simple model of the human eye.

	Surface	Type	Comment	Radius	Thickness	Material	Clear Semi-Dia
0	OBJECT	Standard ▼		Infinity	Infinity		0.000
1		Standard ▼	Offset	Infinity	5.000		3.300
2		Standard ▼	Cornea	7.800	3.600	WATER	3.300
3	STOP	Standard ▼	Lens	10.000	3.600	1.42,47.0 M	2.656
4		Standard ▼	Aqu. Humor	-6.000	16.749 M	WATER	2.580
5	IMAGE	Standard ▼		-10.000	-		3.300 U

Title (OSeye); EPD (6 mm); Field (0°); Wavelength (d-line)

Figure 4.23 Modified LDE for the OSeye.

The first-order values for the modified model eye shown in Fig. 4.22 are on the left side of Table 4.4. In this model, we added an offset surface and curved the image plane to represent a curved retina. Unfortunately, both the offset surface and the curvature of the immersed image plane interfere with the calculation of the FFL and cause an error in the first-order data table. This can be corrected by changing the offset thickness to 0 and setting the radius of the image surface to infinity before running the **FIRST** macro. The corrected values are shown on the right side of Table 4.4.

Armed with the values from the second set of numbers, we can calculate the cardinal points of the eye. From Table 4.4, the EFL of this eye model is 16.5845 mm, the BFL is 16.7492 mm, and its front focal point F is –15.0420 mm (to the left) of the vertex V of the eye, as shown in Fig. 4.24.

Table 4.4 Comparison of the first-order values with both a 5-mm offset surface and a curved image surface and then without.

Infinite Conjugates	5 mm Offset & Curved Image	No Offset & Flat Image
Effective Focal Length	16.5845	16.5845
Back Focal Length	16.7492	16.7492
Front Focal Length	-0.8818	-15.0420
F/#	2.7641	2.7641
Image Distance	16.7492	16.7492
Lens Length	12.2000	7.2000

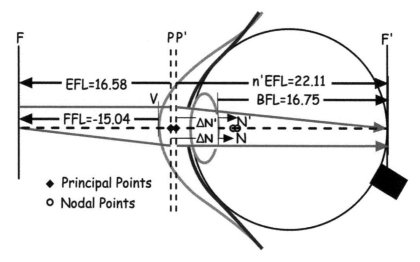

Figure 4.24 Eye model showing the principal, nodal, and focal points.

Substituting these values into the formula for the hiatus [Eq. (4.1)], modified to account for the refractive index of object space and image space, n and n', respectively, we obtain

$$-\text{FFL} + \text{LL} + \text{BFL} = n \cdot \text{EFL} + H + n' \cdot \text{EFL}, \quad (4.3)$$

$$15.04 + 7.2 + 16.75 = 16.58 + H + 1.333 \cdot 16.58.$$

Solving for the hiatus H, we find that the principal planes are only 0.3 mm (300 μm) apart.

The principal planes of the eye are found using the same numbers from Table 4.4. The distance VP is located at EFL + FFL = 16.58 − 15.04 = 1.54 mm from the front vertex. P′ will be located at a distance P′F′ from the retina. That is, the BPP is $n' \cdot \text{EFL} = 1.333 \cdot 16.58 = 22.11$ mm to the left of the retina. Because the length of the eye from the front vertex V to the focal point on the retina F′ is 23.95, P′ will be 23.95 − 22.11 = 1.84 from the front vertex V. This separation between the principal planes (1.84 − 1.54 = 0.3 mm) agrees with the hiatus calculation using Eq. (4.3).

To First Order... 81

Finally, the shift of the nodal points away from the principal points because of immersion can be computed from Eq. (4.2):

$$\Delta N = (n' - n) \bullet \text{EFL} = (1.333 - 1) \bullet 16.58 = +5.52 \text{ mm}.$$

Thus, the nodal points are then located +5.52 mm beyond the principal planes. N is 7.06 mm and N' is 7.36 mm from the front vertex V of the lens (Fig. 4.22). The location of the the nodal points toward the center of the eyeball means that as the eye rotates in its socket, the movement of the image on the retina will be less than if the nodal points were at the eye lens.

4.5 A Concluding Remark

This chapter introduced first-order imaging in a number of basic systems. We showed how any system could be simplified by using its cardinal points (focal points, principal points, and nodal points). Now, to make the modeling of these systems more realistic, the collection of light and the size of objects and lens openings will be addressed. Instead of single rays, bundles of rays will be used to analyze the performance of a real lens.

Exercise Answers

Ex. 4.1

Open the **OSlens** and make the radii changes in the LDE. The BFL for this lens (T2 = 66.634) is the same as the FFL of the OSlens given in Table 4.1 (−66.634 mm), with the exception of a sign change. The BFP and FFP also flip when the lens is reversed.

	Surface	Type	Comment	Radius	Thickness	Material	Clear Semi-Dia
0	OBJECT	Standard ▾		Infinity	Infinity		0.000
1	STOP	Standard ▾		50.000	5.000	N-BK7	10.000
2		Standard ▾		−120.000	66.634 M		9.751
3	IMAGE	Standard ▾		Infinity	-		0.244

Ex. 4.2

Hiatus = −FFL + LL + BFL − 2•EFL.
From Table 4.2:

    Effective Focal Length (EFL)   100.0038
    Back Focal Length (BFL)   61.4875
    Front Focal Length (FFL)   −29.3208
    Lens Length (LL)   77.9674

29.3208 + 77.9674 + 61.4875 − (2 • 100.0038) = −31.2319 mm.

Ex. 4.3
Open the **OSsinglet** and generate a first-order data table (**FIRST**).

Infinite Conjugates	
Effective Focal Length	50.0000
Back Focal Length	49.2902
Front Focal Length	−47.3762
F/#	2.5000
Image Distance	49.2902
Lens Length	5.0000

F and F′ can be located with the FFL (−47.38 mm) and BFL (49.29 mm), respectively.

To find P:
  VP = EFL + FFL = 50 − 47.3762 = 2.6238 mm from the front vertex V.

To find P′:
  V′P′ = −(EFL − BFL) = −(50 − 49.2902) = −0.7098 mm from back vertex V′.

Because the lens is in air, the nodal points N and N′ are located at the principal points P and P′.

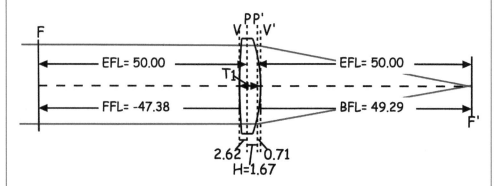

From Eq. (4.1), H = BFL + T1 − FFL − 2EFL;
$H = 49.2902 + 5.0000 + 47.3762 − 2 \cdot 50 = 1.6664$.

Chapter 5
Stops and Pupils and Windows, Oh My!

After a lens has been entered into OpticStudio® and the location of the image plane for an axial object point has been found using a paraxial image thickness solve, we need to determine how the lens performs at off-axis object points. We begin with a small number of points in the object plane, called **fields**, that are used to represent extended objects. Then we examine the passage of light through the lens by introducing the concepts of the aperture stop and field stop of the system, as well as its pupils and windows. These definitions may seem to be a needless complication ("C'mon, you've found the image, haven't you?"). But the purpose of a lens is to transfer radiation (light) from an object to an image. If the image is too dim or the object is not completely imaged, the design is a failure no matter how well resolved the image might be on axis. Students encountering the stops, pupils, and windows of a lens for the first time may find them bewildering. Thus, the title of this chapter. The objective for this chapter is to present stops, pupils, and windows as clearly as possible so that you feel comfortable using them in the construction and evaluation of optical systems.

5.1 Fields

Thus far, the object to be imaged has been a single point on the optical axis of the lens. We could add many points, covering the entire scene, so that the object would be a 2D picture. This type of object is available in OpticStudio, but the approach requires significant computing power and time. By taking advantage of the symmetry of the lens, we could determine the quality of the image by tracing rays from many points along a line in the object plane. But even that may be too complicated. To start, we simplify the evaluation by using three object points (see Fig. 5.1). One is the on-axis point; a second is at the edge of the picture; and the third is a point somewhere between the other two, shown in the figure as light dots on the arrow in the object plane.

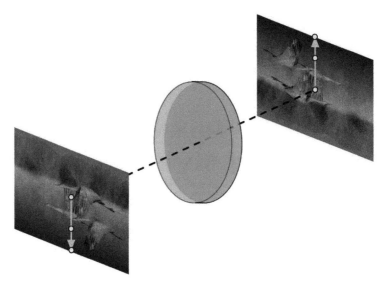

Figure 5.1 An object with three field points.

To demonstrate the entry of object fields in OpticStudio, let's create a 5-mm-thick N-BK7 equiconvex ($R1 = -R2 = 60$) singlet with an object located 120 mm in front of the lens. Start by opening a new lens (File > New) and insert one surface after the stop surface. Give the lens a 20-mm-diameter entrance aperture and change the operating wavelength to the d-line (0.5876 μm). Enter the object distance and the lens prescription data (radius, thickness, and material). Use a paraxial image thickness solve to find the paraxial image location. Figure 5.2 shows the resulting LDE for this lens. Note: For finite object distances (like this one), it is important to change the object field specification in the System Explorer (System Explorer > Fields > Settings > Type) from Angle to Object Height as angular fields of view only make sense for objects at infinity.

	Surface Type	Comment	Radius	Thickness	Material	Clear Semi-Dia
0	OBJEC Standard ▾		Infinity	120.000		0.000
1	STOP Standard ▾		60.000	5.000	N-BK7	10.071
2	Standard ▾		-60.000	112.440 M		10.059
3	IMAGE Standard ▾		Infinity	-		0.657

Title (OSequiconvex); EPD (20 mm); Field (0 mm); Wavelength (d-line)

Figure 5.2 LDE for OSequiconvex lens.

At this point, only an on-axis field point has been defined. You can see how the lens focuses a fan of rays from the on-axis object point to the on-axis image point by opening a Layout window (Crtl-L). As with the OSlens, the marginal rays at the edge of the aperture do not cross at the paraxial image plane but rather at some distance in front of it (Fig. 5.3). This is due to the spherical aberration of the lens and will be discussed in detail in the next chapter.

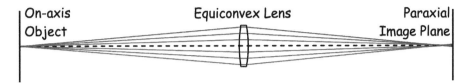

Figure 5.3 OSequiconvex lens.

We can analyze the off-axis performance of the lens by adding more object fields, but how many field points do we need and where should we put them? If only three field points are used, the midfield point is usually set at 0.707 ($= 1/\sqrt{2}$) of the full field h. At this setting, the area inside a circle whose radius equals the midfield distance is half of the total area of the full field (Fig. 5.4). This gives an equal-area weighting to the three field points used to represent the entire object. The three field points can be labeled as Axis (F1), Midfield (F2), and Full Field (F3), which may be shortened to Axis, Mid, and Full in the figures and entered as 0, $0.7h$, and h, where h is the full object height in millimeters. There may be instances where the lens performance needs to be analyzed at more specific points in the field, but if this has not been requested, these three equal-area field points is a good place to start.

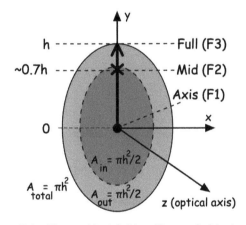

Figure 5.4 Three object fields with equal object areas.

For example, let's use three equal-area object points (0, 14, and 20 mm) to describe a 40-mm *diameter* object ($h = 20$ mm) located 120 mm in front of our equiconvex lens. Open the **Field Data Editor** (FDE) by selecting the seventh icon in the Setup tab (Setup > Field Data Editor) or, even easier, by using the keyboard shortcut Ctrl-F. On the left-hand side of the FDE window is a table where the field values are entered, and to its right is a map of the field points. Two new fields can be quickly added with the Insert key on your keyboard. In the Y (mm) column, enter 14 in Row 2 (Field 2) and 20 in Row 3 (Field 3). The plot on the right automatically updates to show the location of the current field points. Check that the Field Type is set to Object Height using the drop-down arrow next to Field 3 Properties. Your FDE should match Fig. 5.5 before moving on. Save this lens as **OSequiconvex**.

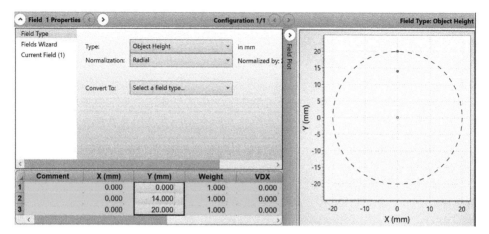

Figure 5.5 Field Data Editor (FDE) for the OSequiconvex lens.

An updated lens cross-section showing all three field points is shown in Fig. 5.6. The performance for the off-axis field points is visibly worse than the on-axis point (note the larger blur sizes on the image plane). You may also notice that we have changed the default colors (blue, green, red) for the rays from different field points to three different shades of blue. The field point colors can be changed in Project Preferences (Setup > Project Preferences > Colors).

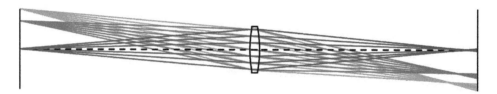

Figure 5.6 OSequiconvex lens with three fields.

To get another idea of what these additional fields look like, we can generate a 3D rendering of the optical system using the 3D viewer on the Setup Tab (Setup > 3D Viewer). Clicking on the 3D icon produces a view of the lens that is the same as the 2D cross-section shown in Fig. 5.6. However, if the rotate icon is active, you can use your left mouse button and drag the cursor around the window to change the orientation of the lens in space. You can also click on the 3D Layout Settings drop-down arrow to specify a specific rotation for the window (Fig. 5.7). Set the Rotation values to X = –20°, Y = 45°, and Z = 27° to get the orthographic view of the lens shown in Fig. 5.8. Note that the Field drop-down menu indicates All. Later on, this will be changed to demonstrate certain fields and other special rays.

Stops and Pupils and Windows, Oh My! 87

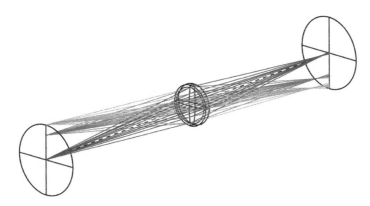

Figure 5.7 The 3D Layout Settings window.

Figure 5.8 Orthographic projection of the OSequiconvex with three fields.

5.2 Special Rays

Once the object fields have been established, the analysis of a system can be further simplified by using sets of specially defined rays. These rays are more complex than the simple focal, center, and parallel rays, defined in Section 2.3 and depicted in Fig. 2.3, that were used to locate images in ray sketches.

5.2.1 Meridional or tangential rays

Restore the **OSequiconvex** lens. Figure 5.9 shows a 3D plot of a single fan of rays from the full-field point (F3) that lies in a plane that contains the object and the optical axis of the lens. This type of ray fan is called a Y-fan and can be generated by changing the 3D Layout Settings for the Field to 3 and the Ray Pattern to Y Fan. The specific X, Y, Z rotation angles are listed at the bottom of the figure. Any ray that is confined to a plane that contains the object point from which the ray originated and the optical axis is a **meridional or tangential ray**.

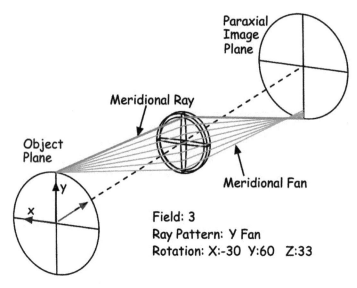

Figure 5.9 Meridional ray and fan.

5.2.2 Sagittal rays

Another fan of rays that is important in understanding lens performance is a **sagittal fan**. This is a fan that shares the chief ray with the meridional fan but is at a right angle to it. If you change the Ray Pattern in Settings from Y Fan to X Fan, you will get a fan of rays from the object to the image perpendicular to the meridional plane, as shown in Fig. 5.10. In poorly corrected lenses, the locations of the focus point for the meridional and sagittal fans from the same object point are not usually in the same plane, leading to a lens error called astigmatism, a concept that will be discussed in Chapter 7.

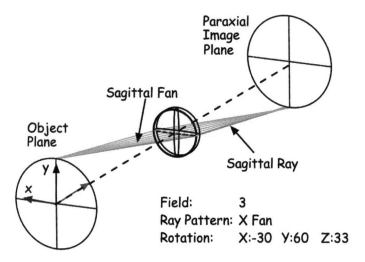

Figure 5.10 Sagittal ray and fan.

5.2.3 Skew rays

Because meridional rays are confined to a plane and that plane must contain both the object point and the optical axis, they are very special rays. There are a great many more rays from an object point that don't meet these requirements. These rays are called **skew rays**. For example, all but one of the rays in the sagittal fan shown in Fig. 5.10 are skew rays. The exception is the central ray in the fan, which is a meridional ray. Do you see why?

5.2.4 Axial rays

Axial rays have already been defined and employed in ray sketching (Section 2.3) and illustrated in Fig. 2.4. Figure 5.11 shows a bundle of axial rays starting at the on-axis point. To generate this plot, the on-axis point is chosen in Settings (Field = 1) and the Ray Pattern is set to Ring. The ray fan in Fig. 5.11 has one ring of 10 rays (Number Of Rays = 10) with a 36° ray separation around the ring.

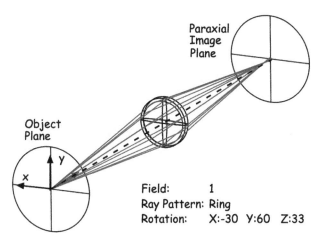

Figure 5.11 An axial ray bundle.

5.2.5 Rays for objects at infinity

In the above equiconvex example, we employed an object at a finite distance from the lens to illustrate the fields and special rays in an optical design. For an object located at infinity, the object height would be enormous. (For example, an object point at an angle of 7° from the axis has an object height of 1.2187×10^{12} mm for an object distance of 1.0×10^{13} mm.) For an object located at infinity, it is easier to describe the location of an object point by the angle that a ray directed at the lens makes with the optical axis (rather than by its height). In OpticStudio, this is accomplished by changing the Field Type from Height to Angle in the FDE (see Fig. 5.12).

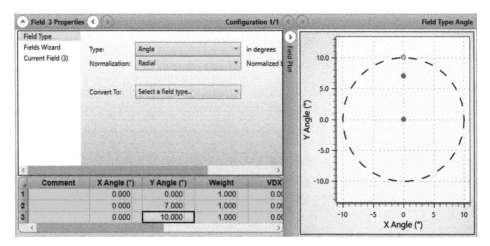

Figure 5.12 FDE settings for an object at infinity with a 10° full-field angle.

Many of our design examples in this text use an object at infinity with a standard set of angles (0°, 7°, 10°). Because these are angles between the axis and a ray from the object point, the full-field angle represents half of the angular coverage by the lens. The full angular coverage is then 20°. This coverage is modest compared to most lenses these days, but it provides a good starting point for exploring the operation of simple lenses.

> **Exercise 5.1 Change distances**
> Restore the OSequiconvex lens. Open the FDE and change the Field Type from Height to Angle. Locate the object fields at 0°, 7°, and 10°. Change the object distance to infinity and then generate 2D and orthographic 3D plots of the lens.

5.3 The Aperture Stop and Marginal Rays

One of the most useful concepts in the design and improvement of an optical system is the size and placement of the system's aperture stop. The **aperture stop** is the opening in the lens assembly that limits the size of the on-axis bundle of rays. As you will see in subsequent chapters, changing the size and location of the aperture stop can often dramatically improve lens performance. However, when the aperture stop is changed, the sizes of the other apertures in the system must change to ensure that the new aperture stop does indeed limit the axial ray bundle. In this section we will use a lens system with two identical lenses and a circular aperture halfway between them to demonstrate this concept.

Open a new lens (File > New) and enter the lens parameters from the LDE shown in Fig. 5.13, where we have included a 10-mm offset surface to draw the incoming rays from infinity. The system values for entrance pupil diameter, wavelength, and field, listed beneath the LDE, are entered by way of the System Explorer. Use a marginal ray height solve on S6 to find the image plane. To make the fourth surface the aperture stop, first click on Surface 4 in the LDE and then open the Surface Properties (for Surface 4) with the drop-down arrow. On the far right of the Type tab, check the checkbox labeled Make Surface Stop (see

Fig. 5.14). Note that for the current lens, the FDE contains only the axial (0°) field. Check that your lens cross-section looks like Fig. 5.15 and the first-order quantities for this system (**FIRST**) match those given in Table 5.1. Title and save this lens as **OSasdoubletAxis**.

	Surface Type	Comment	Radius	Thickness	Material	Clear Semi-Dia
0	OBJECT Standard ▼		Infinity	Infinity		0.000
1	Standard ▼	Offset	Infinity	10.000		5.000
2	Standard ▼		60.000	6.000	N-SK16	5.000
3	Standard ▼		-150.000	18.000		4.817
4	STOP Standard ▼		Infinity	18.000		3.516
5	Standard ▼		150.000	6.000	N-SK16	2.219
6	Standard ▼		-60.000	19.168 M		1.921
7	IMAGE Standard ▼		Infinity	-		0.027

Title (OSasdoubletAxis); EPD (10 mm); Field (0°); Wavelength (d-line)

Figure 5.13 LDE for the OSasdoubletAxis, an air-spaced doublet with an internal stop.

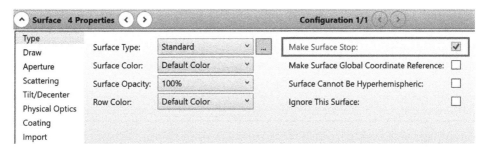

Figure 5.14 Surface Properties settings to change the location of the stop.

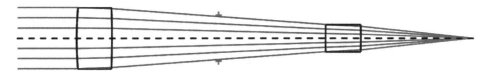

Figure 5.15 Air-spaced doublet OSasdoubletAxis.

Table 5.1 First-order quantities for the OSasdoubletAxis.

```
Infinite Conjugates
   Effective  Focal Length        49.6049
   Back Focal Length              19.1683
   Front Focal Length             -9.1683
   F/#                             4.9605
   Image Distance                 19.1683
   Lens Length                    58.0000
   Paraxial Image
      Height                       0.0000
      Angle                        0.0000
```

For the given EPD of 10 mm, the program automatically calculates the size (semi-aperture) of the stop surface needed to pass all of the rays from an axial object point. This semi-aperture or Clear Semi-Diameter (CSD) is listed in the LDE under the Clear Semi-Dia column as 3.516. Therefore, the stop surface's circular aperture is 7.032 mm in diameter.

> **Exercise 5.2 Determining the stop diameter**
> Change the entrance pupil diameter of the OSasdoubletAxis to 20 mm and determine the diameter of the aperture stop that is needed to pass all rays from an axial object point using the LDE. When you finish this exercise, change the EPD back to 10 mm before you continue the narrative in the text.

OpticStudio also calculates the CSDs needed to pass all of the rays from each field point through all of the surfaces and lists them in the LDE Clear Semi-Dia column. In a cross-section, the lens surfaces are then drawn out to these dimensions. For example, in the air-spaced doublet, shown in Fig. 5.15, the rays are drawn to the very edge of the lens surfaces. However, in practice lenses are typically fabricated to diameters larger than their ray-based clear apertures. This gives room on the surface outside the clear aperture for polishing irregularity errors (referred to as "roll-off"), coating fixtures to coat the surface, and a mounting area for lens assembly. With this in mind, the lens drawing can be improved by adding a small amount to the lens diameters to make it look more like what you would see in practice.

As discussed in the text box "Making Your Lens Cross-Section Look Good" in Chapter 3, add a CSD margin of 10% to the lens surfaces (System Explorer > Aperture > Clear Semi Diameter Margin %). Figure 5.16 shows the new LDE for the lens. Comparing this with the earlier LDE in Fig. 5.13, it is clear that the CSDs for the lens surfaces have been increased by 10% (e.g., 5.5 vs 5.0 on surface 1), while the CSD for both the stop and the image surface are unchanged.

	Surface Type	Comment	Radius	Thickness	Material	Clear Semi-Dia
0	OBJECT Standard ▼		Infinity	Infinity		0.000
1	Standard ▼	Offset	Infinity	10.000		5.500
2	Standard ▼		60.000	6.000	N-SK16	5.500
3	Standard ▼		-150.000	18.000		5.299
4	STOP Standard ▼		Infinity	18.000		3.516
5	Standard ▼		150.000	6.000	N-SK16	2.441
6	Standard ▼		-60.000	19.168 M		2.113
7	IMAGE Standard ▼		Infinity	-		0.027

Figure 5.16 LDE for the OSasdoubletAxis with a 10% CSD margin.

An updated lens cross-section with the increased CSDs is shown in Fig. 5.17. After adding the extra margin to the lens surfaces, it is much clearer that the STOP surface is the surface that restricts the size of the on-axis ray bundle and therefore must be its aperture stop (compare Fig. 5.17 with Fig. 5.15). This clarity is one of the reasons that most experienced designers will typically add some margin to their non-clipping lens surfaces.

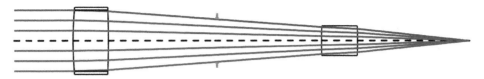

Figure 5.17 Updated lens cross-section for the OSasdoubletAxis with a 10% CSD margin.

Figure 5.17 also shows the **marginal rays** of the optical system. These are the limiting rays of a ray fan (one on each side) from an on-axis object point that go through the edge of the aperture stop.

Now add two off-axis object points with field angles at 7° and 10° using the FDE (as was shown in Fig. 5.12). Again, the CSDs of the lenses in the LDE (Fig. 5.18) automatically increase to transmit the rays from these object points through the optical system. If you click on the small box next to "11.234" in the CSD column for S2, a drop-down menu titled, "Clear Semi-Diameter solve on surface 2" is disclosed with the current Solve Type, Automatic. This indicates that the CSD for S2 has automatically been calculated. Other solve types are Fixed, Pickup, Maximum, and ZPL Macro. The Fixed solve type lets the user set the CSD, overriding the automatic solve. The use of this type of solve will be discussed later in this chapter. Title this lens OSasdoublet.

	Surface Type		Comment	Radius	Thickness	Material	Clear Semi-Dia
0	OBJECT	Standard ▾		Infinity	Infinity		Infinity
1		Standard ▾	Offset	Infinity	10.000		13.343
2		Standard ▾		60.000	6.000	N-SK16	11.234
3		Standard ▾		-150.000	18.000		10.304
4	STOP	Standard ▾		Infinity	18.000		3.517
5		Standard ▾		150.000	6.000	N-SK16	7.444
6		Standard ▾		-60.000	19.168 M		7.989
7	IMAGE	Standard ▾		Infinity	-		9.333

Figure 5.18 LDE for the OSasdoublet showing the updated clear semi-diameters with three field points.

Figure 5.19 shows a cross-section of the OSasdoublet, where the number of rays for each field has been set to three, two marginal rays and a center ray. Note that the central ray for each field intersects at the center of the aperture stop. These are the **chief rays** for the three field points and will be discussed in more detail in the next section. It is also clear from Fig. 5.19 (compare it to Fig. 5.17) that the diameters of the lenses have increased to allow the rays from the off-axis field points to make it through the optical system.

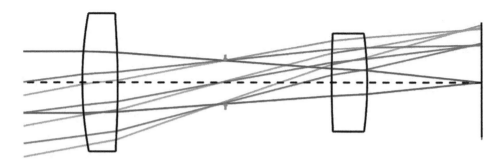

Figure 5.19 Lens cross-section for the OSasdoublet with three field points.

By definition, the aperture stop is the single aperture in the lens system that limits light from an on-axis field. If there are no other clipping apertures in the optical system, then the aperture stop also limits the amount of light collected from the off-axis field points, and rays from these field points should also fill the stop. At first glance it would appear this is true for our lens, but a closer look at the stop reveals that the marginal rays the from off-axis fields do not completely fill the lower portion of the stop surface. This is due to pupil aberrations, an advanced topic in lens design outside the scope of this text. For now, OpticStudio uses an algorithm, called ray aiming, to fix this error by determining the rays at the object that correctly fill the stop surface for each field for a given stop diameter. Generally, ray aiming is required when the image of the stop as seen from object space is considerably aberrated, shifted, or tilted. However, it is turned off in the default settings to increase ray tracing speed.

To turn ray aiming on, go to the Ray Aiming tab of the system explorer (System Explorer > Ray Aiming > Paraxial) and select Paraxial from the drop-down menu (Fig. 5.20). Figure 5.21 shows a magnified region around the aperture stop comparing the marginal rays for ray aiming Off [Fig. 5.21(a)] and ray aiming Paraxial [Fig. 5.21(b)]. The lower marginal rays that are above the lower edge of the stop in Fig. 5.21(a) are fixed so that they properly fill the stop in Fig. 5.21(b) when ray aiming is turned on. Paraxial ray aiming will be used for all subsequent examples and demonstrations in this text. If you see small differences in your LDEs and lens cross-sections, check your ray aiming—it should be turned on and it's easy to forget!

When ray aiming is turned on, the clear semi-diameters for each lens surface change to accommodate the new ray paths through the lens system. The modified LDE is shown in Fig. 5.22. Save this lens as **OSasdoublet**. It will be used later in this chapter.

Although the aperture stop for the air-spaced doublet is currently set at a surface halfway between the two lenses, there is no requirement that this surface must be the aperture stop. Any other surface in the lens system could be assigned that function. However, for that to occur, the diameters of all lenses must change so that the new aperture stop becomes the surface that limits the on-axis ray bundle. Provided that the sizes of the apertures have not been fixed by the user, OpticStudio will vary the semi-diameters of the surfaces to accommodate all of the rays for all of the object points to achieve this condition.

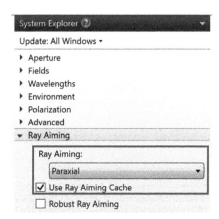

Figure 5.20 System Explorer settings to turn on paraxial ray aiming.

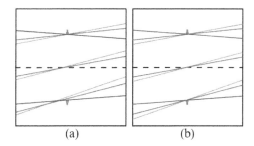

Figure 5.21 The effect of ray aiming at the aperture stop: (a) Ray Aiming: Off; (b) Ray Aiming: Paraxial.

	Surface Type	Comment	Radius	Thickness	Material	Clear Semi-Dia
0	OBJECT Standard ▼		Infinity	Infinity		Infinity
1	Standard ▼	Offset	Infinity	10.000		13.902
2	Standard ▼		60.000	6.000	N-SK16	11.776
3	Standard ▼		-150.000	18.000		10.852
4	STOP Standard ▼		Infinity	18.000		3.520
5	Standard ▼		150.000	6.000	N-SK16	7.446
6	Standard ▼		-60.000	19.168 M		7.990
7	IMAGE Standard ▼		Infinity	-		9.397

Figure 5.22 Clear Semi-Dia changes for the OSasdoublet after Ray Aiming: Paraxial is selected.

Let's see how the program modifies the air-spaced doublet when the first surface of the first lens is specified as the aperture stop. Select the first lens surface (S2) in the LDE, open the **Surface 2 Properties** window, and check the **Make Surface Stop** box. Immediately, the STOP designation in the Surface Type column will move to S2 (Fig. 5.23), and the Layout window (Fig. 5.24) shows that the lens apertures for S5 and S6 have been increased (to eliminate clipping of the 10° off-axis bundle of rays at the second element), whereas the size of the ray bundle at the first element has been reduced (compare Fig. 5.24 with Fig. 5.19). Note that the chief ray from each field point now intersects at the center of the first surface.

	Surface Type	Comment	Radius	Thickness	Material	Clear Semi-Dia
0	OBJECT Standard ▾		Infinity	Infinity		Infinity
1	Standard ▾	Offset	Infinity	10.000		7.480
2	STOP Standard ▾		60.000	6.000	N-SK16	5.000
3	Standard ▾		-150.000	18.000		5.968
4	Standard ▾		Infinity	18.000		7.937
5	Standard ▾		150.000	6.000	N-SK16	9.925
6	Standard ▾		-60.000	19.168 M		10.132
7	IMAGE Standard ▾		Infinity	-		8.885

Figure 5.23 LDE for the OSasdoublet with the stop at the first surface of lens 1 (S2).

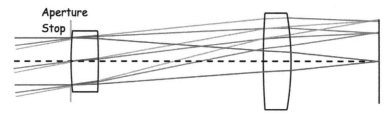

Figure 5.24 Marginal and chief rays for the OSasdoublet with the aperture stop at the first surface of lens 1 (S2).

> **Exercise 5.3 Aperture stop at S5**
> Restore the OSasdoublet and locate the aperture stop at S5. Set the field angles to 0°, 3.5°, and 5° and examine the LDE and Layout of the changed lens. What is the aperture stop diameter now? What is the minimum clear aperture for lens #1 that is needed to pass all of the rays? If the first element is only 25 mm in diameter, is it large enough to pass all of the rays for these fields? What happens if you try to maintain the original full 10° field with this new stop position? After this exercise, do not save this lens.

The symmetry of a design like the OSasdoublet with the stop at S4 can be useful in improving lens performance, but it is also a very practical feature for manufacture and assembly. Because of its symmetry, this design can be fabricated using two identical lenses that have the same diameter. This makes sense because only one type of lens is required, and the lenses can be inserted into the optical system in either position.

To illustrate this, move the stop back to surface 4 (or simply reload the OSasdoublet file). We can make the lenses identical by changing the element diameters to 25 mm for both elements. Enter a value of 12.5 into the Clear Semi-Dia cell for the first lens surface (S2). If this is repeated for surfaces S3, S5, and S6, the two lenses are now the same size. Each of the semi-diameters that we assigned values to have a "U" in the box next to it, indicating that semi-diameters are no longer automatically calculated and have been set by the user (Fig. 5.25). A new lens layout (Analyze > Cross-Section) will display the symmetrical design (Fig. 5.26). Save this lens as **OSasdoubletSym.**

Stops and Pupils and Windows, Oh My! 97

	Surface Type	Comment	Radius	Thickness	Material	Clear Semi-Dia
0	OBJECT Standard ▾		Infinity	Infinity		Infinity
1	(aper) Standard ▾		60.000	6.000	N-SK16	12.500 U
2	(aper) Standard ▾		-150.000	18.000		12.500 U
3	STOP Standard ▾		Infinity	18.000		3.520
4	(aper) Standard ▾		150.000	6.000	N-SK16	12.500 U
5	(aper) Standard ▾		-60.000	19.168 M		12.500 U
6	IMAGE Standard ▾		Infinity	-		16.609

Figure 5.25 LDE for the OSasdoubletSym with user-defined apertures.

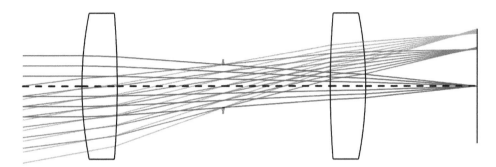

Figure 5.26 Lens cross-section for the symmetrical OSasdoubletSym.

5.4 Chief Rays and Pupils of a Lens

The analysis of the collection and transmission of light through an optical system depends on the location of the system's aperture stop. If you have closed OpticStudio, open it and restore the symmetric air-spaced doublet **OSasdoubletSym**. However, instead of displaying rays from all object points, show only the maximum field point F3 (10°) by changing the field from All to 3 in the layout settings. The result is shown in Fig. 5.27. Although any of the rays might be representative of the bundle, it is the central ray or chief ray of the bundle that can be distinguished and easily defined. A **chief ray** is a ray that goes through the center of the aperture stop.

For any lens, there can be many chief rays, one for each object point. All of the chief rays intersect at the center of the aperture stop, as shown in Fig. 5.26. (What is the chief ray for the axial object point?) Each field also has its own set of associated marginal rays. However, you will often find in the optical literature a reference to a single chief ray and a single marginal ray for the entire system. By convention, this is the chief ray from the largest field point, while the marginal ray is from the on-axis field point that goes through the top of the aperture stop.

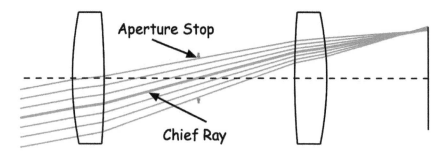

Figure 5.27 Chief ray for the maximum field point (F3) through the OSasdoubletSym.

A chief ray may be considered as the axis for a ray bundle from some object point. Each ray bundle represents the light falling on the image point from its corresponding object point. The size of that bundle is determined by the size of the aperture stop and any vignetting, a term that will be discussed in Section 5.8. When the aperture stop is at the first lens surface (Fig. 5.24), the size of the incoming bundle that will fill the lens system is easily determined. It equals the diameter of the aperture on the first lens surface. But how can the size of the incoming bundle be determined if the aperture stop is buried within the lens?

At first, this may seem like an unimportant question. However, the size of the light bundle that can be collected by a lens can determine how well it will work in low light or high-speed situations, or if it can be used in high-resolution applications. The feature that is used to assess the light collection capacity of a lens is its entrance pupil.

The **entrance pupil** is defined as the image of the aperture stop as seen from object space. If the first surface of a lens is the aperture stop, as in Fig. 5.24, then it is also the entrance pupil of the lens. But when the aperture stop is buried inside the lens as it is in Fig. 5.26, it becomes an object to be imaged. That is, if you consider the aperture stop to be an object and you calculate where its image will be and what the magnification would be due to all the elements between the aperture stop and the object space, then you will have determined the size and location of the entrance pupil of the lens. The importance of the entrance pupil is that it represents the aperture that limits the amount of light entering the optical system. When it is filled by a ray bundle from the on-axis object point, rays traced from the entrance pupil to the aperture stop will fill the aperture stop with light. The entrance pupil tells a user how much light a lens can collect and deliver to the image plane.

Delete the offset surface in the OSasdoubletSym (for clarity) and run the **FIRST** macro (Table 5.2). The entrance (and exit) pupil location and diameter are listed at the end of the output. For this lens, the entrance pupil (EnP) is the image of the circular aperture stop as seen through the first lens and is located 30.4366 mm from the first surface vertex (a positive value indicates that it is to the right of the first vertex, whereas a negative value would indicate that the pupil is to the left of the first surface).

Table 5.2 First-order quantities for the OSasdoubletSym (offset surface deleted).

```
Infinite Conjugates
    Effective Focal Length          49.6049
    Back Focal Length               19.1683
    Front Focal Length             -19.1683
    F/#                              4.9605
    Image Distance                  19.1683
    Lens Length                     48.0000
    Paraxial Image
        Height                       8.7467
        Angle                       10.0000
    Entrance Pupil
        Diameter                    10.0000
        Location                    30.4366
    Exit Pupil
        Diameter                    10.0000
        Thickness                  -30.4366
```

The locations of the pupils are demonstrated graphically in Fig. 5.28, where only the chief ray and the two marginal rays are plotted for the full field, field F3. The chief ray crosses the optical axis at the aperture stop S3, the third surface in the lens, located 24 mm (T1 + T2) beyond S1. But it is *aimed* at a point 30.4366 mm beyond S1 where its extension crosses the axis. This is where the image of surface S3, the entrance pupil, is located. The listing also indicates that the entrance pupil diameter is 10 mm, which was the value we assigned to this lens when we set it up. This value is larger than the diameter of the aperture stop calculated in the LDE in Fig. 5.25. Twice the semi-aperture for the STOP surface is 7.040 (2 × 3.520); therefore, the first lens magnifies the aperture stop.

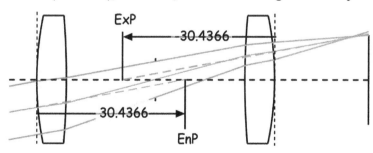

Figure 5.28 The pupil locations of the air-spaced doublet.

By now, you may have noticed that there is also a lens feature called an exit pupil (ExP), which is the image space counterpart to the entrance pupil because the **exit pupil** is the image of the aperture stop in image space. Because this lens is symmetrical about the aperture stop, the exit pupil is located the same distance in front of the last lens surface as the entrance pupil was behind the first lens surface (again, a positive value in the first-order table indicates that the distance is measured to the right of the last surface, whereas a negative value would indicate that the pupil is to the left of the last surface). A simple example illustrating these concepts using ray sketching is shown in the text box, "Ray Sketching Some Pupils."

Just as the entrance pupil represents the hole into which an axial ray bundle enters, the exit pupil represents the hole out of which that axial ray bundle exits. So, when it comes to light control, a collection of lens elements can be replaced by two holes with the diameters and locations of the entrance and exit pupils. These simplifications, along with the EFL of a lens, allow designers to compare lenses of different construction in the prediction of lens performance. This will become clearer when we look at the light-collecting ability of a lens in Section 5.7.

Ray Sketching Some Pupils

We will start with two lenses, Lens1 and Lens2, with a circular aperture stop between them. The stop is inside the front focal point F_2 of the second lens Lens2. To find an image of the aperture in image space, we trace two rays from the edge of the stop into image space, as shown in the following ray sketch (see Chapter 2). By extending the two rays from the stop (one parallel, the other through the center of the second lens) after they exit the lens backward to where they cross, the location of its image is found. So, there is a magnified virtual image of the aperture stop, which corresponds to the exit pupil (ExP).

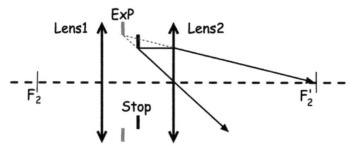

In a similar manner, by tracing two rays from the top of the stop into object space, the location of the image is found from extending the two rays after they exit Lens1 forward to where they cross. This locates the image of the aperture stop as seen from object space and is therefore the entrance pupil (EnP). The entrance pupil is a magnified, virtual image of the stop that is located farther from the Lens1 than the stop.

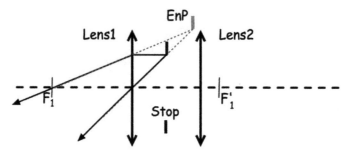

In the air-spaced doublet example used here, we located the aperture stop between the two lenses. But where is the stop supposed to go? In many optical designs, the stop position is constrained by system requirements. For example, for an eyepiece, the aperture stop needs to be out in front of the lens. For infrared systems, the stop is cooled and located near the infrared sensor to shield it from the warm housing. In addition, as will be discussed in Chapters 7 and 8, the size and location of the stop are features of a lens that a designer can use to optimize its performance. In most cases, the stop location for many designs has been determined from years of practice.

5.5 The Field Stop and Its Windows

After the limits on the light collection of an optical system have been determined from the sizes and locations of the aperture stop and the pupils, the next item to be considered is the maximum size of the image that can be accommodated by the system. Because the chief ray represents the central ray of a bundle of rays from a particular point in the object field, the edges of the image are determined by the chief ray at the largest field angle (or largest object point) that can be transmitted by the optical system. This is the full-field chief ray of the system, and the system component that limits this ray is the **field stop** of the system. In most cases, the field stop of a lens system is the detector, be it a digital CMOS array, photographic film frame, or some other light sensor. The elements are then oversized so that they do not clip the chief ray, as shown in Fig. 5.29.

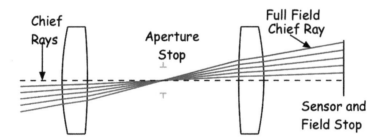

Figure 5.29 A fan of chief rays through the OSasdoubletSym.

In summary, the aperture stop is the surface that limits the size of a bundle of rays from an axial object point that can be transmitted by the lens system, whereas the field stop is the surface in the lens system that limits the system's maximum field size.

As with the aperture stop, the images of the field stop (called windows) can also be located. In the case of this doublet, the **exit window** of the system is easy to locate because the field stop is located in image space, so it is also the exit window. Although it takes a little more effort to locate the **entrance window** of the system, the image of the field stop, as seen from object space through the air-spaced doublet, can be computed. If the sensor is the field stop and it is located at the BFP, then the entrance window is at infinity. For objects at finite distances, if the sensor is the field stop, then the entrance window is located at the object plane.

5.6 Tracing General Rays

In the previous sections we described various special rays that propagate through an optical system. But how can we trace a general ray? In OpticStudio, there is a straightforward way to describe any ray in any system. The rays are defined by two specific locations: where the ray came from (an object point) and where it's going (a point in the entrance pupil). To keep things general for all systems, the locations of these points are normalized from 0 to 1 and given as a ratio of the current point to its maximum possible value.

In the case of object points, a normalized field coordinate h is the ratio of a field point to its maximum value. If a lens with its object at infinity has three fields (0°, 7°, 10°), the intermediate field point would be $h = 0.7$, with a 10° field being $h = 1.0$. Because this point can have any location in the object plane relative to the optical axis, the coordinate has a subscript, x or y. Thus, any point in the object plane can then be described by a pair of coordinates, (h_x, h_y). Similarly, any point in the entrance pupil is designated by normalized pupil coordinates (p_x, p_y). Then any ray propagating through the optical system can be defined (and traced) by these two sets of coordinates.

Examples of this approach for two special rays, the marginal ray and the chief ray, are shown in Fig. 5.30. On the top, a marginal ray is traced from the bottom of the object ($h_x = 0$, $h_y = 0$) to the top of the entrance pupil ($p_x = 0$, $p_y = 1$). On the bottom, the chief ray starts at the top of the object ($h_x = 0$, $h_y = 1$) and, by definition, is traced through the center of the entrance pupil ($p_x = 0$, $p_y = 0$).

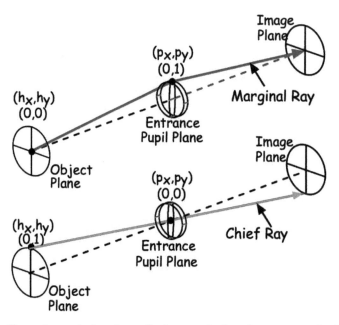

Figure 5.30 Example ray designations: (top) a marginal ray from an on-axis object point to the edge of the entrance pupil and (bottom) a chief ray from an off-axis (full field) object point through the center of the entrance pupil.

5.7 Field and Pupil Specifications

For many modern devices, there is usually a set of numbers that specify their size and performance. For example, high-definition digital displays are characterized by their diagonal dimension (size) and their vertical pixel count (performance). Of course, other specifications, such as image contrast, might be added to provide information to justify the cost and establish the benefits of purchasing one display over another. What other values might be used to evaluate a high-definition television?

In the case of lenses, the two primary numbers that are used to initially sort and rank lenses are a field specification (typically given as an angular field of view) and a pupil specification, such as an $f/\#$ or numerical aperture (NA). When searching the lens catalogs and patent lists for a likely candidate as a starting point for a new design, the values for the aperture and the field of view of the lens (or ranges of numbers about some nominal value) are entered into the search filters to reduce the number of potential candidates. A third specification, the EFL of the lens (defined and described in Section 4.2), is usually added to the list to help constrain the overall package size of the lens system. In most design problems, these three quantities are specified long before work begins.

5.7.1 Field of view

How large of a view can you record with a particular lens? If you are standing with a handheld camera next to a professional sports photographer in a baseball park, your iPhone camera will be able to record a fair amount of the field and the stands, but its wide-angle lens will not be able to record the on-field action with any detail. The camera of the photographer next to you has a narrow-angle (telephoto) lens that permits incredible close-ups of the action despite his position on the sidelines. The specification of the angular coverage of a scene by a lens is its field of view. In the case of the air-spaced doublet shown in Fig. 5.29, the angle between the entering full-field chief rays is a measure of the field of view for this system. Many cameras have a built-in zoom capability, so you can change the field of view to take a range of pictures, from close-ups to landscapes.

In Fig. 5.31, the full field-of-view angle is shown as 2θ. If the full field of view is 20°, this layout can be created in OpticStudio for the OSasdoubletSym lens by entering two fields +10° and –10° in the FDE and specifying one ray in the Layout settings. The axis, labels, and the extensions of the entering rays have been added to the figure.

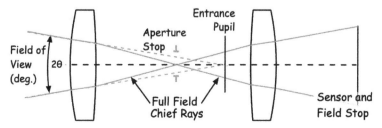

Figure 5.31 Field of view (2θ) of the air-spaced doublet lens.

For some systems, the field of view is determined by the size of the detector. In such cases, the field size is specified in image space using either the paraxial image height or the real image height. The difference between the two lies in whether distortion (see Section 8.5) is taken into account. For an object at infinity, we showed in Chapter 2 how to find the paraxial image height h' given the half field angle θ and the focal length of a lens, using Eq. (2.1):

$$h' = f \tan \theta. \tag{2.1}$$

In practice, this equation is extremely useful for checking your first-order specifications. For example, given a sensor size and field of view, you can calculate the focal length of the lens needed for mapping object space to image space. You can also use this equation to convert spatial resolution at the sensor into angular resolution at the object. A full discussion of sensor sizes and formats is available at the beginning of Chapter 11 on analyzing lens performance.

Five different field specification options in OpticStudio can be displayed by clicking on the drop-down arrow for **Type** (see Fig. 5.32) in the FDE under the field type tab. We've already used two of them, **Angle** and **Object Height**. If the detector is the field stop, the **Field Type** can be set to the **Paraxial** or **Real Image Height** and one of the dimensions of the detector can be entered as the y height. Because many detectors are rectangularly shaped, both dimensions can be entered to provide a field point in a corner of the detector by entering both the x and y coordinates of the field point. However, this breaks the circular symmetry of the optical system. It is much simpler and faster to find the radius of the smallest circle that encompasses the entire detector and use that for the full-field y image height (see Exercise 5.4). One must remember that the field dimensions and angles are measured from the optical axis, which intersects the center of the detector.

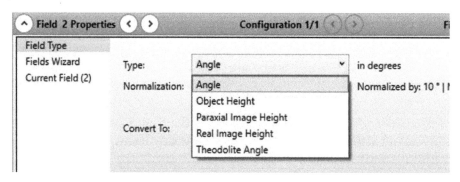

Figure 5.32 FDE window showing available options for field type.

Exercise 5.4 Angular field of view
The OSasdoubletSym is used to image an object at infinity onto a Canon APS-C sensor whose dimensions are 22.2×14.8 mm. Use Eq. (2.1) to find the three equal-area angular field points (axis, 0.7 field, and full field) that should be used to ensure that every point on the sensor is covered by the design.

5.7.2 F-number and numerical aperture

To this point, only the EPD has been used to specify the aperture of a lens. However, a more common way to express how compactly a lens can focus a ray bundle is the **F-number** of the lens, which is the ratio of the EFL to the EPD:

$$f/\# = \frac{\text{EFL}}{\text{EPD}}. \tag{5.1}$$

Note: This definition only applies to lenses when imaging objects at infinity. It is also the value that appears on the side of many camera lenses.

The $f/\#$ of the lens is a measure of the size of the EPD relative to the focal length. So, an $f/8$ lens (the value of the $f/\#$ replaces the pound sign (#) in the symbol) has an EPD that is one-eighth (1/8) of the EFL of the lens. The $f/\#$ designation is then a relative measure of the light collection of a lens that is "independent" of the specific EFL of a lens. This fractional value is also referred to as the **speed** of the lens. The term speed reflects the fact that the smaller the $f/\#$ the larger the EPD relative to the EFL. When more light is collected, it is easier to capture objects moving at a high speed. So, an $f/2$ lens, which has an EPD that is four times larger than that of an $f/8$ lens with the same EFL, will record a moving object with reduced blur. This is because the $f/2$ lens will produce acceptably exposed images with shorter exposure times, which is why photographers refer to a lens with a lower $f/\#$ as a "faster" lens.

Figure 5.33 shows three lenses with the same 40° field of view but very different F-numbers (arrayed from slowest to fastest). As the $f/\#$ decreases, the number of elements increases, the axial marginal ray angle U' at the image surface increases, and the surfaces become strongly curved. A large-diameter lens with a short focal length, such as the $f/2$ lens, is usually a high-performance lens that delivers bright, crisp images and may be quite expensive. In contrast, the performance of the other two lenses in Fig. 5.33 have lower performance over the same-size field. To be useful, their field sizes must be reduced, or elements added, to meet high-performance requirements.

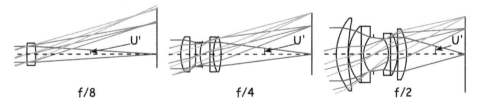

Figure 5.33 Sample lenses of differing F-numbers but the same 40° field of view.

Restore the **OSasdoublet** from Section 5.3. The **Aperture Type** in the **Aperture** tab of the System Explorer was chosen as **Entrance Pupil Diameter** and assigned a value of 10 mm. According to Table 5.1, its $f/\#$ is 4.96. Now, suppose you wish to "speed up" the lens a bit to, say, $f/4$. Instead of setting the EPD, we can enter the $f/\#$ directly in the Aperture tab of the System Explorer by selecting

Image Space F/# in the drop-down menu for Aperture Type and entering 4.0 for the Aperture Value (Fig. 5.34).

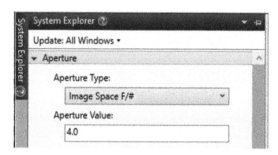

Figure 5.34 Changing the aperture type to Image Space F/#.

Figure 5.35 shows a comparison of the LDE (top) before and (bottom) after the $f/\#$ change. The basic lens parameters (e.g., radii, element thicknesses) have not changed, while all of the Clear Semi-Diameters have increased so that the lens is now operating at $f/4$. The controlling feature, the aperture stop CSD, has increased from 3.520 to 4.365 mm. In practice, if the interior stop diameter (e.g., the iris diaphragm) can be varied in an optical system such as this one, a user can control the amount of light reaching a sensor or sensing surface. Many camera systems include a control ring around the lens marked with its F-stop settings to open and close its aperture.

	Surface Type	Comment	Radius	Thickness	Material	Clear Semi-Dia
0	OBJEC Standard ▾		Infinity	Infinity		Infinity
1	Standard ▾		Infinity	10.000		13.902
2	Standard ▾		60.000	6.000	N-SK16	11.776
3	Standard ▾		-150.000	18.000		10.852
4	STOP Standard ▾		Infinity	18.000		3.520
5	Standard ▾		150.000	6.000	N-SK16	7.446
6	Standard ▾		-60.000	19.168 M		7.990
7	IMAGE Standard ▾		Infinity	-		9.397

	Surface Type	Comment	Radius	Thickness	Material	Clear Semi-Dia
0	OBJECT Standard ▾		Infinity	Infinity		Infinity
1	Standard ▾		Infinity	10.000		15.357
2	Standard ▾		60.000	6.000	N-SK16	13.183
3	Standard ▾		-150.000	18.000		12.284
4	STOP Standard ▾		Infinity	18.000		4.365
5	Standard ▾		150.000	6.000	N-SK16	8.011
6	Standard ▾		-60.000	19.168 M		8.475
7	IMAGE Standard ▾		Infinity	-		9.584

Figure 5.35 LDE comparison for the OSasdoublet with an internal stop (top) at the original $f/4.96$ and (bottom) at $f/4$.

Figure 5.36 shows the six different aperture specification options available in OpticStudio (System Explorer > Aperture > Aperture Type). We've already covered two of them, **Entrance Pupil Diameter** and **Image Space F/#**. These two settings are typically used for optical systems with an object at infinity. Two other aperture specifications, **Paraxial Working F/#** and **Object Space NA**, are more appropriate for lenses when the object distance is not at infinity. For example, the working $f/\#$ is a more general definition of $f/\#$ that works for finite object distances and is discussed in more detail in the text box, "Working $f/\#$." The other two aperture options, **Float by Stop Size** and **Object Cone Angle**, are not covered in this text.

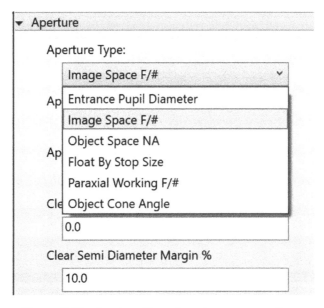

Figure 5.36 Aperture Type drop-down-menu options.

Numerical aperture is another measure of the light-gathering capability of a lens. In image space, the **numerical aperture** is defined as

$$\text{NA} = n' \sin U', \quad (5.2)$$

where angle U' is the real angle of the marginal ray for the axial object point in image space [Fig. 5.37(a)]. (Angle U' is not to be confused with the paraxial angle u'.) For lenses used in air, the refractive index is unity, but for lenses immersed in water or oil, the refractive index of the image space medium n' multiplies $\sin U'$. For a finite object distance, there is a corresponding numerical aperture in object space. The numerical aperture in object space (NAO) is given as

$$\text{NAO} = n \sin U, \quad (5.3)$$

where n is the refractive index of object space.

Working f/#

Although we have just provided a simple relationship between the $f/\#$ and NA, it represents the special case of an object at infinity. However, this constraint is rarely stated when this relation is applied. For example, it is common for the $f/\#$ to be displayed on the barrel of most fixed-focal-length camera lenses even though these lenses are intended to be used for both objects at infinity and close-up objects.

One reason for some confusion is that the $f/\#$ is considered to be a constant of the lens because it is the ratio of two fixed quantities, the EFL and EPD, that do not change as the object distance is changed (this assumes that the aperture stop diameter is also fixed). In contrast, the angle between the marginal ray and the optical axis U' decreases as the object approaches the lens. Therefore, for magnifications other than zero (object at infinity), the NA changes with the magnification of the optical system.

This means that as the NA decreases, the effective $f/\#$, or "working" $f/\#$, of the lens increases. A working $f/\#$ is defined as

$$f/\#_{working} \equiv \frac{1}{2\text{NA}}.$$

It can be shown that the working $f/\#$ is related to the $f/\#$ of the lens according to $f/\#_{working} = (1 - m) f/\#$, where m is the magnification.

For example, for a single thin lens, if an object is located $2f$ in front of the lens ($t = -2f$), its image appears at the same distance on the other side of the lens ($t' = 2f$). The magnification ($m = t'/t = 2f/(-2f) = -1$) is minus one, the minus sign indicating image inversion. The $f/\#_{working}$ is then $[1 - (-1)]$ $f/\#$, or 2 times the stated $f/\#$ of the lens. We can demonstrate how this works in OpticStudio using the 50-mm-focal-length OSsinglet lens from Section 3.5. The top figure shows rays from an axial object point at infinity converging to the focal point. The bottom figure shows rays from an object 100 mm (twice the focal length) from the lens.

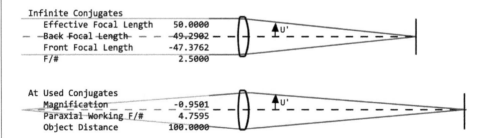

For an object at infinity, the $f/\#$ ($= 50/20$) is 2.5. When the object is located 100 mm in front of the lens, the $f/\#_{working}$ is listed in the output of the FIRST macro under At Used Conjugates as 4.7595. The small discrepancy between this value and the thin lens expectation of 5 is due to the 5-mm thickness of the lens.

If the object is located at the front focal point, the range of angles between the axis and the edge of the EPD over which light is collected is given as U, as shown in Fig. 5.37(b). In OpticStudio, only the object space NA can be specified. If an image space metric is needed, then the $f/\#$ or working $f/\#$ can be used.

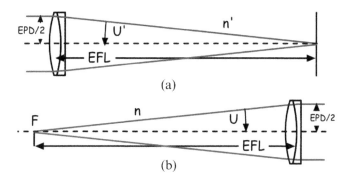

Figure 5.37 (a) Numerical aperture in image space defined by the n' in image space and the maximum marginal ray angle U'. (b) Numerical aperture in object space defined by the n in object space and the maximum marginal ray angle U.

For an object at infinity, both the image space NA and the $f/\#$ describe the axial marginal ray angle $U = \tan(\text{EPD}/2\text{EFL})$ at which a lens focuses light. In the case where U is small, the approximation for NA is given by

$$\text{NA} = \sin\arctan\left(\frac{\text{EPD}}{2\text{EFL}}\right) \approx \left(\frac{\text{EPD}}{2\text{EFL}}\right). \quad (5.4)$$

Combining Eqs. (5.1) and (5.4), the relationship between NA and $f/\#$ is given by

$$\text{NA} = \frac{1}{2\,f/\#}. \quad (5.5)$$

In the cases of the three lenses pictured in Fig. 5.33, their NAs are 0.0625, 0.125, and 0.25, respectively. So, as the lens gets faster, the NA increases and the $f/\#$ decreases. Microscope objectives are typically labeled by their NA.

5.8 Vignetting

As we noted in Section 5.3, when a field point is specified, as long as the Clear Semi-Diameters (CSDs) are set to automatic, OpticStudio computes the sizes of the lens elements needed to ensure that the ray bundle from that field point will be transmitted by the lens. But when an actual lens (whose apertures are predetermined by how it was made and the lens barrel that holds it) is entered into the program for analysis, the aperture stop may not be the limiting aperture for field points other than the axial object point. Rays entering the lens from larger field angles may be blocked by one of the components, thereby reducing the amount of light falling in the image plane for those image points. This reduction in light transmission due to the sizes of the components is called **vignetting**.

For example, restore the saved **OSasdoubleSym** with the 25-mm-diameter user-defined lens apertures from Fig. 5.25. Add a fourth field to the FDE with a value of 16°. The rays from the first three field points (0°, 7°, and 10°) have already been shown to completely fill the aperture stop (see Fig. 5.26). Adjust the layout settings to plot only the rays for the 16° field (Field 4), with 7 as the setting for Number of Rays to be plotted.

The result is shown in Fig. 5.38. Whereas the earlier layout (Fig. 5.26) showed seven rays for each field point, for this larger field point, only five rays are transmitted by the lenses as the bottom two rays do not make it through the first lens aperture. This is a simple example of vignetting, the reduction of light transmission for field points at the edge of an object field.

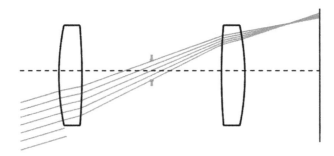

Figure 5.38 Ray fan for 16° field of the OSasdoubletSym showing vignetting.

In optical system design, the vignetted off-axis fields are modeled, measured, and addressed by replacing the circular entrance pupil with a vignetted pupil for each field. The shape of this pupil can be seen in OpticStudio by looking at a Footprint Diagram (Analyze > Rays & Spots > Footprint Diagram) at the stop surface (remember that the aperture stop is an image of the entrance pupil). When first opened, the footprint diagram shows the plot of rays for all four fields on the Image surface. Change the displayed surface to the aperture stop (S4) by clicking on the Settings tab and change Surface to 4. Note the Ray Density is set to Ring. Click OK to get a plot of rays around the edge of the stop for each of the field points (Fig. 5.39). Because the rays from the first three field points fill the aperture stop, they all lie on the same circle, but the shape for the 16° field point looks like a squashed circle that is missing its bottom edge, indicating that not all of the rays from that field point get through the lens. Note: The Footprint Diagram from OpticStudio has been rearranged to save space, and its font sizes have been increased to improve legibility.

Now open the Footprint Settings and change the Field from All to 4 to isolate the effect of vignetting on the 16° field. (Or uncheck the top three fields in the box in the upper right corner of the footprint diagram.) Then change the Ray Density to 20 and click OK to see the footprint for a bundle of rays at the 16° field with 20 rays across the half pupil (Fig. 5.40). The bottom of the text tab for the footprint diagram shows that only 79.66% of the rays traced for field 4 pass through the system. In many ways this is not very efficient—why trace rays that you know won't get through the system?

Stops and Pupils and Windows, Oh My! 111

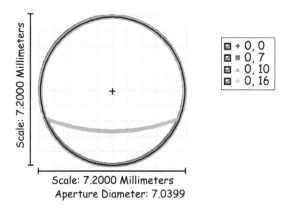

Figure 5.39 Footprint diagram for all fields, showing vignetting for 16° field, with the ray density set to Ring.

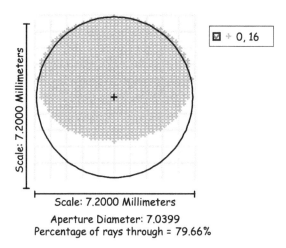

Figure 5.40 Footprint diagram for field 4, showing vignetting for a 16° field, with the ray density at 20.

To fix this, return to the FDE. In the banner above the field entries, there are two icons: ⊖ ⊖. The one on the left sets the vignetting and the one on the right clears all vignetting. Note that all four vignetting columns (VDX, VDY, VCX, VCY) have zero values to start. Click the left icon to set the vignetting. The result is shown in Fig. 5.41. Now three of the four vignetting factors are no longer zero for the 16° field point, and the footprint plot (Fig. 5.42) shows that 100% of the rays traced get through! The meaning and application of these vignetting factors are described in the text box "Vignetting Parameters."

	Comn	X Angle (°)	Y Angle (°)	Weight	VDX	VDY	VCX	VCY	TAN
1		0.000	0.000	1.000	0.000	0.000	0.000	0.000	0.000
2		0.000	7.000	1.000	0.000	0.000	0.000	0.000	0.000
3		0.000	10.000	1.000	0.000	0.000	0.000	0.000	0.000
4		0.000	16.000	1.000	0.000	0.225	0.015	0.225	0.000

Figure 5.41 FDE for the OSasdoubletSym with a 16° field with vignetting factors set.

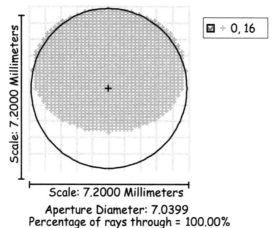

Aperture Diameter: 7.0399
Percentage of rays through = 100.00%

Figure 5.42 Footprint diagram for the 16° field with vignetting factors set.

Vignetting Parameters

There are four vignetting parameters displayed in the FDE (VDX, VDY, VCX, VCY) for each field. As shown below, VDX and VDY represent a shift of the vignetted pupil in x and y, respectively. A positive value shifts the pupil up or to the right, while a negative value shifts the pupil down or to the left. VCX and VCY give the amount of compression (positive value) or expansion (negative value) applied to the pupil in each dimension. These values are all normalized to 1 such that a 0.5 value represents a 50% shift (for VDX or VDY) or a 50% compression/expansion (for VCX or VCY) of the pupil's actual dimensions.

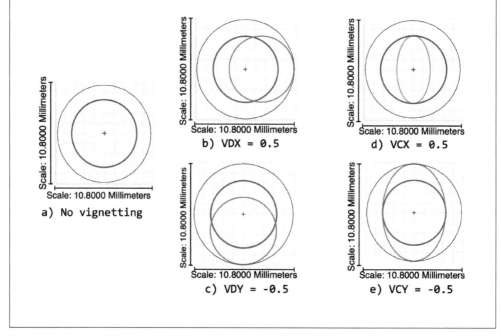

In many lens designs (e.g., projection lenses, lithography lenses, and most microscope objectives), vignetting is to be avoided. However, there are many other designs (e.g., almost all camera lenses on the market) where vignetting is used to control aberrations at the edges of an image. Vignetting is also used to reduce the size and weight of these lenses because the image format is rectangular, and the field angle required to cover the corners of the sensor demands much larger element diameters. In such cases, reducing the element diameters increases the vignetting in the corners of the image, making the corners of the sensor darker than the center, but this is balanced "in software," and users never notice.

5.9 A Final Comment

After the type of lens has been chosen and the EFL, field of view, and $f/\#$ (or the corresponding EPD) are selected, the design can be analyzed to see if it provides the required performance. If the analysis shows that the cost and performance of the lens is satisfactory, the design process is complete. Most of the time, some aspects of the lens need to be modified and improved, so the design work continues. To find out how well the lens performs, OpticStudio provides analytical tools to determine the lens errors so that it can be improved. We will look at a set of imaging errors in Chapters 6–8. Perhaps the easiest errors to understand are color errors or chromatic aberrations, which we will explore in Chapter 9. With information from these results, we can begin to analyze lens performance (Chapter 10) and then reduce the imaging errors in lenses (Chapter 11).

Exercise Answers

Ex. 5.1

Two different views of the **OSequiconvex** lens with an object at infinity and object fields of 0°, 7°, and 10° are provided:

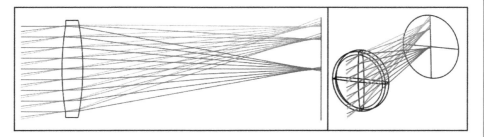

Note: An offset surface (with a thickness of 10 mm) has been added to show the rays before the first lens surface with 10% CSD margin on the lens surfaces. (For the 3D plot Y-fan, $X = -25$, $Y = 60$, and $Z = 27$.)

Ex. 5.2
Change the EPD of the **OSasdoubletAxis** in the System Explorer to 20. The stop is now approximately 7 mm in radius and 14 mm in diameter, as shown in the LDE.

	Surface Type	Comment	Radius	Thickness	Material	Clear Semi-Dia
0	OBJECT Standard ▾		Infinity	Infinity		0.000
1	Standard ▾	Offset	Infinity	10.000		10.000
2	Standard ▾		60.000	6.000	N-SK16	10.000
3	Standard ▾		-150.000	18.000		9.687
4	STOP Standard ▾		Infinity	18.000		7.005
5	Standard ▾		150.000	6.000	N-SK16	4.359
6	Standard ▾		-60.000	19.168 M		3.771
7	IMAGE Standard ▾		Infinity	-		0.225

Ex. 5.3
Change the stop surface of the **OSasdoublet** to surface 5 by clicking on the Make Surface Stop checkbox in the surface 5 properties dialog box. Change the y field angles to 0°, 3.5°, and 5° in the FDE. The result is shown below.

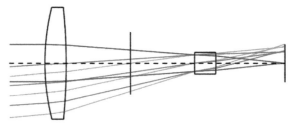

The aperture stop diameter is now 4.4 mm, and the minimum CSD (with the 10% margin and ray aiming activated) for lens #1 is 15.15 mm. This means that a 25-mm-diameter lens (with a semi-diameter of 12.5 mm) would not be large enough to pass all of the rays from the full-field point. Increasing the field further to 10° results in some very odd-looking ray traces as the first lens is not large enough (or thick enough) to accommodate this large of a field with the stop in this location.

Ex. 5.4
The diagonal width of the 22.2 × 14.8 mm sensor is

$$\sqrt{22.2^2 + 14.8^2} = 26.7 \, \text{mm}.$$

The distance from the optical axis to the corner is then $h' = 26.7/2 = 13.3$ mm. Therefore, the angle can be found from $\tan(\theta) = h'/f$. From Table 5.1, the EFL is 49.6; therefore, $\tan(\theta) = 13.3/49.6 = 0.270$, and $\theta = 15.0°$. The three equal-area field angles are then 0, 10.5, and 15°.

Second Hiatus
Rays and Waves

Beginning in classical times, light patterns were investigated as occurring from light traveling in straight lines (e.g., objects casting shadows), and the operation of the simplest optics was described in terms of light rays (see Chapter 1). In the next few chapters, we will also examine the performance of lenses using the ray model, but there are situations where the evaluation of lens performance using rays fails to provide a realistic assessment of the design. This is because light consists of electromagnetic wavefronts moving through space, and its wave properties must be included in the evaluation of lens performance. Because light can be modeled as both a ray and a wave, this dual aspect of light must be addressed to correctly model an optical design. When a light wave encounters an aperture, it can bend around the corner of the aperture and enter the region of the geometrical shadow. While a ray-based model would argue that this is impossible, this wave-based phenomenon is known as diffraction and depends on the wavelength of the light and the size of the aperture in the optical design.

H2.1 Rayleigh Criterion

The first demonstration of the diffraction of light is usually that of light diffracted by a narrow slit. Instead of a narrow band of light beyond the slit, the light pattern is a bright central band flanked by dark and bright bands that diminish away from the central peak. The distribution of light [Fig. H2.1(a)] is described by the function $\sin^2\alpha/\alpha^2$. The variable α depends on the wavelength of the light and the width of the slit. For a lens with a circular aperture, the diffraction profile has circular symmetry and can be described with a circular function called a first-order Bessel function $J_1(x)$. Just as in the case of the narrow slit, the diffraction pattern for the circular aperture is the square of the function over its argument, $4J_1(x)^2/x^2$. This function is known as an Airy function, and the resulting pattern is called an Airy diffraction pattern [Fig. H2.1(b)]. As in the case of the narrow slit, the dimensions of the pattern are determined by the diameter of the diffracting aperture D and the wavelength of the light λ. The angular separation θ between the peak of the pattern and the first dark ring is given by

$$\theta = 1.22\frac{\lambda}{D}. \tag{H2.1}$$

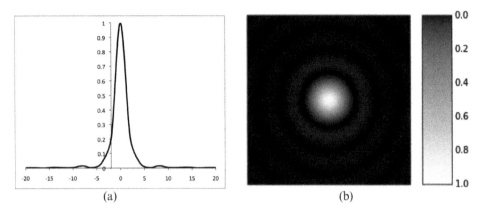

Figure H2.1 (a) Diffraction pattern for a single-slit aperture and (b) Airy diffraction pattern for a circular aperture.

A diffraction-limited lens is a well-corrected lens wherein diffraction dominates any ray-based aberrations in the lens. This type of lens will focus a plane wave from a distant point source to a converging spherical wave, and because of diffraction, the point source image will be an Airy pattern. An example of this is the imaging of a star field by a telescope. Because stars are point sources for most telescopes, the quality of a telescopic system is its ability to determine if an astronomical object is one star or two. An English scientist, Lord Rayleigh, proposed that two point sources of equal brightness are resolved if the peak of the diffraction pattern of one source falls on the first dark ring of the diffraction pattern of the other source. In Fig. H2.2, two point source images are shown with differing separations in the image plane. In Fig. H2.2(a), the images are completely resolved. In the center picture [Fig. H2.2(b)], the sources are just resolved by the **Rayleigh criterion**. In the last panel [Fig. H2.2(c)], the sources are no longer resolved. This criterion is somewhat artificial because it is based on the fact that the two sources are equally bright.

As it is given in Eq. (H2.1), the Rayleigh criterion is not a practical expression for use with an optical design. However, if we multiply the angular separation θ by the image distance, the separation $\Delta\ell$ between the resolved spots in the image plane can be calculated. For an object at infinity, this distance is the EFL. Because the diffracting aperture is essentially the EPD, the Rayleigh criterion can be rewritten as

$$\Delta\ell = \theta \cdot \text{EFL} = 1.22 \frac{\lambda}{\text{EPD}} \cdot \text{EFL}. \tag{H2.2}$$

As expressed in Eq. (5.1), the $f/\#$ of a lens is defined as EFL/EPD. Therefore, the separation between resolved spots can be recast as

$$\Delta\ell = 1.22\lambda \cdot f/\#. \tag{H2.3}$$

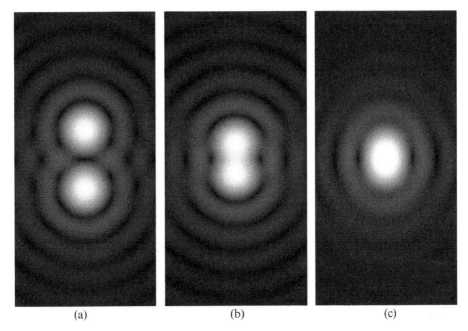

(a) (b) (c)

Figure H2.2 Two point source images separated by (a) the diameter of the first dark ring, (b) the radius of the first dark ring (just resolved according to the Rayleigh criterion), and (c) a distance too small to be resolved.

From this, we see that in the absence of aberrations, smaller details can be imaged with shorter wavelengths and smaller $f/\#$s.

One of the most elaborate and costly imaging lenses is the lithographic stepper lens. This lens is used for imaging patterns onto a photoresist coating on a silicon wafer that forms the basis for integrated circuit chips. By exposing specific areas on the wafer and then using chemical deposition and etching techniques, the electronic elements of the integrated circuits are formed. Because the dimensions of the areas are critical, the lens must have diffraction-limited performance over the area of the chip. By stepping the lens from one chip location to another, an array of integrated circuits can be manufactured. The Rayleigh criterion as expressed in Eq. (H2.3) shows why a high-resolution semiconductor stepper lens operates using ultraviolet lasers near 200 nm and a $f/\#$ less than 1.

H2.2 The Pinhole Camera

As discussed in Section H2.1, the Airy diffraction pattern represents the fundamental limit of the focus of a perfect optical system when analyzed using wave optics. But when do we need to use wave optics versus ray optics to evaluate the performance of a lens? The experiment described below helps explain the ray/wave dual nature of light and gives some insight into when wave optics is needed to completely describe the resolution of an optical system.

To keep things simple, we use a pinhole camera, a dark box with a small pinhole on one side and a translucent sheet (or photographic film) on the opposite side. Two point sources, whose separation can be varied, are imaged by the optical system. The angle at which the overlap of the images of these sources can be just resolved serves as a measure of the resolution of the optical system. Because the ray from each object point goes through the center of the pinhole, all of the rays are chief rays. Thus, the pinhole camera image contains no ray errors, off-axis or on. Geometrically, if the point source is at infinity, the spot size will be nearly the same as the pinhole size. As the pinhole size is reduced, the spot size gets smaller, and the two sources will be resolved when the centers of the spots are separated by their common radius. Therefore, the geometric resolution angle is $\theta_{ray} = D/2$ and is linear with the pinhole diameter D.

However, in Section H2.1, we showed that due to diffraction and the wave nature of light, the resolution angle using the Rayleigh criterion is $\theta_{diff} = 1.22 \lambda / D$. In this case, as D gets smaller, diffraction becomes evident, and the spot begins to grow in size. A plot of these two resolution angles (Fig. H2.3) as a function of pinhole diameter D shows two overlapping curves. So, for large pinholes, the diffraction of light is small, and the ray model is sufficient to model the resolution. But as the size of the pinhole is reduced, the diffraction of the light by the pinhole determines the resolution.

In the next few chapters, we will be using standard ray-tracing techniques to evaluate the performance of a design and then optimize the lens parameters to find a solution with reduced spot sizes. In practice, a designer can keep pushing on the design until the ray-based spot diameter is very small (much smaller than the diffraction limit), but if the final evaluation does not incorporate the wavelike features of light, the design will not perform as expected when built. The designer should be aware of these limitations when evaluating performance with ray-based calculations, such as spot diagrams.

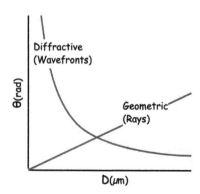

Figure H2.3 Angular resolution θ of an aperture of diameter D.

Chapter 6
Spherical Aberration

Once upon a time, the evaluation of an optical design required a great deal of ingenuity. With no high-speed computers to aid them, designers were forced to find shortcuts and clever approximations to assess the performance of a lens. They developed analytical tools to provide insight from limited input data in the shortest computational time. A paraxial ray trace, accurate for a small area about the optical axis, was used as a baseline measure of a perfect image. If the lens provided a perfect image, the areas on the object that were farther from the axis would also be imaged exactly with the correct magnification. To the extent that a lens fails to do this, the differences between the rays of a perfectly imaged object and those directed by the actual lens are a measure of the optical errors, or **aberrations**, in the lens.

6.1 Propagating Real Rays

Figure 6.1 illustrates the propagation of a ray through an optical system. It compares two rays. One ray is propagated by a lens with no aberrations (a "perfect" lens); the other is an aberrated skew ray propagated by a real lens. To keep the situation as general as possible, all quantities in the figure are normalized to their maximum values. Thus, at a point in the object plane, the fraction of the field between the full field and the axis is designated as h, a proper fraction between one and zero ($1 \geq h \geq 0$).

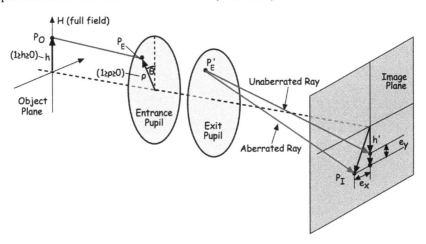

Figure 6.1 Propagation of an aberrated skew ray through an optical system compared to an unaberrated ray.

For systems that are rotationally symmetric about the optical axis, the object point P_O can be located anywhere in the object plane, but for convenience and by convention, the object is located in the *y*-*z* plane, as shown in Fig. 6.1. Starting at P_O in the meridional plane, the ray is launched to a point in the entrance pupil P_E that is not in the plane, making it a skew ray (Section 5.2.3). The location of the ray in the pupil is typically described with normalized polar coordinates (see the box, "Conventions and Coordinate Systems"), where ρ is the fractional distance of the entrance pupil radius (EPD/2), and θ is its angle with respect to the *y* axis. One of the benefits of using the pupils of the system rather than the lens surfaces and apertures is that the operation of the lens can be discussed without needing to provide a specific lens prescription, just the pupil locations and sizes.

As shown in the figure, the skew rays are launched to the same point in the entrance pupil for the real lens and its perfect version. Both travel through the system and emerge from the exit pupil plane at P_E'. For a perfect lens, its unaberrated ray intersects the paraxial image plane at a point in the meridional plane with a paraxial image height h' that is the magnified object height mh. (Because it is a perfect lens, *any* ray from P_O will end in the meridional plane at h'.) In comparison, the aberrated ray rattles through the optical system, obeying the laws of optics, and emerges from a point on the exit pupil P_E' and intersects the image plane at a point P_I that is not in the meridional plane. The *x* distance between the perfect image point and the aberrated image point is the sagittal ray error e_x, whereas the *y* distance between the perfect image point and the aberrated image point is the tangential ray error e_y, as shown in Fig. 6.1.

Conventions and Coordinate Systems

Two coordinate systems (rectangular and polar) may be used to describe a ray intersection with a pupil plane. For example, the location of the intersection of the ray with the entrance pupil P_E can be expressed as either $P_E(\rho_x, \rho_y)$ or $P_E(\rho, \theta)$, where $\rho_x = \rho \sin\theta$ and $\rho_y = \rho \cos\theta$. Which of these is used when discussing aberrations depends on which more easily demonstrates the concept. Also, in the description of ray propagation in Fig. 6.1, we used uppercase symbols for exact quantities and lowercase symbols for relative quantities. You may find that other notations reverse this convention.

6.2 Third-Order Aberrations

In 1857, Phillip Ludwig von Seidel devised a polynomial expansion of the aberrations of a lens in terms of the odd-order products $h^a \rho^b$ of entrance pupil ray location ρ and object height h. (The even-order terms in the expansion are all zero for a rotationally symmetric system.) The lowest-order non-trivial terms, those with third-order products $(a + b = 3)$, are the five third-order aberrations: spherical aberration, coma, astigmatism, Petzval curvature, and distortion. One reason that Seidel's expansion was such an important analytical tool was that the coefficients of each term in the expansion could be computed using only two paraxial rays (the marginal ray and the chief ray). Another was that the Seidel aberration coefficients for each individual surface could be computed separately. This meant that a designer could identify which surfaces contributed most to the errors, which surface contributions canceled each other, and what strategies might be useful for optimizing lens performance. Although the reasons for using this approach grew out of the limited computational power available at the time, these Seidel or third-order aberrations remain useful in this age of high-speed calculations because they can still provide a designer with insight during the design of an optical system.

In a third-order approximation, the transverse ray errors (e_y and e_x, shown in Fig. 6.1) are computed by adding the contribution from each of the five third-order transverse aberrations. These aberrations and their dependence on the pupil ρ and the field h are listed in Table 6.1. We see that four of the five aberrations depend on the field coordinate h, while spherical aberration depends only on the position of the ray in the pupil ρ. So, the spherical aberration of a lens is the same for off-axis object points as it is for the axial object point. Therefore, we will start to understand the evaluation and identification of aberrations by examining the spherical aberration of a lens for an on-axis object point in this chapter. After that, we will look at the third-order ray errors at off-axis points.

Another approach to analyzing the aberrations in an optical system looks at wavefronts in the exit pupil. This model consists of considering each object point as the source of an expanding spherical wave. In a perfect lens, the wavefront propagates through the lens and emerges from the exit pupil as a spherical wavefront converging on the ideal image point. In a real lens, the propagation through the system causes deformations to the wavefront. These deformations from the ideal spherical surface represent the aberrations in a lens, and they can be expressed as a wavefront aberration polynomial whose coefficients can be directly related to the coefficients of the Seidel aberrations.

Table 6.1 Pupil and field dependence for the third-order transverse aberrations, where $\rho_x = \rho \sin\theta$ and $\rho_y = \rho \cos\theta$.

Aberration	ρ-h dependence
Spherical aberration	ρ^3
Coma	$\rho^2 h$
Astigmatism (tangential)	$\rho_y h^2$
Astigmatism (sagittal)	$\rho_x h^2$
Petzval curvature	ρh^2
Distortion	h^3

6.3 On-Axis Ray Errors for a Singlet Lens

For an on-axis object point ($h = 0$), the only third-order aberration for a rotationally symmetric lens will be spherical aberration. This is most easily demonstrated by launching a fan of rays from the axial object point that is spread across the breadth of the entrance pupil. For example, if we locate the object at infinity and the lens exhibited no spherical aberration, all of the rays in the fan would intersect on the axis in the focal plane of the lens. To see what happens in a real lens, we restore the **OSsinglet** (File > Open). If you can't retrieve it, its LDE is shown in Fig. 6.2.

	Surface	Type	Comment	Radius	Thickness	Material	Clear Semi-Dia
0	OBJECT	Standard ▼		Infinity	Infinity		0.000
1	STOP	Standard ▼		120.000	5.000	N-BK7	10.000
2		Standard ▼		-32.464	49.290 M		9.914
3	IMAGE	Standard ▼		Infinity	-		1.350

Title (OSsinglet); EPD (20 mm); Field (0°); Wavelength (d-line)

Figure 6.2 LDE of the OSsinglet.

We can see what happens if we trace a fan of 15 on-axis rays through the lens by selecting Analyze > Cross-Section and changing the **Number of Rays** in the Layout > Settings to 15. As shown in Fig. 6.3, the rays do not all focus in the paraxial image plane. Instead, they cross the axis at different distances from the plane. Because of the large number of rays in a small region, the focal region is magnified in the inset box. Parallel rays close to the axis cross at the paraxial focal point (PF). The pair of marginal rays that enter at the edges of the entrance pupil cross at a point on the axis (MargF) much closer to the lens. There is a point on the axis where the focusing ray bundle is smallest (MinF). This is the location where all of the rays in the fan are contained in the smallest area.

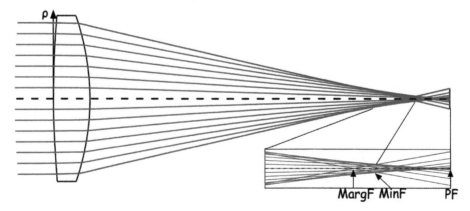

Figure 6.3 Spherical aberration of a singlet with a magnified focal region. MargF is the marginal focal plane; MinF is the plane with the smallest spot containing all the rays; and PF is the paraxial focal plane.

6.4 Displaying Spherical Aberration

As can be seen from the inset in Fig. 6.3, the focal point for a pair of rays varies with their distance from the axis in the entrance pupil. A plot of the focal points for pairs of rays as a function of their normalized ray height in the entrance pupil ρ is shown in Fig. 6.4. To get this plot in OpticStudio®, go to the Analyze tab, choose the Aberrations icon, and then choose Longitudinal Aberration (Analyze > Aberration > Longitudinal Aberration). This plot displays the **longitudinal spherical aberration** (LSA) of the lens when the paraxial focal point is located at 0 (the right-hand portion of the plot has been omitted for clarity in Fig. 6.4).

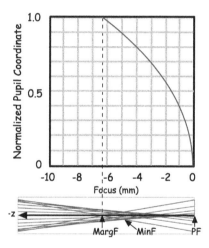

Figure 6.4 Ray bundle exhibiting spherical aberration of a singlet (from Fig. 6.3) and a plot of its longitudinal spherical aberration.

Another way to show the amount of aberration in a lens is to plot the intersection with the image plane of a large number of rays distributed in a 2D array across the entrance pupil from a single object point. All of the rays that get through the optical system are plotted on the image plane. The result is a **spot diagram** for that object point. In OpticStudio, select the Analyze icon, and under Rays & Spots, select Standard Spot Diagram (Analyze > Rays & Spots > Standard Spot Diagram). Go to Settings and change the Ray Density to 11 rays, change the Pattern to Square, and uncheck "Use Symbols."

The resulting spot diagram is shown in Fig. 6.5 (with the legend removed to simplify the figure). A spot diagram gives a designer an idea of how compact the spot is in the evaluation plane, which, in this case, is the paraxial focal plane. The extent of the ray pattern can be estimated with the scale bar in the plot. The root-mean-square (RMS) radius for the spot and the radius of the circle (GEO for geometric) that encloses 100% of the traced rays (typically found in the legend of the plot) have been overlaid on the bottom of the figure.

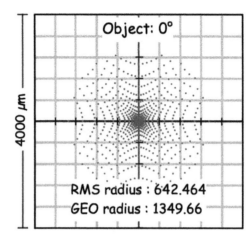

Figure 6.5 Spot diagram for the OSsinglet in the paraxial focal plane.

Figure 6.4 showed that the size of the spot changes with image plane location and that the paraxial image plane is not the location with the smallest spot. Moving the image plane from paraxial focus to another distance along the optical axis is referred to as **defocusing** the lens, and the distance moved is the **defocus**. To defocus our singlet, insert a new surface in the LDE after S2 and label it "Defocus" in the Comment column (see Fig. 6.6). The focus position with the minimum 100% (GEO) diameter that encloses all transmitted rays can be found by manually changing the thickness of the defocus surface (T3) and checking the values on the spot diagram for each change. For this lens we found that a defocus of -4.6 mm yields the smallest GEO radius.

	Surface Type	Comment	Radius	Thickness	Material	Clear Semi-Dia
0	OBJECT Standard ▼		Infinity	Infinity		0.000
1	STOP Standard ▼		120.000	5.000	N-BK7	10.000
2	Standard ▼		-32.464	49.290 M		9.914
3	Standard ▼	Defocus	Infinity	-4.600		1.350
4	IMAGE Standard ▼		Infinity	-		0.331

Figure 6.6 LDE of the OSsinglet with a defocus surface (S3) inserted.

OpticStudio also has a quick optimization feature that finds the optimum focus of a lens based on the greatest amount of energy delivered to the smallest area. Open the Optimize tab and click the Quick Focus icon (Optimize > Quick Focus). This opens a window to select the type of optimal focus point. Choose Spot Size Radial from the drop-down menu. When clicked, OpticStudio calculates the location of the best-focus image plane (for that criteria) and replaces the thickness of surface 3 with that value (in this case, -4.059 mm). The spot diagrams at the three different focus positions (paraxial, minimum GEO, and optimum are shown in Fig. 6.7 for comparison.

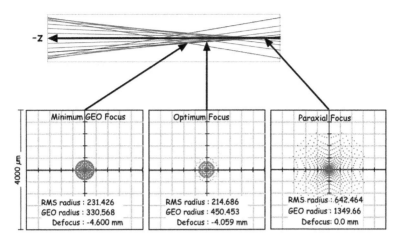

Figure 6.7 Spot diagrams for the OSsinglet for three defocus distances.

One of the applications for spot diagrams is to determine the size of the radiation pattern from a single object point relative to the sensor elements in a detector array as well as the resolution requirements of the system. However, this geometrical calculation does not include diffraction effects (see the Second Hiatus and Chapter 11 for more information). This can be misleading if the lens is diffraction limited and very well corrected for aberrations. For these types of systems, it is better to use the diffraction-based calculation of enclosed energy (Analyze > Enclosed Energy).

6.5 Transverse Ray Plots

Although spot diagrams provide some insight into the amount of error in a lens, it can be difficult to use them to determine which aberrations are causing the error, especially, when multiple aberrations are present. One approach to distinguishing specific aberrations in a lens is to plot the ray errors (e_x and e_y) for fans of rays through the entrance pupil (both tangential and sagittal). The resulting figure is called a **transverse ray plot**. A description of how this plot is generated is given on the next page in the box, "Constructing a Transverse Ray Plot."

Reset the focus position for the OSsinglet back to paraxial focus (T3 = 0) and create its transverse ray fan plots by selecting Analyze > Rays & Spots > Ray Aberration. This generates a window with two tabs. The "Graph" tab displays the transverse ray plots (Fig. 6.8). The left-hand plot is for the tangential fan (e_y vs ρ_y when $\rho_x = 0$), and the right-hand plot is for the sagittal fan (e_x vs ρ_x when $\rho_y = 0$). These plots have been automatically scaled to a value of ± 2000 μm [see a portion of the plot legend in Fig. 6.9(a)] so that all points on the curves are shown. The "Text" tab lists the data used to generate these curves for 41 rays across the pupil from the lowest marginal ray ($\rho = -1$) to the highest ($\rho = +1$). Scrolling down to 1.000, the ray error e_y for a ray at the edge of the pupil ($\rho = +1$) is listed as −1349.656178 μm [Fig. 6.9(b)] or −1.349656 mm.

Constructing a Transverse Ray Plot

To show how a transverse ray plot is generated, we will propagate a fan of rays in the *y-z* plane through a singlet lens. A 3D depiction of the lens with a fan of tangential rays (another term for meridional rays) is generated using the 3D viewer to produce a fan of 15 rays into a 3D wireframe of the lens. In the following figure, the 3D rendering includes graphic elements that have been added to explain the generation of a transverse ray plot, which is displayed beyond the grid in the image plane.

If not for spherical aberration, all rays in the fan would cross the optical axis at the image plane. Because of spherical aberration, the error for each ray in the fan will be the distance e_y of the ray from the optical axis. The transverse ray plot always shows the ray's transverse error (the plot's *y* axis) versus the ray's pupil coordinates (the plot's *x* axis). It is important to understand that the ray's pupil coordinates can be along the *x* or *y* axis of the lens but will be shown as *x* values on the plot; and the ray's transverse error can be e_x or e_y but will be shown as *y* values on the plot. In the following illustration, the ray error e_y for each of the rays in the tangential fan is plotted on the *y* axis as a function of the location of that ray in the entrance pupil (from $+\rho_y$ to $-\rho_y$) along the *x* axis. (If there were no ray errors, all rays would cross the image plane at the axis, and the plot would be a straight line along the *x* axis.) A complete transverse ray plot also includes a (sagittal) plot of the e_x ray error as a function of pupil coordinate ρ_x and gives both a tangential and sagittal plot for every field point *h* defined in the lens.

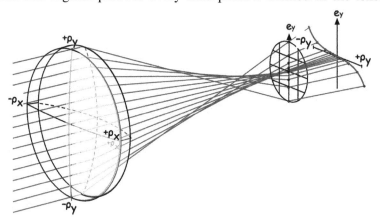

[Note: To demonstrate the construction of this transverse ray plot, the singlet was opened up to 30 mm ($f/1.67$) and its thickness was increased so that the more-aberrated marginal rays farther off the axis could be included. However, all calculations for the OSsinglet in the surrounding text are derived from the $f/2.5$ OSsinglet we have been using.]

Spherical Aberration

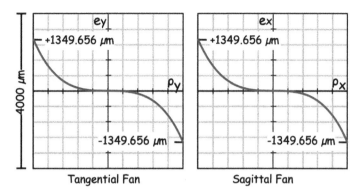

Figure 6.8 Transverse ray curves for the OSsinglet evaluated at the paraxial image plane.

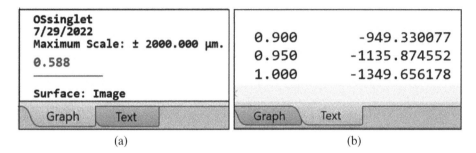

Figure 6.9 Segments of the Ray Aberration window (a) Graph tab (b) Text tab.

The transverse ray curves provide the designer with a means to identify the aberrations in a lens. For this lens, the S-shaped curve indicates that third-order spherical aberration is present. As listed in Table 6.1, spherical aberration is the only third-order aberration to show up on axis (at $h = 0$), and it is the only third-order aberration with a cubic dependence in ρ, hence, the S-shaped curve. The sagittal fan (at a right angle to the tangential fan) shows exactly the same curve as the tangential fan. This is because the object is on the optical axis, and the sagittal fan will be focused by the lens in the same way. In the next chapter, when we look at off-axis points, this will not be the case.

One of the attractions of the transverse ray curve is that when the evaluation plane location is changed by defocusing, the shape of the curve does not change. It simply rotates about the origin, so the aberration curve is recognizable for any evaluation plane. For example, in Fig. 6.10, the curves are plotted for the paraxial focus plane (PF), the marginal focus plane (MargF), and the minimum 100% diameter plane (MinF). Even when additional aberrations are present, this will still be true. (Note that this plot also uses the faster and slightly thicker lens that was used in the box, "Constructing a Transverse Ray Plot" to create more spherical error for a better display.)

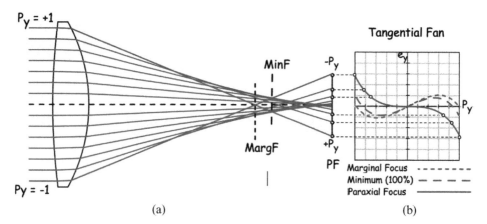

Figure 6.10 Generation of transverse ray curves: (a) tangential ray fan and (b) transverse curves for the corresponding evaluation planes.

Two other features of these curves should also be pointed out. First, when the evaluation plane is the paraxial focus, those rays entering close to the axis focus close to the paraxial focus, so the ray errors are very small, and the transverse ray curve is tangent to the x axis at the origin, as can be seen in Fig. 6.8. Therefore, any slope at the origin in these curves indicates that the lens is not being evaluated at the paraxial image plane, and defocus is present. Second, the diameter of the aberration blur spot can be estimated from these plots by drawing a horizontal line through the maximum point on the curve and a similar line through the minimum point of the curve. Together these lines form a band that encloses all points on the curve and indicates an estimate of the spot diameter.

Figure 6.11 shows an example of this type of calculation for the OSsinglet at the minimum GEO focus plane (use a defocus distance of −4.6 mm and a plot scale of ±2000 μm). The width of the curve, estimated as a fraction of the e_y axis, provides a rough measure of the minimum 100% spot diameter. From the figure, the band containing the curve is about 16% of the 4000-μm axis length, or about 0.640 mm, which is very close to the 0.66-mm 100% spot diameter given in Fig. 6.7 for this evaluation plane.

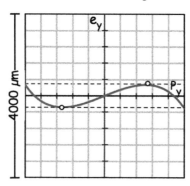

Figure 6.11 Construction to estimate the minimum 100% spot diameter from the transverse ray curve.

6.6 Third-Order Aberration Coefficients

As shown in the previous section, for the case of spherical aberration, the transverse ray error (e_y or e_x) is dependent on the height of the ray in the entrance pupil and goes as the cube of the relative entrance pupil ray position ρ^3. The magnitude of the total error at $\rho = 1$ is then determined by the transverse spherical aberration (TSPH) coefficient:

$$e_y(\rho, h) = \text{TSPH}\rho^3, \qquad (6.1)$$

which can be calculated using a paraxial marginal ray and its values (height, angle of incidence, etc.) at each surface in the lens. The formulas for the calculation of the aberration coefficients can be quite complex and are beyond the scope of this text, but they can be found in many optical design references.

OpticStudio automatically calculates these aberration coefficients for each surface in the lens in the analysis option Analyze > Aberrations > Seidel Coefficients. There you will find five listings of the aberration coefficients in different formats and units. We have chosen to use the third set of coefficients, the Transverse Aberration Coefficients, because they are given in terms of ray error (in millimeters) in the image plane and relate directly to the transverse ray plots discussed in the last section. All of the labels for transverse errors begin with a "T" for transverse. Table 6.2 gives a full summary of the transverse aberration coefficients, their labels, and their definitions. Two of the coefficients (TSFC and TTFC) are combinations of two other coefficients (TAST and TPFC) and are useful when discussing field curvature (Chapter 8).

As with the **FIRST** macro, we have created a **THIRD** macro that quickly computes and displays only the transverse aberration coefficients to simplify the aberration analysis output. One of the coefficients, the sagittal coma (TSCO) is equal to one-third of the tangential coma (TTCO). TSCO is not needed for the text, so it is omitted and not reported by THIRD. The full path to access this macro is Programming > Macro List > THIRD.ZPL. [Note: If your macros list does not include the **THIRD** macro, you can create one using the listing and instructions in the Appendix.]

Table 6.2 Transverse aberration coefficients.

Labels	Aberrations	
TSPH	**SPH**erical aberration	
TSCO	**S**agittal **CO**ma	(not used in THIRD)
TTCO	**T**angential **CO**ma	
TAST	Average **AST**igmatism	
TPFC	**P**etzval **F**ield **C**urvature	
TSFC	**S**agittal **F**ield **C**urvature	(= TAST/2 + TPFC)
TTFC	**T**angential **F**ield **C**urvature	(= 3TAST/2 + TPFC)
TDIS	**DIS**tortion	
TAXC	**AX**ial **C**olor	
TLAC	**LA**teral **C**olor	

Restore the **OSsinglet**. Table 6.3 lists the third-order aberration coefficients as given by the output of the **THIRD** macro where the axial and lateral color errors, TAXC and TLAC, were omitted. All coefficients except TSPH are zero. This is because there is only a single, axial field, and with the exception of spherical aberration, all other values represent off-axis contributions. When additional fields are added in Chapter 7, the table will be filled with non-zero values. Note: These values are computed for the full aperture (when $\rho = 1$) and full field (when $h = 1$).

Table 6.3 Third-order aberration coefficients for the OSsinglet (THIRD).

Surf	TSPH	TTCO	TAST	TPFC	TSFC	TTFC	TDIS
STO	-0.0032	-0.0000	-0.0000	-0.0000	-0.0000	-0.0000	-0.0000
2	-1.1333	0.0000	-0.0000	-0.0000	-0.0000	-0.0000	0.0000
IMA	-0.0000	-0.0000	-0.0000	-0.0000	-0.0000	-0.0000	-0.0000
TOT	-1.1366	0.0000	0.0000	0.0000	0.0000	0.0000	0.0000

In the TSPH column of Table 6.3, there is a contribution to the spherical aberration of −0.0032 for the stop/first surface of the lens and −1.1333 for the second surface of the lens. So, the strongly curved back surface is responsible for almost all of the spherical aberration in this lens. The total spherical aberration error (TOT) is their sum, −1.1366. The units of the coefficients are millimeters, and the value for the total, −1.1366 mm, is slightly less than the exact value displayed in the transverse ray plot (when converted to millimeters), −1.349656 mm. This discrepancy occurs because the TSPH in Table 6.2 includes only the third-order term in the polynomial expansion of the aberrations, and the higher-order contributions (e.g., fifth- and seventh-order spherical aberration) are ignored.

One method of reducing spherical aberration is to stop down the lens, i.e., to reduce the EPD so that it cuts off the more aberrant rays at the edge of the pupil. This also significantly reduces the higher-order spherical aberration contributions. For example, the effect of reducing the EPD from 20 mm to 10 mm on the transverse ray plot for the OSsinglet can be seen in Fig. 6.12.

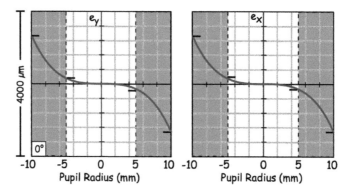

Figure 6.12 A modified transverse ray plot for the OSsinglet. The unshaded area encloses the curve for the lens when it is stopped down to a 10-mm EPD.

The unshaded area in Fig. 6.12 contains the part of the transverse curve for the OSsinglet when it is stopped down to a 10-mm EPD. The transverse ray error is reduced dramatically—from -1349.656 μm to -147.797 μm, and the third-order value for TSPH (-0.1421 mm = 142.1 μm) matches the transverse ray aberration curve to within a few microns.

6.7 Lens Bending

A simple optimization of a lens design with the goal of reducing aberrations can be demonstrated using the current OSsinglet by reversing its surfaces. It is easy to flip a lens over in your hand. It is almost as easy to do so in OpticStudio. Restore the **OSsinglet**. In the LDE, add a defocus surface before the image plane and then swap the radii of curvatures and signs: $R1 = +32.464$; $R2 = -120$. You can also do so by selecting the R1 and R2 cells and clicking the Reverse Elements icon (circled at the top of the LDE in Fig. 6.13). Save this lens as **OSsingletRev**.

	Surface Type	Comment	Radius	Thickness	Material	Clear Semi-Dia
0	OBJECT Standard ▼		Infinity	Infinity		0.000
1	STOP Standard ▼		32.464	5.000	N-BK7	10.000
2	Standard ▼		-120.000	47.376 M		9.669
3	Standard ▼	Defocus	Infinity	0.000		0.450
4	IMAGE Standard ▼		Infinity	-		0.450

Figure 6.13 LDE for a reversed OSsinglet (OSsingletRev).

Comparing the lens cross-section (Fig. 6.14) of the reversed OSsinglet (with 15 rays) to that of the earlier version (Fig. 6.3), it is evident that spherical aberration has been considerably reduced, and this new orientation of the same lens yields considerably better performance. This is also reflected in the reduction of its third-order spherical aberration (TSPH) coefficient (Table 6.4).

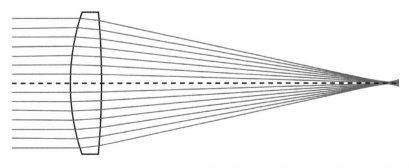

Figure 6.14 Reduced spherical aberration of the "flipped" OSsinglet (OSsingletRev) lens. Compare this to Fig. 6.3.

Table 6.4 Third-order aberration coefficients for the reversed OSsinglet.

```
Surf   TSPH    TTCO    TAST    TPFC    TSFC    TTFC    TDIS
STO   -0.1641 -0.0000 -0.0000 -0.0000 -0.0000 -0.0000 -0.0000
  2   -0.2411  0.0000 -0.0000 -0.0000 -0.0000 -0.0000  0.0000
  3    0.0000 -0.0000  0.0000  0.0000  0.0000  0.0000 -0.0000
IMA   -0.0000 -0.0000 -0.0000 -0.0000 -0.0000 -0.0000 -0.0000
TOT   -0.4053  0.0000  0.0000  0.0000  0.0000  0.0000  0.0000
```

As we did with the original lens in Section 6.4, we can display spot diagrams for the lens at three different focus positions: the minimum 100% spot diameter, the quick focus, and the paraxial focus (Fig. 6.15). Certainly, the most dramatic effect of flipping the lens is the large reduction in spherical aberration. Compared to the spot sizes listed in Fig. 6.7 for the original lens orientation, the values for the flipped lens shown in Fig. 6.15 are about one-third as large. For example, the minimum 100% spot radius (GEO) was originally 330 μm, whereas for the reversed lens it is 111 μm. The spot diagrams are plotted at one-quarter of the scale of the unflipped lens (Settings: ray density of 11 rays with square pattern). This reduction is also evident in transverse ray aberration curves for the flipped lens (Fig. 6.16) plotted at one-quarter of the scale of the unflipped lens (Fig. 6.8).

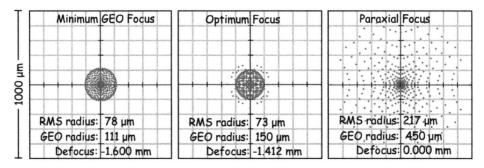

Figure 6.15 Spot diagrams for the OSsingletRev.

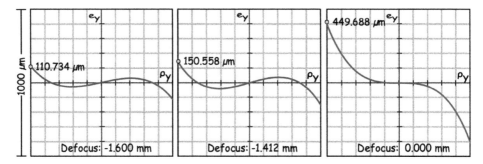

Figure 6.16 Tangential transverse ray curves for the OSsingletRev evaluated at three different defocus planes.

Another way to approach reversing the singlet (flipping the lens) is to think of it as if it were made of a flexible material. Instead of looking at the reversed singlet as a lens that was simply rotated about an axis perpendicular to its optical axis, its shape was formed by pulling on the center of the front surface to get a larger curvature c_1 (shorter radius of curvature) and pushing on the second surface to get a smaller curvature c_2 (flatter surface). If the difference between the curvatures of the two surfaces $(c_1 - c_2)$ remains the same, the optical power (and focal length) doesn't change [Eqs. (1.9) and (1.10)]. Thus, there is a range of combinations of curvatures that will generate a 50-mm-EFL lens.

By bending the lens through a range of combinations, the spherical aberration of the lens can be examined to determine which lens shape will result in the smallest error. One way to keep track of the lens bending is to use a shape factor q, which is defined as

$$q = \frac{R_2 + R_1}{R_2 - R_1}. \qquad (6.2)$$

Figure 6.17 illustrates the shape factor for five positive lens shapes located along the q axis.

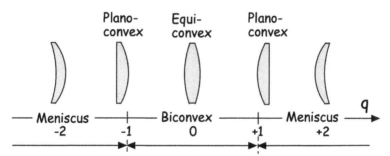

Figure 6.17 Shape factor illustrated.

Specifically, for a 50-mm-EFL thin lens (optical power $\phi = 1/50$ mm $= 0.02$ mm^{-1}) made of N-BK7 ($n = 1.517$) glass, we can solve for the bending factor $\beta = (c_1 - c_2)$, using $\phi = (n-1)\beta$ [Eq. (1.10)] to get 0.03868 mm^{-1}. Thus, any N-BK7 lens with this bending factor is a 50-mm-EFL lens. For the unreversed OSsinglet, the shape factor q ($R_1 = +120$ mm and $R_2 = -32.5$ mm) is -0.57. When the singlet is reversed, its shape factor has the same magnitude, but its sign is reversed, $+0.57$.

Lens bending is most easily done using a computer-based approach by tracing rays through a lens while varying the curvature of the first surface. The lens is "bent" by manually changing the radius of curvature of the first surface while using an angle solve to determine the curvature of the second surface that maintains the focal length of the lens (as was described in Section 3.5 and illustrated in Fig. 3.14 for a 50-mm EFL). Thus far we have determined the TSPH for two shape factors, the original lens and its reverse. But we don't know the optimum lens shape, i.e., the lens with the smallest TSPH.

It is possible to perform lens bending by manually changing R1 in the LDE over a range of values and then checking to see when the TSPH has reached a minimum. However, OpticStudio has a nifty feature, a Slider, which makes it easier to find the optimum shape for the 50-mm OSsinglet. To use it, first, we must set up the LDE. Restore the **OSsingletRev** and put a marginal ray angle solve of −0.2 on the radius of S2 to maintain a 50-mm EFL. If there is a defocus thickness, set it to zero so that the back focal distance is at the paraxial focus plane. The LDE for the lens should now look like the LDE shown in Fig. 6.18.

	Surface Type	Comment	Radius		Thickness		Material	Clear Semi-Dia
0	OBJECT Standard ▾		Infinity		Infinity			0.000
1	STOP Standard ▾		32.464		5.000		N-BK7	10.000
2	Standard ▾		-120.000	M	47.376	M		9.669
3	Standard ▾	Defocus	Infinity		0.000			0.450
4	IMAGE Standard ▾		Infinity		-			0.450

Figure 6.18 LDE for the OSsingletRev to be used with a Slider to optimize the lens for minimum TSPH using lens bending.

Slider is the third icon on the Optimize banner (Optimize > Slider). Click on its icon, and the Slider window will open. The radius of surface 1 should be preselected as the variable parameter (if it is not, don't worry, we're going to change it). For this next exercise, it is easier to work in curvature space ($c = 1/R$), so change the Parameter drop-down menu from Radius to Curvature and check that the Surface number is 1. A range of curvatures to explore can then be specified in the Start and Stop cells. In this case, we will permit the curvature to vary between 0.0 (R1 = infinity) and 0.05 (R1 = 20 mm). As we change the lens shape, we can monitor the changes in any Analysis window (the default setting is one where all open windows will be updated). If we check the box, Update ZPL Windows, the changes in the transverse ray coefficients in the **THIRD** macro are also updated as we "slide." Your Slider window should now look like Fig. 6.19.

Figure 6.19 Slider to vary c_1 between 0 (R1 = Infinity) and 0.05 (R1 = 20 mm).

Of the various plots offered in the Analyze tab, the one that provides the best feedback for this exercise, is the Ray Fan diagram (Analyze > Aberrations > Ray Aberration). As you move the slider, you can see the Ray Fan curve grow larger and smaller as the lens shape is changed. However, there will be jumps in the plot because the Ray Fan plot autoscales—as the error curve gets flatter, the plot scale is reduced. To eliminate these jumps, open the Ray Fan Settings and change the Plot Scale to 1000. This will fix the size of ray fan e_y axis and permit us to watch the changes in the Ray Fan curve as the lens shape is changed, without any jumps in the plot.

When you click the Animate button in Slider, you will see the ray fan curve changes as Slider cycles c_1 between the values of 0.0 and 0.05. The animation won't tell you much—except that somewhere in the 0 to 0.05 range there is a lens shape with the smallest TSPH. The question is, where is it? The answer is arrived at by slowly moving the Slider bar through its range and looking at both the values on the Text tab on the Ray Fan plot and the TSPH value in **THIRD**.

For example, we know the OSsingletRev was an improvement over the OSsinglet. We can scan the performance of the singlet around this value ($c_1 = 1/R_1 = 1/32.464 = 0.031$) using the slider. If you click on the slider bar in the Slider window, pressing the left or right key will then change c_1 by 0.001. Examining the Ray Fan data and the **THIRD** text window, for each c_1 value, a table of the performance as a function of curvature can be generated (see Table 6.5). The minimum value for the tangential ray error and the TSPH are in blue. The lens shape providing the best performance is approximately $q = 0.692$, which is not far from the $q = 0.59$ for the reversed OSsinglet. There, $R1 = 1/c_1 = 1/0.033 = 30.3$ mm.

Table 6.5 Tangential Ray Errors and TSPH for a range of lens shapes for a 50-mm EFL singlet.

C1	q	Tan. Ray Err.(μm)	TSPH
0.031	0.590	-448.527	-0.4042
0.032	0.641	-444.256	-0.3999
0.033	0.692	-442.735	-0.3980
0.034	0.744	-443.962	-0.3984
0.035	0.795	-447.942	-0.4011

Using **THIRD** to calculate TSPH for a range of values of R1 from 120 mm to 20 mm and calculating the q values for each R1, pairs of q and TSPH values can be entered in a spreadsheet or plotting program to produce a plot of spherical aberration versus q like the one shown in Fig. 6.20. (Note that the values of R1 used for this plot were 120, 110, 100, 90, 80, 70, 60, 50, 40, 35, 31, 30, 29, 25, 22, and 20.) The minimum absolute value of TSPH for the lens is close to $q = 0.0.692$ and c_1 (R1 = 30 mm). This is not far from the R1 value 32.464 mm for the flipped lens (Fig. 6.18).

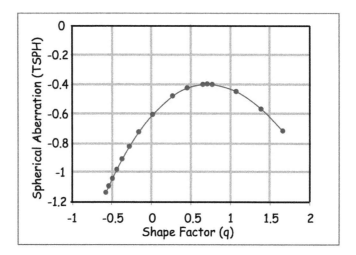

Figure 6.20 Plot of spherical aberration as a function of the shape factor q of the N-BK7 singlet. The minimum absolute value of TSPH is in the vicinity of $q = 0.7$ (R1 = 30 mm).

6.8 Going Off-Axis

In this chapter, we demonstrated the properties of spherical aberration and explored its behavior for an on-axis object point using transverse ray curves and spot diagrams. If spherical aberration were the only aberration in the lens, the performance would be the same across the entire field. However, at field points away from the axis, additional errors are added to the spherical aberration that is present. These were listed in Table 6.1. In the next chapter, we examine the next two third-order transverse errors: coma ($\rho^2 h$) and astigmatism ($\rho_y h^2$ and $\rho_x h^2$). Then in Chapter 8, the effects of field curvature, or Petzval curvature (ρh^2), and distortion (h^3) will be addressed. As we have seen in this chapter, third-order aberrations and transverse ray curves for a lens can provide the designer with a basic evaluation of its design at a glance, proving useful even in this age of high-speed calculations.

Exercises

Exercise 6.1 Biconvex 50-mm lens
Start with the **OSsinglet** and add a marginal ray angle solve to the radius of the second surface to maintain a 50-mm EFL as you change the radius of curvature of the first surface. Manually change R1 (or use a slider) to find the lens solution that results in a biconvex lens ($R_2 = -R_1$). Determine the TSPH of the new lens.

Exercise 6.2 Using a plano-convex lens
Start with the **OSsinglet**, make the first surface flat, and then add a marginal ray angle solve to the radius of the second surface to yield a 50-mm EFL. What is the radius of curvature of the second surface? Determine the TSPH for this lens. Flip (reverse) the lens and reevaluate the TSPH of the lens. Considering that many catalog singlet lenses found in research labs are plano-convex lenses, what rule might be useful when placing a plano-convex lens in an optical system with a collimated beam?

Exercise 6.3 Singlet with a different glass
Start with the **OSsinglet**, change the material from N-BK7 to F2, and then add a marginal ray angle solve to the radius of the second surface to yield a 50-mm EFL. Try using the slider technique from Section 6.7 to find the lens shape that gives the smallest spherical aberration. What is R_1 and shape factor q for the smallest TSPH? How does the minimum TSPH value for F2 compare to the minimum TSPH value and lens shape for N-BK7? Spherical aberration is highly dependent on the index of refraction of a material. What is the index of refraction of the two materials at the design wavelength?

Exercise 6.4 Germanium lens
Start with the **OSsinglet**, change the wavelength to 10 μm and the material to germanium (located in the INFRARED material catalog). Add a marginal ray angle solve to the radius of the second surface to yield a 50-mm EFL. Find the lens shape that gives the smallest spherical aberration. What is R_1 for the smallest TSPH? What is the index of refraction of germanium at the design wavelength? How do the minimum TSPH value and lens shape for germanium compare with the minimum TSPH value and lens shape for N-BK7?

Exercise Answers

Ex. 6.1

Open the **OSsinglet** and add a −0.2 marginal ray angle solve to the radius of the second surface to maintain a 50-mm EFL. You can change the R1 radius manually to find a biconvex ($R_2 = -R_1$) lens solution or use a slider. For example, an R1 radius of 60 mm yields an R2 radius of −44.098 mm. Changing the radius to be halfway between 60 mm and 44 mm gets close to equal radii of curvature:

    R1 = 50 gave R2 = −51.655;
    R1 = 50.8 gave R2 = −50.827;
    R1 = 50.814 gave R2 = −50.813

Close enough.
Running the THIRD macro gives a TSPH SUM = −0.6159.

Ex. 6.2

Open the **OSsinglet**, make the first surface flat (R1 = infinity), and add a −0.2 marginal ray angle solve to the radius of the second surface to maintain a 50-mm EFL. The R2 radius is now −25.84. The TSPH for this lens is −1.7228, while it is almost 4× lower (−0.4311) for the flipped lens. The rule for using plano-convex lenses with large object distances / collimated beams is to orient the lens with the flat side to the focus. This results in much lower spherical aberration.

Ex. 6.3

A plot of the spherical aberration as a function of shape factor q for the new F2 material (shown below) shows that the lens with minimum spherical aberration has $R_1 = 33$ mm, $q = 0.872$, and TSPH = −0.3101.

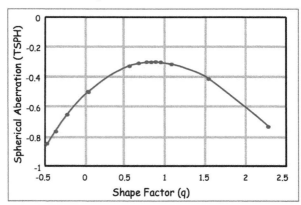

The index of refraction of N-BK7 at the design wavelength (Analyze > Reports > Prescription Data) is 1.5168, while the index of refraction of F2 is 1.62004. As the index of refraction of the lens material increases, the minimum spherical aberration (SA) decreases.

Ex. 6.4
For a germanium singlet at a 10-μm wavelength, the lens shape that results in minimal SA has $R_1 = 51$ mm, $q = 5.963$, and TSPH $= -0.0541$.

The index of refraction of germanium at the design wavelength is 4.0044 (much larger than the index of refraction of N-BK7). Due to the relatively high index of refraction, the minimum TSPH for the germanium lens is significantly smaller (~8×), and the shape of the lens is much more meniscus ($q > 1$).

Chapter 7
Coma and Astigmatism

In the previous chapter, we demonstrated the properties of spherical aberration and explored its behavior along the optical axis using transverse ray curves and spot diagrams. If spherical aberration were the only aberration in the lens, the performance would be the same across the entire field. However, at field points away from the axis, additional errors add to spherical aberration. In this chapter, we examine the next two third-order transverse errors: coma ($\rho^2 h$) and astigmatism ($\rho_y h^2$ and $\rho_x h^2$).

7.1 Coma

Restore the $f/2.5$ **OSsinglet** with an object at infinity. In the FDE (Setup > Field Data Editor), enter our standard set of fields (0°, 7° and 10°) as described in Section 5.2.5. For more complex lenses, these angles are fairly modest, but for a singlet, they are large enough to demonstrate two off-axis errors, coma and astigmatism. As is indicated in Table 6.1, coma, or comatic error, is quadratic in ray height in the pupil ρ and linear in ray height in the field h. The maximum coma occurs for rays from the largest field value $h = 1$ and at the edge of the pupil $\rho = 1$.

To examine the off-axis errors in the singlet, create a lens cross-section (Analyze > Cross-Section) with an off-axis fan of 15 rays from the 7° field (F2) as shown in Fig. 7.1. This is generated by changing the Number of Rays to 15 and the Field to 2 in the Layout Settings window. For reference, we have thickened the 7° chief ray. If you want to see the area near the focal region in more detail, you can use the magnifying glass icon on the Layout toolbar to draw a box around the area you want to zoom in on. To reset the zoom of the layout, click the black icon with the counter-clockwise circular arrow at the end of the Layout toolbar.

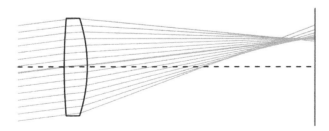

Figure 7.1 Meridional ray trace of the OSsinglet for a 7° object field.

The off-axis performance of the lens can be compared to the on-axis performance (Sections 6.4–6.5) by plotting the transverse ray curves (Analyze > Rays & Spots > Ray Aberration) and spot diagrams (Analyze > Rays & Spots > Standard Spot Diagram) for all three fields. In Fig. 7.2, the transverse ray plots and spot diagrams for the different field points have been stacked on top of each other and then placed side by side for an easy comparison. The plots scale to a neat 10 mm (10,000 μm) where the spot diagrams were generated by changing the Ray Density to 11 rays, the Pattern to Square, and "Use Symbols" was unchecked.

The on-axis curves at the top of the transverse ray plots in Fig. 7.2(a) are the same as the curves in Fig. 6.8 in the last chapter. The off-axis curves are also S-shaped curves, but they are asymmetric about the origin because of coma. The off-axis spot diagrams [Fig 7.2(b)] also show this asymmetry, while the on-axis spot diagram, the top pattern, is the same as Fig. 6.5. As discussed in Chapter 6, a spot diagram gives a good idea of the size of the overall image blur, but it is much easier to identify aberrations by the shape of the curve in a transverse ray plot.

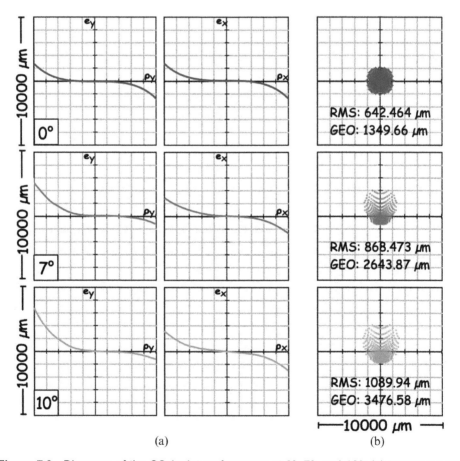

(a)   (b)

Figure 7.2 Diagrams of the OSsinglet performance at 0°, 7°, and 10°: (a) transverse ray curves and (b) spot diagrams.

Running the **THIRD** macro introduced in Chapter 6 for the OSsinglet with a 10° field generates the third-order aberration coefficients listed in Table 7.1. At the largest field angle, 10°, the transverse coma (TTCO) is 0.98636 and is approximately the same amount as the third-order spherical aberration (−1.1366) but opposite in sign. This leads to an approximate cancellation of error in the 10° tangential transverse ray plot in Fig. 7.2(a) for $+\rho$ (the right side of the curve) and an approximate doubling of the error for $-\rho$ (the left side of the curve).

Table 7.1 Third-order errors for the OSsinglet (EPD = 20 mm, Max. Field = 10°).

Surf	TSPH	TTCO	TAST	TPFC	TSFC	TTFC	TDIS
STO	-0.0032	-0.0206	-0.0291	-0.0221	-0.0366	-0.0657	-0.0775
2	-1.1333	1.0070	-0.1988	-0.0816	-0.1810	-0.3798	0.0536
IMA	-0.0000	-0.0000	-0.0000	-0.0000	-0.0000	-0.0000	-0.0000
TOT	-1.1366	0.9864	-0.2279	-0.1036	-0.2176	-0.4455	-0.0239

Exercise 7.1 Third-order coma variation with pupil diameter
Change the EPD for the **OSsinglet** (with a 10° full field) to 5 mm and 10 mm, and show that the third-order coma error (TTCO) varies quadratically with aperture ρ^2, as given in Table 6.1. Include in the results the TOT line for the 20-mm EPD from Table 7.1.

Exercise 7.2 Third-order coma variation with field
Change the maximum field for the **OSsinglet** to 5° in the FDE and find the third-order coma error (TTCO). Do the same for 20°. Including the values for the 10° full field given in Table 7.1, show that the coma error varies linearly with field h, as given in Table 6.1. Note that the field coordinate h varies as the tangent of the field angle [see Eq. (2.1) in Chapter 2].

7.1.1 Aberration contributions

Because spherical aberration and coma are both present in this lens, it is not easy to see the effect of pure coma on the transverse ray curves or spot diagrams. However, we can use the aperture and field dependence of each aberration given in Table 6.1 to simulate what coma ($\rho^2 h$) or other lens errors would like on the transverse ray curves. These plots can be created in Excel or any other graphing program.

For example, Fig. 7.3 displays a set of simulated transverse ray curves for a lens with a 10° field and −100 μm of spherical aberration. As expected, we see the same cubic curve plotted for all three fields because spherical aberration is independent of field h, and the maximum value for $\rho = +1$ is −100 μm.

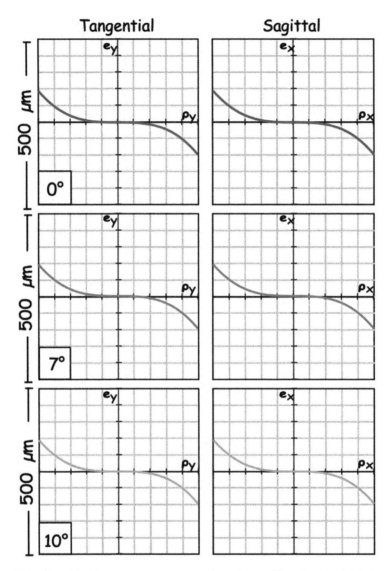

Figure 7.3 Simulated transverse ray curves for a lens with only spherical aberration.

We can also simulate a lens that has pure coma for fields of 0°, 7°, and 10°. The transverse ray curves and spot diagrams for these fields are shown in Fig. 7.4. What is obvious from the transverse ray plot for a lens with pure coma is that the curve is U-shaped [Fig. 7.4(a)], i.e., quadratic in pupil coordinate ρ. In contrast, the spherical aberration curve is S-shaped (Fig. 7.3 above), i.e., cubic in pupil coordinate ρ.

Coma and Astigmatism

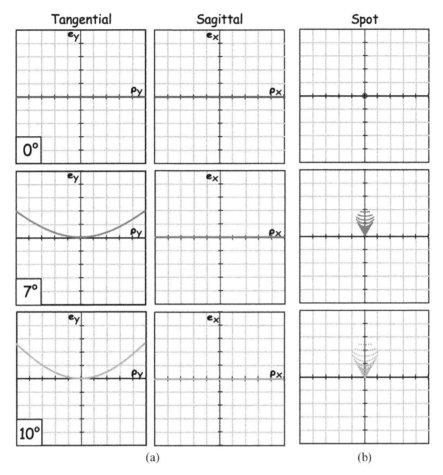

Figure 7.4 Simulated lens plots for pure coma at 0°, 7°, and 10°: (a) transverse ray curves and (b) spot diagrams.

The spot diagram for a lens with pure coma is shown in Fig. 7.4(b). For off-axis point sources, such as stars seen through a telescope, their images are shaped like a comet. It is this resemblance that gives the aberration its name. (The term "comet" comes from the Greek word *kometes*, a long-haired star, which derives from the Greek word for hair, *kome*.) It can be shown that the comatic image of a star is confined to a V-shaped area with a 60° angle.

When both spherical aberration and coma are present in approximately equal amounts and they are summed, the resulting curve shows a larger amount of error for rays on one side of the axis and near cancellation on the other, as shown in the 10° curve in Fig. 7.2(a).

7.1.2 Coma and lens bending

Like spherical aberration, coma changes with the shape of the lens (lens bending). To see this, you can restore the reversed OSsinglet, **OSsingletRev**, and establish three fields (0°, 7°, and 10°) to examine the off-axis performance of the lens. The transverse ray aberration curves and spot diagrams for this reversed lens are shown in Fig. 7.5. Note that the scale of the plots is 2.5×smaller (4 mm versus 10 mm) than that for the unreversed lens in Fig. 7.2(b).

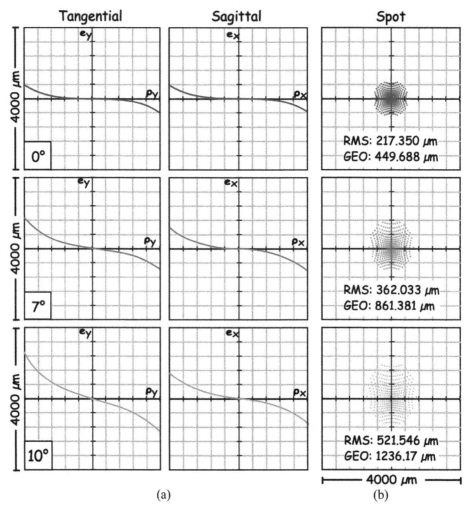

Figure 7.5 Performance of the reversed OSsinglet at 0°, 7°, and 10°: (a) transverse ray curves and (b) spot diagrams.

The aberration coefficients (**THIRD**) for the reversed lens, shown in Table 7.2, are also markedly smaller. The TSPH decreased by almost two-thirds (from −1.1366 to −0.4053), and the TTCO decreased by six-fold from 0.9864 to 0.1563. It is possible to further change the bending of the lens and find a shape of the lens that will give you zero coma (see Exercise 7.3 at the end of this chapter). Do not save OSsingletRev because off-axis fields have been added; it will be used later.

Table 7.2 Third-order errors for the OSsingletRev.

```
Surf    TSPH     TTCO     TAST     TPFC     TSFC     TTFC     TDIS
STO    -0.1641  -0.2819  -0.1076  -0.0816  -0.1354  -0.2429  -0.0775
  2    -0.2411   0.4382  -0.1770  -0.0221  -0.1105  -0.2875   0.0670
  3     0.0000  -0.0000   0.0000   0.0000   0.0000   0.0000  -0.0000
IMA    -0.0000  -0.0000  -0.0000  -0.0000  -0.0000  -0.0000  -0.0000
TOT    -0.4053   0.1563  -0.2845  -0.1036  -0.2459  -0.5304  -0.0105
```

7.2 Aplanatic Lenses

Design programs such as OpticStudio® are used to find lens solutions with reduced aberrations. Currently, lenses are designed on computers, but earlier lenses, such as the (narrowly) air-spaced doublet shown in Fig. 7.6, were designed with far more primitive tools. This lens is called an **aplanat** because it almost eliminates both spherical aberration and coma. The design was taken from Rudolph Kingslake's text, *Lens Design Fundamentals* (Academic Press, 1978) and scaled to produce a 50-mm EFL. (Much of the discussion of this lens in the Kingslake text concerns strategies for improving this lens based on graphical solutions. This approach has been supplanted by computer optimization.)

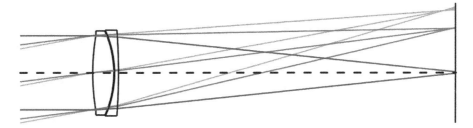

Figure 7.6 The OSaplanat.

The prescription for the lens, OSaplanat, is given in Fig. 7.7. Open a new lens (File > New). In the System Explorer, set the EPD to 10 mm and change the system wavelength to the d-line (0.5876 μm). (Note: While you are in the System Explorer, check that ray aiming is set to Paraxial). Open the FDE, insert two more fields, and set the field angles to 0°, 7°, and 10°. Enter the radii, thickness, and material values from Fig. 7.7 into the LDE. As was done in Section 4.4.2, the model glass for the first lens is entered in the Material column of S1 by clicking on the box to the right of the cell and from the Solve Type menu by selecting Model then entering 1.523 in the Index N_d space and 58.6 in the Abbe V_d space. For the second lens, the values are 1.617 and 36.6, respectively. Put a marginal ray height thickness solve on S4 to find the paraxial image plane and save the lens as **OSaplanat**.

	Surface Type	Comment	Radius	Thickness	Material	Clear Semi-Dia
0	OBJECT Standard ▼		Infinity	Infinity		Infinity
1	STOP Standard ▼		30.000	3.000	1.52,58.6 M	5.000
2	Standard ▼		-14.717	0.050		5.091
3	Standard ▼		-14.930	0.750	1.62,36.6 M	5.084
4	Standard ▼		-73.596	47.930 M		5.186
5	IMAGE Standard ▼		Infinity	-		9.139

Title (OSaplanat); EPD (10 mm); Fields (0° 7° 10°); Wavelength (d-line)

Figure 7.7 Lens prescription for the OSaplanat.

This lens performance cannot be compared directly with the reversed OSsinglet because it is a doublet lens operating at $f/5$, whereas the reversed OSsinglet is a much faster $f/2.5$ lens and is expected to have larger aberrations. However, the shape of the aberrations of the OSaplanat shown in a transverse ray plot [Fig. 7.8(a)] and a spot diagram [Fig. 7.8(b)] can be compared to Fig. 7.5. The transverse ray curves for the aplanat are almost straight, indicating that spherical aberration and coma have been nearly eliminated. This is also shown by the near-zero aberration coefficients TSPH and TTCO for the lens given in Table 7.3.

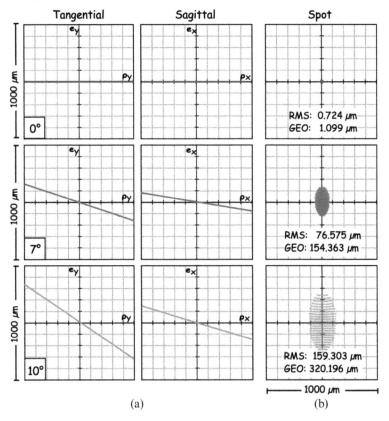

Figure 7.8 Performance of the OSaplanat at 0°, 7°, and 10°: (a) transverse ray curves and (b) spot diagrams.

Coma and Astigmatism

Table 7.3 Third-order errors for the OSaplanat.

```
Surf     TSPH     TTCO     TAST     TPFC     TSFC     TTFC     TDIS
STO    -0.0261  -0.0829  -0.0585  -0.0445  -0.0738  -0.1323  -0.0781
  2    -1.8408   1.3211  -0.2107  -0.0908  -0.1961  -0.4068   0.0469
  3     1.9213  -1.3899   0.2234   0.0994   0.2112   0.4346  -0.0509
  4    -0.0502   0.1544  -0.1055  -0.0202  -0.0729  -0.1784   0.0747
IMA    -0.0000  -0.0000  -0.0000  -0.0000  -0.0000  -0.0000  -0.0000
TOT     0.0041   0.0027  -0.1512  -0.0561  -0.1317  -0.2829  -0.0073
```

Although the on-axis image for the aplanat is nearly perfect and the coma is very close to zero, the spot sizes for the two off-axis field points still look terrible. It is obvious from the spot diagram in Fig. 7.8(b) that the aplanat is not the end of the search for a better lens. The 7° and 10° spot diagrams are much broader than the on-axis spot, and they are asymmetrical, the tangential width being larger than the sagittal width. These asymmetries are also evident in the differing slopes of the tangential and sagittal transverse ray curves for each of the field points. Having slain two dragons, spherical aberration and coma, we find that there is a smaller dragon lying in the weeds. Now that the twins have been dispatched, a dragonette, astigmatism, raises its head.

7.3 Astigmatism

In contrast to coma, **astigmatism** is linear in the pupil coordinate ρ and quadratic in the field coordinate h. By specifying the object angle or object height in y only (in the FDE the choice of Field Type is either Angle or Object Height), the field points h are confined to the y-z plane, but the orientation of the ray in the entrance pupil need not be confined to that plane. There are two entrance pupil locations, one in the y-z plane (tangential) and the other at right angles to it in the x-z plane (sagittal), that are important when describing astigmatic error. If we plot two ray fans through the aplanat with one fan consisting of blue rays in the tangential (y-z) plane and the other fan of red rays perpendicular to it in the sagittal (x-z) plane (as shown in Fig. 7.9), we see that the fans are focused at different points along the chief ray. For this lens, the tangential ray focus is located on the chief ray closer to the lens than the sagittal ray focus. This is indicated more clearly in the figure inset.

An examination of the transverse ray curves [Fig. 7.8(a)] shows that the nearly flat curves for the off-axis object points make an angle with the x axis, suggesting that the image plane for each curve is "out of focus." (Remember that an object point is focused in the evaluation plane when its transverse ray curves are coincident with the x axis of the transverse ray plots.) However, the angle that the tangential ray curve makes with the x axis is different than the sagittal ray curve for that object point. This means that there are image plane locations where rays from an off-axis object point will focus to a point in one dimension but spread out in the other dimension, producing a line focus. The fact that the tangential and sagittal curves for the off-axis field points have different slopes is an indication that astigmatism is present. Also, the larger difference in slopes for the 10° field compared to the 7° field tells you that the astigmatism increases with field.

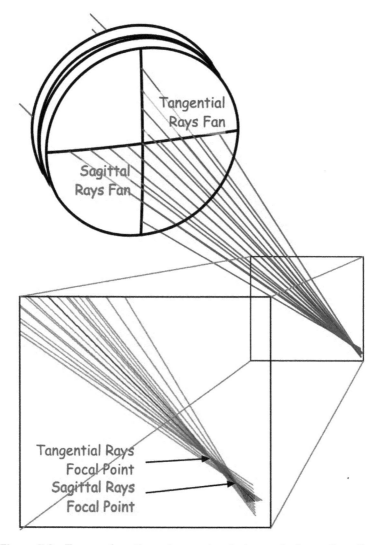

Figure 7.9 Two ray fans through an aplanat, demonstrating astigmatism.

To illustrate the utility of the transverse ray curves, we can use the Slider feature (Optimize > Slider) that was introduced at the end of Chapter 6 for lens bending to change the image evaluation plane by defocusing the lens. First, a defocus surface needs to be inserted in the LDE before the Image surface as S5. Then, in the Slider window, change the parameter to the thickness of surface 5 and change the start/stop range to be ±1.5 mm. Move the slider along the total range. This causes the transverse ray curves [shown in Fig. 7.8(a)] to be rotated about the origin. As you move the evaluation plane from the paraxial focal plane toward the lens, stop at about −1.30 mm. The new transverse ray curves and spot diagrams are shown in Fig. 7.10. This amount of defocus causes the tangential curve for the 7° off-axis point to be coincident with the x-axis [Fig. 7.10(a)], so the rays come to a focus in the tangential plane but are spread in the sagittal plane, leading to an elliptical spot [Fig. 7.10(b)]. Coincidentally, this −1.3-mm defocus also focuses the 10° off-axis point in the sagittal plane. The spot diagram [Fig. 7.10(b)] shows the line focus for this object point.

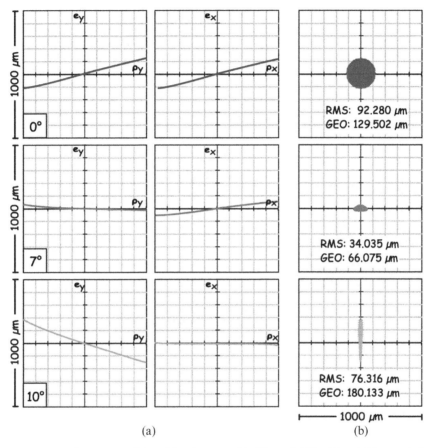

Figure 7.10 Performance of the OSaplanat at 0°, 7°, and 10° with −1.30-mm defocus: (a) transverse ray curves and (b) spot diagrams.

We can quantify the amount of astigmatism in a lens by looking at the third-order astigmatism coefficient (TAST = −0.1513) in Table 7.3. However, if we compare this number with the tangential transverse ray error when $h = 1$ and $\rho = 1$ in Fig. 7.8(a), it is only about half of the total error. This is because the OSaplanat also suffers from another error that is linear in pupil coordinate, Petzval curvature (ρh^2). Since both astigmatism (TAST) and Petzval curvature (TPFC) are linear in pupil coordinate, they are often lumped together into a field curvature "sum" with a tangential component (TTFC) and a sagittal component (TSFC). In Fig. 7.8(a), the transverse ray errors on the x and y axes at $\rho = 1$ and $h = 1$ are approximately equal to these field curvature coefficients for the tangential curve (TTFC = −0.2829) and sagittal curve (TSFC = −0.1317). Petzval curvature and its impact on the total field curvature will be discussed in more detail in the next chapter along with the last of the third-order aberrations, distortion (h^3).

Additional Exercises

Exercise 7.3. Zero coma using lens bending
Start with the **OSsinglet** (with a 10° max field) and add a marginal ray angle solve to R2 to hold a 50-mm EFL. Change R1 and make a plot of both the spherical (TSPH) and coma (TTCO) for a range of lens shapes. What shape factor q results in zero TCO? How close is this to the minimum TSPH found in Section 6.7?

Exercise 7.4 Linear foci for the aplanat
In Fig 7.10, a defocus of -1.30 mm produced a linear sagittal focus at 10° and a tangential focus at 7° for the **OSaplanat**. What defocus value will produce a sagittal focus for the 7° point? What defocus will produce a tangential focus for the 10° point?

Exercise 7.5 Through-focus of 10° field for the aplanat
The spot diagram for the **OSaplanat** at 10° goes from a horizontal line at a -2.8-mm defocus to a vertical line at a -1.3-mm defocus. Use a slider to find the defocus amount where the spot size for the 10° field is the smallest (based on the RMS spot diameter).

Exercise Answers

Ex. 7.1
Start with the **OSsinglet** with a 10° max. field and change the EPD from 5 mm to 10 mm to 20 mm, and evaluate THIRD each time.

EPD	TSPH	TTCO	TAST	TPFC	TSFC	TTFC	TDIS
5mm	-0.0178	0.0616	-0.0570	-0.0259	-0.0544	-0.1114	-0.0239
10mm	-0.1421	0.2466	-0.1140	-0.0518	-0.1088	-0.2228	-0.0239
20mm	-1.1366	0.9864	-0.2279	-0.1036	-0.2176	-0.4455	-0.0239

As indicated in Table 6.1, coma is quadratic in ray height in the pupil ρ. If the EPD were increased by a factor of 4 (5 mm to 20 mm), the TTCO value should increase by a factor of 16 (or 4^2). This agrees with the above results, where the TTCO value for a 20-mm pupil (0.9864) divided by the TTCO value for a 5-mm pupil (0.0616) equals 16.

Ex. 7.2
Start with the **OSsinglet** with a 20-mm EPD and change the max. field angle from 5° to 10° to 20° and evaluate THIRD each time.

FOV	TSPH	TTCO	TAST	TPFC	TSFC	TTFC	TDIS
5°	-1.1366	0.4894	-0.0561	-0.0255	-0.0536	-0.1097	-0.0029
10°	-1.1366	0.9864	-0.2279	-0.1036	-0.2176	-0.4455	-0.0239
20°	-1.1366	2.0360	-0.9712	-0.4416	-0.9272	-1.8984	-0.2100

As indicated in Table 6.1, coma is linear in field height h. If the field angle is increased by a factor of 4 (5° to 20°), the TTCO value should increase by a factor of 4.16 (or tan 20°/tan 5°). This agrees with the above results, where the TTCO value for a 20° field (2.0360) divided by the TTCO value for a 5° field (0.4894) equals 4.16.

Ex. 7.3
A range of R1s were entered into the LDE. For each R1, the shape factor q was calculated and THIRD was used to evaluate TSPH and TTCO. The values are listed, and a chart is plotted below. The values for the minimum TSPH and closest-to-zero TTCO are bold in the list.

R1	q (shape factor)	TSPH	TTCO
120	-0.57415	-1.13660	0.986358
100	-0.48976	-1.04033	0.930133
80	-0.36338	-0.90881	0.844004
60	-0.15276	-0.72364	0.695905
40	0.27197	-0.47875	0.384176
35	0.45706	-0.42419	0.244908
31	0.65111	-0.39918	0.097987
30	0.70840	**-0.39784**	0.054572
29	0.77003	-0.39937	**0.007896**
25	1.07229	-0.44914	-0.219698
20	1.65622	-0.71714	-0.644665

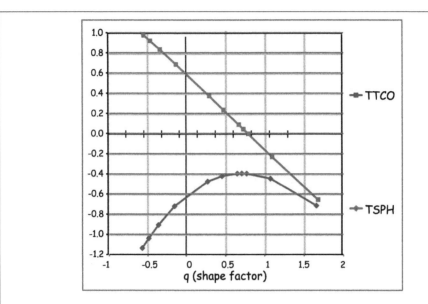

As shown here, the lens shape ($q = 0.77$) that gives zero coma is very close to the lens shape ($q = 0.70$) that provides minimum spherical aberration.

Ex. 7.4

A defocus value of -2.8 mm is needed for the 10° tangential focus, and a defocus value of -0.65 mm is needed for the 7° sagittal focus.

Ex. 7.5
The defocus amount is about halfway, i.e., -2.05 mm.

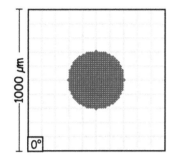 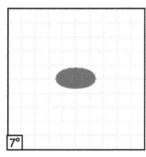 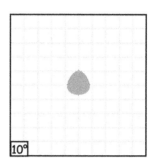

Chapter 8
Aberrations of the Image Surface

In most cases, when you design an optical system such as camera lens, you require that the lens form an image of a scene on a flat surface. For example, Fig. 8.1 shows the imaging of an object at infinity onto a flat paraxial image plane by a perfect lens, i.e., one with no aberrations. However, the image surface for a real lens is not a plane but a curved surface whose z-axis departure from a flat paraxial image plane represents an image surface error for the lens called Petzval curvature. If the image plane is a curved surface (consider the shape of your retina), Petzval curvature may not cause any problem. However, because most modern sensors are flat, not curved, Petzval curvature must be managed when designing a lens.

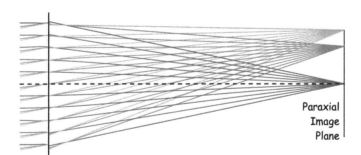

Figure 8.1 Imaging through a perfect lens.

This chapter explores Petzval curvature and a second image-surface aberration, distortion, in more detail. Distortion is an image-surface error that does not blur the image but causes errors in the chief ray locations in the x-y plane, resulting in a change in magnification with image height. This results in a stretched (or compressed) image of the object.

8.1 Field Curves

As was shown in the previous chapter, an aplanatic lens exhibits little spherical aberration or coma, and the focal images reside on curved surfaces. Restore the **OSaplanat**, shown in Fig. 7.6, either from the saved lens or from its prescription in Fig. 7.7. Trace a fan of 9 rays for each field in the Layout window (Analyze > Cross-Section). Figure 8.2 shows that the image points for the two off-axis object points focus at a distance in front of the paraxial image plane defined by the on-axis ray fan. Although you could trace additional fans at other angles to get a better picture of the image-surface error, they won't add much to what is already demonstrated in Fig. 8.2. This same information was presented in a different manner by the transverse ray curves for the lens shown in Fig. 7.8. In those curves, the slope of the tangential curves increases with an increasing field. This indicates that the lens focuses farther away from the paraxial image plane with increasing angles. The same is true for the sagittal image surface, although the amount of focus shift is smaller for the sagittal surface than for the tangential surface for a given field point.

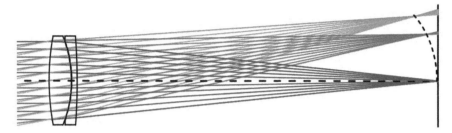

Figure 8.2 OSaplanat cross-section for three object angles (0°, 7°, and 10°).

Although the transverse ray curves provide information only at a few field points, another plot, called a field curvature plot, provides a more compact and easily understandable presentation of the lens performance over its entire field. The field curves for the OSaplanat, shown in Fig. 8.3, are created by clicking on the Aberrations icon in the Analyze tab and selecting Field Curvature and Distortion in the drop-down menu (Analyze > Aberrations > Field Curvature and Distortion). The plot in Fig. 8.3(a) shows both the tangential and sagittal field curves; the plot in Fig. 8.3(b) is the curve for image distortion. Because distortion will be addressed later, the field plots in the next few sections will show only the left plot, labeled as field curvature.

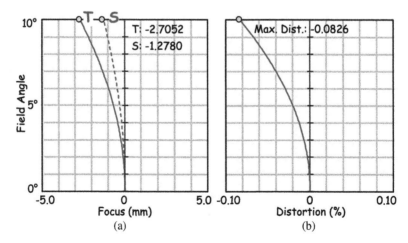

Figure 8.3 Image surface errors for the OSaplanat (a) field curvature and (b) distortion.

The two labeled field curves in Fig. 8.3(a) show the variation of the tangential T and sagittal S image locations as a function of the field angle. The values for the plots can be examined by selecting the Text tab at the bottom of the Field Curv/Dist window. At the bottom of the listing are the focus shift values at full field (10°): −1.2780 for the sagittal ray fan and −2.7052 for the tangential ray fan. These are the actual values in millimeters of the z-axis distance for each of the image surfaces from the paraxial image plane. Because the lens is symmetrical about the optical axis, the image surface curves can be reflected about the axis. This would not provide any additional information, but it is useful to think of the field curves as the top of a cross-sectional cut through two tangential, umbrella-shaped surfaces with the optical axis as their common handle.

As we noted at the end of the previous chapter, the error shown by the field curves is not a single third-order aberration but a combination of two transverse third-order aberrations (astigmatism and Petzval curvature) that share a ρh^2 dependence, as listed in Table 6.1. As we will show, astigmatism causes the separation between the tangential and sagittal curved image surfaces, whereas Petzval curvature is a basic field curvature that remains when a lens is corrected for astigmatism. These third-order aberrations must be considered separately because they require different correction strategies to achieve a flat field.

8.2 Petzval Curvature

Because the OSaplanat still has a significant amount of astigmatism, it is difficult to see the effect of pure Petzval curvature. The simplest way to demonstrate a system with only Petzval curvature is with a pair of mirrors, one concave and one convex, such as the Schwarzchild system that was given as a problem in Exercise 3.8. If you finished the exercise and saved it as **OSschwarzchild**, you can open it again from your lens folder. If you have not done the exercise, the prescription for the lens is given in Fig. 8.4 and can be entered from a new lens (File > New). In the FDE (Setup > Field Data Editor), enter our standard set of fields (0°, 7°, and 10°) and check that paraxial ray aiming is on. A cross-section (Analyze > Cross-Section) of the mirror system should look like Fig. 8.5, and its first-order quantities from **FIRST**, given in Table 8.1, can be used to check your entry.

	Surface Type	Comment	Radius	Thickness	Material	Clear Semi-Dia
0	OBJEC Standard ▼		Infinity	Infinity		Infinity
1	Standard ▼		Infinity	150.000		31.485
2	STOP Standard ▼		61.803	−100.000	MIRROR	5.000
3	Standard ▼		161.803	211.803 M	MIRROR	38.433
4	IMAGE Standard ▼		Infinity	-		8.727

Title (OSDschwarzchild); EPD (10 mm); Field (0°, 7°, 10°); Wavelength (d-line)

Figure 8.4 Lens prescription for the OSschwarzchild mirror system.

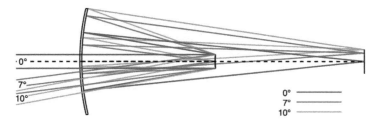

Figure 8.5 Cross-section of the OSschwarzchild mirror system.

Table 8.1 First-order quantities for the OSschwarzchild.

Infinite Conjugates	
Effective Focal Length	49.9996
Back Focal Length	211.8026
Front Focal Length	161.8034
F/#	5.0000
Image Distance	211.8026
Lens Length	50.0000

The third-order aberration coefficients for this system (**THIRD**) are shown in Table 8.2. (We have removed the color errors, TAXC and TLAC, to fit the quantities under discussion onto single lines.) Note that the system has no spherical aberration (TSPH) or coma (TCOM). The astigmatism coefficient (TAST) is also zero, indicating that the astigmatism of the lens is zero. Therefore, the amount of sagittal and tangential field curvature (TSFC and TTFC) is due only to Petzval curvature (TPFC), and all three coefficients have exactly the same value = –0.0777.

Table 8.2 Third-order errors for the OSschwarzchild.

Surf	TSPH	TTCO	TAST	TPFC	TSFC	TTFC	TDIS
1	0.0000	0.0000	0.0000	0.0000	0.0000	0.0000	-0.0000
STO	0.0265	0.1731	0.2515	-0.1258	0.0000	0.2515	0.0000
3	-0.0265	-0.1731	-0.2515	0.0480	-0.0777	-0.3293	-0.1694
IMA	-0.0000	-0.0000	-0.0000	-0.0000	-0.0000	-0.0000	-0.0000
TOT	0.0000	0.0000	0.0000	-0.0777	-0.0777	-0.0777	-0.1694

The absence of astigmatism is also evidenced in the Ray Aberration plots (Analyze > Rays & Spots > Ray Aberration) shown in Fig. 8.6(a), where the tangential and sagittal curves have the same slope for each field. This absence is also demonstrated by the field curves (Analyze > Aberrations > Field Curvature and Distortion) plotted in Fig 8.6(b). Since there is no separation between the S and T field curves, the field curvature in the Schwarzchild mirror system is due to Petzval curvature alone.

A check from the text tab for Fig. 8.6(a) shows that the maximum errors in the transverse ray curves for 10° are –0.0745 mm (tangential) and –0.0760 mm (sagittal) (converted from microns in the table). These values are slightly different than the third-order values given in Table 8.2, –0.0777. This is because this lens has a small amount of higher-order aberration and the transverse ray errors are calculated from real ray traces.

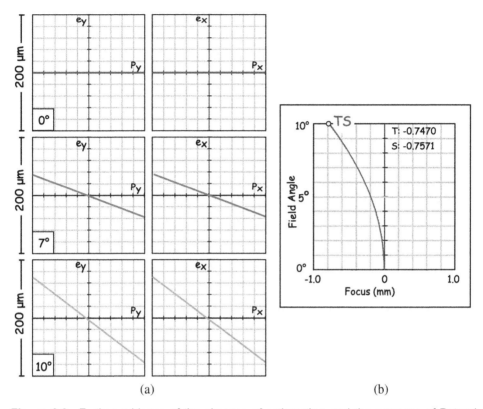

Figure 8.6 Further evidence of the absence of astigmatism and the presence of Petzval curvature in the OSschwarzchild: (a) transverse ray curves with equal sagittal and tangential slopes that increase with field angle and (b) a field curvature plot with identical S and T curves.

8.3 Field Curvature and Third-Order Coefficients

In OpticStudio®, the T and S curves in the field curvature plot are the **sum** of astigmatism and Petzval curvature. Neither the plots nor the listing of the values for the S and T curves provide data for astigmatism and Petzval individually, so how can you separate these field curvature components? Luckily, there is a simple set of relations between these aberrations and the field curvature values that can be used to determine them.

The first relation is that, per aberration theory, the tangential astigmatism is three times larger than the sagittal astigmatism. To simplify the following calculations and graphics, the symbol Δ will be assigned to the maximum value for the sagittal astigmatism measured longitudinally (along the optical axis). Because the T and S curve values are the sum of the astigmatism and Petzval third-order contributions, the relations between T and S in the plots are

$$T = 3\Delta + P, \tag{8.1}$$
$$S = \Delta + P, \tag{8.2}$$

where P is the longitudinal Petzval error.

By substitution, we can eliminate Δ to arrive at a value for P in terms of the field curvature values T and S that are given on the Text tab at the bottom of the Field Curv/Dist window:

$$P = (3S - T)/2. \tag{8.3}$$

In the case of the OSaplanat in Fig. 8.3, the longitudinal field curvature errors at the 10° field were T = −2.7052 mm and S = −1.2780 mm. Applying Eq. (8.3) gives P = −0.5644 mm for full field.

In Fig. 8.7, the Petzval surface P for the OSaplanat is plotted along with the astigmatic fields T and S from Fig. 8.3(a). This shows the astigmatic and Petzval contributions to field curvature. In effect, the Petzval surface can be considered the base curve for the field curvature plots. And the separation between the S curve and the P curve is a measure of the sagittal astigmatism at any field angle. The same is true for tangential astigmatism. If astigmatism is eliminated, the residual field curvature is due to only the Petzval curvature.

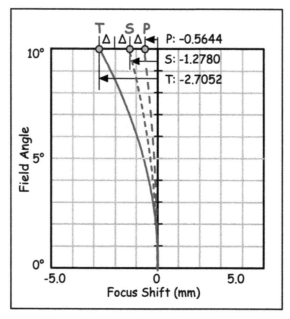

Figure 8.7 Astigmatic field curves for the OSaplanat with an added curve P to show the Petzval contribution.

The field curve values can also be related to the third-order aberration coefficients generated using the **THIRD** macro. For example, the third-order values for the OSaplanat are shown again in Table 8.3. The relevant third-order values are TTFC, TSFC, and TPFC, where TTFC = −0.2829, TSFC = −0.1317, and TPFC = −0.0561. However, these are transverse ray errors, whereas the field curve values from the field curvature plot are longitudinal (z-axis) errors. To understand how they are connected, we need to first know how to convert from a transverse ray error in y to a longitudinal ray error in z.

Aberrations of the Image Surface 163

Table 8.3 Third-order aberration values for the OSaplanat.

Surf	TSPH	TTCO	TAST	TPFC	TSFC	TTFC	TDIS
STO	-0.0261	-0.0829	-0.0585	-0.0445	-0.0738	-0.1323	-0.0781
2	-1.8408	1.3211	-0.2107	-0.0908	-0.1961	-0.4068	0.0469
3	1.9213	-1.3899	0.2234	0.0994	0.2112	0.4346	-0.0509
4	-0.0502	0.1544	-0.1055	-0.0202	-0.0729	-0.1784	0.0747
IMA	-0.0000	-0.0000	-0.0000	-0.0000	-0.0000	-0.0000	-0.0000
TOT	0.0041	0.0027	-0.1512	-0.0561	-0.1317	-0.2829	-0.0073

Figure 8.8 shows the relation between the transverse and longitudinal errors. The transverse ray error on the paraxial image plane is shown as Δy. The longitudinal ray error Δz is related to the transverse ray error Δy through the angle u between the marginal and chief rays for the full-field point according to $\tan u = \Delta y / \Delta z$. Using a small-angle approximation, $\tan u \approx u$,

$$\Delta z = \Delta y / u. \tag{8.4}$$

Thus, the third-order values TTFC, TSFC, and TPFC can be divided by u to determine the distance Δz to the tangential and sagittal foci and to the Petzval surface.

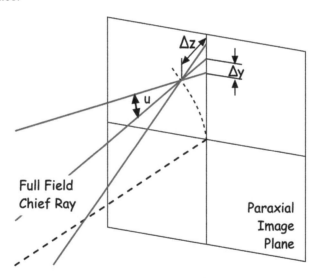

Figure 8.8 Geometry between the transverse ray error Δy, the longitudinal image-surface error Δz, and u, the angle between the chief ray and the marginal ray for the full-field point.

For the OSaplanat, the angle u between the marginal and chief ray is (EPD/2)/EFL = 5 mm/50 mm or 0.1 rad (radians), so the third-order aberration values in Table 8.4 should be multiplied by 10 for comparison with the longitudinal image surface errors given in Table 8.1 for the field plots. When we do so, we see that the full-field value of the T curve is –2.7052 mm, which is close to $10 \cdot \text{TTFC} = -2.829$. For the S curve, its full-field value is –1.2780, which is also close to $10 \cdot \text{TSFC} = -1.317$. The longitudinal measure of the third-order astigmatism coefficient ($10 \cdot \text{TAST} = -1.512$) is also close to the separation between the T and S curves at full field ($-2.7052 + 1.2780 = -1.4272$).

The differences between these field curves are the difference between third-order approximations and the field plots that use real ray traces.

Third-order aberration theory shows that the Petzval aberration coefficient for each surface depends only on the optical power of the surface and the product of the refractive indices enclosing the surface. For a single surface with a curvature of c enclosed by refractive indices n and n', the power of the surface is given by

$$\phi = c(n' - n), \tag{1.7}$$

and the curvature of the Petzval surface (PC) is

$$\mathrm{PC} = c\left(\frac{1}{n'} - \frac{1}{n}\right) = c\frac{n - n'}{n'n} = -\frac{\phi}{n'n}. \tag{8.5}$$

Similarly, it can be shown that the Petzval curvature of a single *thin* lens depends **only** on the lens power and its index of refraction and is therefore independent of its shape.

Exercise 8.1 Petzval blur with changing lens shape

Starting with the **OSsinglet**, introduce off-axis fields (say, 0°, 7°, and 10°). Change R1 in the LDE several times between 120 and 60. What is the Petzval field curvature coefficient (TPFC) as the shape of the lens is changed? R2 must be given an angle solve so that the lens retains a 50-mm focal length, as R1 is used to change the shape. How does the Petzval field curvature change with lens shape?

8.4 An Anastigmatic Lens

The Schwarzchild mirror system shown in Fig. 8.4 is a special mirror configuration that has no astigmatism across the entire field. For lenses, what is needed to combat astigmatism? It turns out that it is not possible to improve the aplanatic doublet much beyond what was described earlier with just two lenses, a small air space, and a stop in contact with them. The problem is that with only four surfaces, two glasses, and two lens thicknesses, there are not enough levers to pull to eliminate spherical, coma, and astigmatism. Adding a lens, using the air spaces as additional degrees of freedom, and allowing the stop to move can create an anastigmat, a lens with reduced astigmatism. One of the best examples of this is an air-spaced triplet, also known as a Cooke triplet (Fig. 8.9).

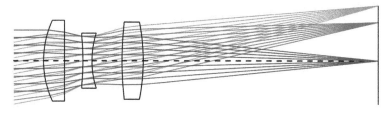

Figure 8.9 OStriplet operating at f/5.

We will use the **OStriplet** from the exercise on lens entry in Chapter 3 (Exercise 3.5). If you have not done the exercise, the prescription for the lens is given in Fig. 8.10 and can be entered from a new lens (File > New). In the FDE (Setup > Field Data Editor), enter our standard set of fields (0°, 7°, and 10°) and check that paraxial ray aiming is on. This lens has the same EFL (50 mm) and EPD (10 mm), and therefore operates at the same $f/\#$ ($f/5$) as the OSaplanat.

	Surface Type	Comment	Radius	Thickness	Material	Clear Semi-
0	OBJEC Standard ▼		Infinity	Infinity		Infinity
1	Standard ▼		15.000	3.500	N-SK16	5.873
2	Standard ▼		Infinity	3.100		5.254
3	STOP Standard ▼		-37.530	1.500	N-F2	3.912
4	Standard ▼		14.080	5.000		4.007
5	Standard ▼		50.260	3.500	N-SK16	5.279
6	Standard ▼		-31.380	38.515 M		5.605
7	IMAGE Standard ▼		Infinity	-		8.826

Title (OStriplet); EPD (10 mm); Field (0°, 7°, 10°); Wavelength (d-line)

Figure 8.10 LDE for the OStriplet operating at $f/5$.

The field curves (Analyze > Aberrations > Field Curvature and Distortion) for the OStriplet are shown in Fig. 8.11 with a maximum scale of ±0.25 mm. If these curves were plotted on the same ±5-mm scale as the OSaplanat (Fig. 8.3), they would be almost flat and coincident with the y axis.

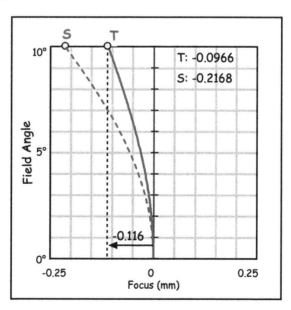

Figure 8.11 Field curves for the OStriplet over a 10° field. Also indicated is the evaluation plane for a −0.116-mm defocus.

Although this triplet doesn't eliminate astigmatism, the separation between the S and T surfaces has been substantially reduced. At full field, the curves for the OSaplanat were separated by 1.43 mm at full field (Fig. 8.6). The values for the S and T curves at full field for the OStriplet are –0.2168 mm and –0.0966 mm, respectively. Thus, the separation between the image surfaces for the OStriplet is 0.12 mm, indicating that the amount of astigmatism across the field is much less for the triplet than for the OSaplanat. The additional lens, stop position, and lens spacings available in the triplet provide the additional degrees of freedom needed to control the astigmatism across the field.

Note that the performance of this lens at the edge of the field can be improved by simply defocusing the lens (Section 6.4). Deleting the paraxial marginal ray solve on the image distance and then using Quick Focus in the Optimize tab (Optimize > Quick Focus), the evaluation plane is shifted from 38.515 mm to 38.399 mm or by –0.116 mm. As shown in Fig. 8.11, the departures of the S and T surfaces from the defocus plane will now be smaller at the edge of the field (however, some of the on-axis performance has been sacrificed to gain this advantage).

A comparison of the field curves (Fig. 8.12) for the three optical systems discussed earlier demonstrates the reduction in astigmatism and Petzval curvature with the increase in elements and geometry. The values for the OSsinglet and OSsingletRev are comparable to those of the OSaplanat over the same field.

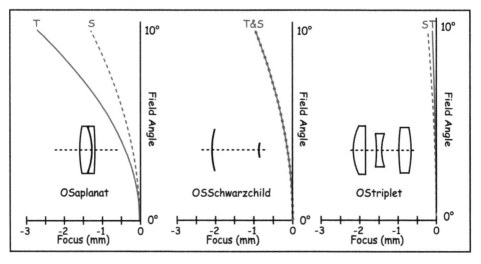

Figure 8.12 Field curves (all plotted on the same scale) for the three optical systems discussed in this chapter.

8.5 Distortion

The last of the five third-order aberrations, distortion, is unique because it does not blur the image (a point object is imaged to a perfect point). Rather, distortion causes a shift of the ideal location of off-axis field points. More specifically, it is a measure of the size of the chief ray errors in the image plane as a function of field. This is illustrated schematically by mapping the chief rays from a grid of points in the object plane onto a grid of image points in the paraxial image plane [Fig. 8.13(a)]. If there were no distortion, the object plane grid would be mapped to the image plane with a magnification equal to the paraxial magnification. Distortion, however, introduces a change in magnification with distance from the optical axis (the center of the grid). For example, if the magnification decreases with distance from the axis, the object grid will be imaged as shown in Fig. 8.13(b). The corners of the grid will be closer to the center than they would be in a rectangular grid. This is referred to as "barrel" distortion. If the magnification of the lens increases with distance from the axis, the grid will display "pincushion" distortion, as shown in Fig. 8.13(c).

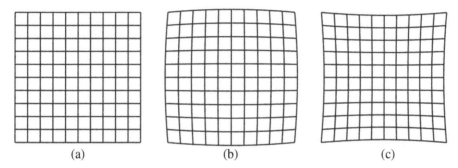

Figure 8.13 Types of distortion: (a) object grid, (b) barrel distortion, and (c) pincushion distortion.

The calculation of the third-order distortion coefficient is a little different from the other aberrations. This is because distortion is defined as the difference between where the chief ray *should* be located in the image plane and where it really ends up. Where it should go is determined from the paraxial ray trace values for a chief ray, while where it really ends up is determined from the real ray trace values for that same chief ray (see the text box "Tracing Single Rays" for more information on how to trace and evaluate both a paraxial ray and a real ray in OpticStudio).

Table 8.4 shows the single-ray-trace data needed to calculate the distortion coefficient for the OSsinglet with a 10° full field. The (truncated) output shows a real ray image height (Y-coordinate value at surface 3) of 8.79251 versus a paraxial ray image height of 8.81634. The difference (8.79251 – 8.81634) is –0.02383, which very closely matches the third-order distortion SUM of –0.0239 from the THIRD output given in Table 7.1.

The distortion of a lens is usually specified as a percentage, where the percent distortion is calculated by taking the difference between the real and paraxial chief ray heights and dividing it by the paraxial chief ray height (×100). The right plot in the field curves (Analyze > Aberrations > Field Curvature and Distortion), which has been ignored until now, shows the distortion percentage as a function of field. Figure 8.14 shows the field curves for the OSsinglet with a 10° maximum field.

Table 8.4 Single Ray Trace output for a paraxial and real chief ray for the OSsinglet.

```
Normalized X Field Coord (Hx) :        0.00000
Normalized Y Field Coord (Hy) :        1.00000
Normalized X Pupil Coord (Px) :        0.00000
Normalized Y Pupil Coord (Py) :        0.00000

Real Ray Trace Data:
Surf           X-coordinate        Y-coordinate        Z-coordinate

OBJ            Infinity            Infinity            Infinity
 1             0.00000             0.00000             0.00000
 2             0.00000             0.57562            -0.00510
 3             0.00000             8.79251             0.00000

Paraxial Ray Trace Data:
Surf           X-coordinate        Y-coordinate        Z-coordinate
OBJ            Infinity            Infinity            Infinity
 1             0.00000             0.00000             0.00000
 2             0.00000             0.58125             0.00000
 3             0.00000             8.81634             0.00000
```

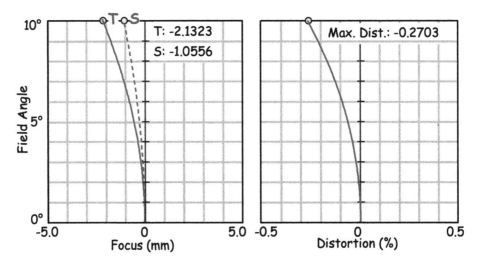

Figure 8.14 Field curves for the OSsinglet.

We can also find specific values for distortion on the Text tab of the field curves. As discussed in Section 8.1, the Text tab lists the tangential (T) and sagittal (S) focal shifts under the columns TAN shift and SAG shift. It also lists the Real Height and Reference Height (i.e., paraxial height) along with a Distortion (%) for each field point. The data from the text tab for the 10° full-field point for the OSsinglet are given in Table 8.5 to four-place accuracy and agree with the Single Ray Trace output in Table 8.4. The negative sign on the distortion % indicates that this is barrel distortion.

Aberrations of the Image Surface

Tracing Single Rays

Under Analyze > Rays & Spots > Single Ray Trace is the option to trace and evaluate a single ray. The output includes the X, Y, and Z coordinates of a ray at each surface along with the direction cosines, angle of incidence, and path length of the ray. Two tables of data are listed for each ray traced; the first is for the real ray and the second is for a paraxial ray. A truncated example of the output ray data that were used to calculate the distortion of the OSsinglet can be found in Table 8.5. This option is an extremely valuable tool when trying to diagnose both aberration errors and lens setup errors (especially for tilted and decentered systems).

The two most common rays that are traced are the marginal ray and the chief ray (see Section 5.2.6). The settings default to trace a marginal ray for the axial field point ($H_x = H_y = 0$) with the ray directed at the top of the pupil ($P_x = 0$, $P_y = 1$).

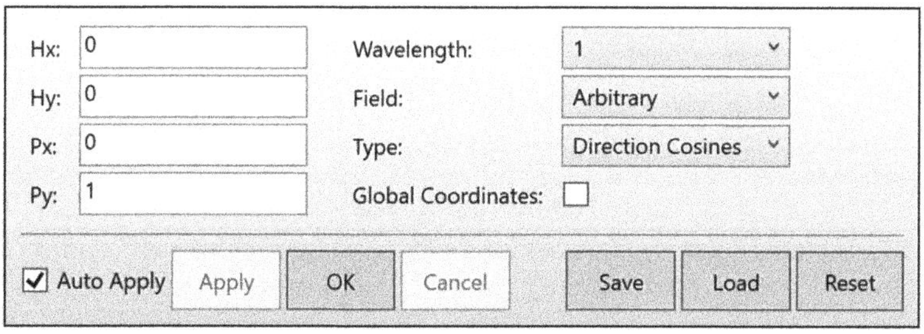

Single Ray Trace settings for a marginal ray

The settings can be easily changed to trace a chief ray for the full-field point ($H_x = 0$; $H_y = 1$) with the ray directed at the center of the pupil ($P_x = P_y = 0$), as shown below.

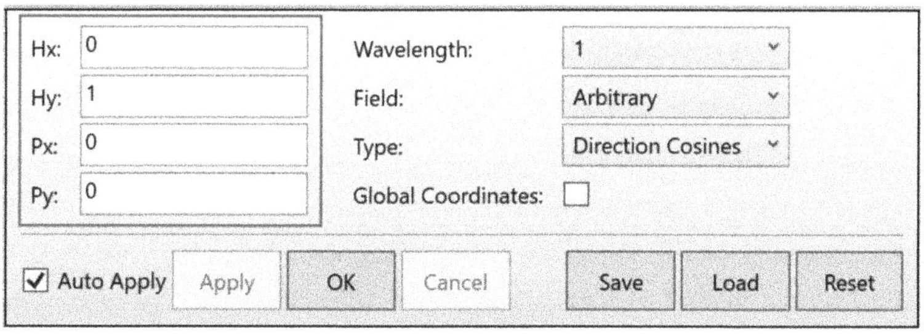

Single Ray Trace settings for a chief ray

Table 8.5 Field curve values for a 10° field for the OSsinglet.

Angle	Tan Shift	Sag Shift	Real Height	Ref. Height	Distortion
10.0000	-2.1323	-1.0556	8.7925	8.8163	-0.2703 %

A distortion grid displaying the difference between a grid of paraxial ray intersections in the image plane and one for real rays can also be plotted in OpticStudio using Analyze > Aberrations > Grid Distortion. The result is two superimposed grids: one is a regular grid of lines that represents the case of no distortion and the other is an array of blue crosses that locates the line intersections of the distorted field caused by the lens. Figure 8.15(a) shows these grids for the OSsinglet for a 20° field. The upper left-hand corner of this plot of the grid is displayed in Fig. 8.15(b) to show more detail. Note: The distortion in the 10° lens is modest, so the full field for the OSsinglet was increased to 20° to show more error.

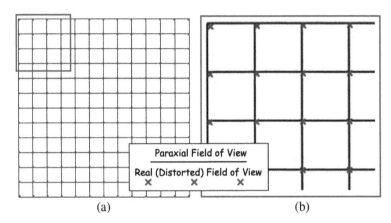

Figure 8.15 Distortion grid for the OSsinglet with a 20° maximum field. An array of blue x's overlays a rectangular grid shown for (a) the entire field and (b) a magnified corner of the grid. The inward displacement of the x's indicates a slight barrel distortion.

> **Exercise 8.2 Distortion for an aplanat**
> Restore the **OSaplanat** and add three field points with a maximum field angle of 10°. Follow the procedure discussed earlier to trace a paraxial chief ray and a real chief ray and then calculate the % distortion in this lens. Compare this value with the % distortion listed in the text output for the field curves. What type of distortion is it? How does the amount of distortion compare to the OSsinglet (over the same field angle)?

Almost all images we view (from television screens to digital photos on our phones and tablets) contain some amount of distortion. It is only when the aberration begins to dominate the image that it becomes noticeable and, therefore, annoying. Some examples are shown in the text box, "Examples of Distortion" along with images that show various amounts of distortion. The first image (a) from the camera lens of a Panasonic DMC-FZ20 camera has very little distortion. The next two images show (b) strong pincushion distortion and (c) strong barrel distortion. The last two images demonstrate that (d) distortion below 3% is difficult to detect by the unaided eye but (e) becomes more obvious when it reaches 10%. The best indications of distortion are the tilt of the windows on the left side of the image.

Examples of Distortion

These images of Incan walls in Cuzco, Peru demonstrate the types of distortion and varying amounts of distortion. The best indications are the tilt of the windows on the left side of the image: (a) undistorted image; (b) strong pincushion distortion; (c) strong barrel distortion; (d) ~3% pincushion distortion at corners (image cropped to a square); and (e) ~10% pincushion distortion at corners (image cropped to a square).

(a)

(b)

(c)

(d)

(e)

Even if the distortion in a system might be so modest that it could not be detected when viewing an image, there are still instances where its presence can be a problem. For example, a series of stellar images or land survey photos cannot be stitched into a panorama because the distortion at the edges of an image does not permit an exact overlap. While this may be addressed using an image editor, such as Adobe Photoshop®, the actions needed to correct for the distortion could in some manner modify other features of the image. So, lenses that are used for metrology and for panoramic images are designed to have virtually no distortion.

Other applications require that the lenses have a specific amount of distortion. For example, the lens in a laser scanner or a laser printer may be designed so that the location of the focused laser beam on a surface is proportional to the angle to the optical axis of the lens so that the velocity of the beam across the surface is constant. Such lenses with this built-in distortion are referred to as $f\theta$ lenses.

Exercise 8.3 Distortion for an aplanat at twice the field

Restore the **OSaplanat** and add three field points with a maximum paraxial image height of 10 mm. Evaluate the third-order distortion coefficient. Now double the maximum image height to 20 mm and evaluate the third-order distortion coefficient. How does the distortion coefficient vary with field size? Is this what you would expect based on Table 6.1?

Exercise Answers

Ex. 8.1

Restore the **OSsinglet** and introduce three fields (0°, 7°, and 10°). Put a −0.2 marginal ray angle solve on R2 to hold a 50-mm focal length as you change R1 in the LDE and then evaluate the Petzval field curvature coefficient (TPFC) as the shape of the lens is changed.

Radius R1	TPFC
120	-0.10365
100	-0.10381
80	-0.10400
60	-0.10419

As discussed in Section 8.3, the Petzval curvature for a single thin lens should remain constant for any lens shape. Even though the OSsinglet is not a thin lens (it has a thickness of 5 mm), we see that the values for TPFC vary with shape by only a very small amount. Try this same exercise after setting the thickness to zero; the value of TPFC should be −0.10249 for all lens shapes. The point is that the variation in Petzval curvature with lens bending is small even when bending a lens with a realistic thickness.

Aberrations of the Image Surface

Ex. 8.2
Restore the **OSaplanat**, add three field points with a maximum field angle of 10°, and then trace a single chief ray ($H_x = 0$; $H_y = 1$) and ($P_x = P_y = 0$).

```
Real Ray Trace Data:
Surf     X-coordinate         Y-coordinate         Z-coordinate
OBJ         Infinity             Infinity             Infinity
 5          0.00000              8.81896              0.00000

Paraxial Ray Trace Data:
Surf     X-coordinate         Y-coordinate         Z-coordinate
OBJ         Infinity             Infinity             Infinity
 5          0.00000              8.82624              0.00000
```

Paraxial chief ray height = 8.82624; real chief ray height = 8.81896. %Distortion = [(8.81896 − 8.82624) / 8.82624] · 100 = −0.0826%. This is in agreement with the −0.0826% listed in the field curve text data. The distortion is negative and therefore barrel, and it is about 1/3 the distortion of the OSsinglet.

Ex. 8.3
Restore the **OSaplanat** and add three field points with a maximum paraxial image height of 10 mm.

```
THIRD
Surf  TSPH    TTCO    TAST    TPFC    TSFC    TTFC    TDIS
TOT   0.0041  0.0031  -0.1942 -0.0720 -0.1690 -0.3632 -0.0106
```

Double the maximum image height to 20 mm.

```
THIRD
Surf  TSPH    TTCO    TAST    TPFC    TSFC    TTFC    TDIS
TOT   0.0041  0.0061  -0.7766 -0.2878 -0.6761 -1.4527 -0.0851
```

If the field size is doubled, the distortion coefficient increases by a factor of 8 (−0.08507 / −0.01063 = 8). This agrees with the h^3 ($2^3 = 8$) field dependence of distortion listed in Table 8.2.

Chapter 9
Chromatic Aberration

When the early astronomers looked at the sky through their telescopes, the images of the stars were surrounded by colored halos caused by an optical error in the telescope, chromatic aberration. The error was so large that it severely limited their telescopes' resolution. During the Great Plague of 1666, Isaac Newton retreated to his family home at Woolsthorpe, where he began a series of experiments on the colors of light. To investigate the problem, Newton ground and polished a triangular prism. He directed a beam of light through the prism onto a wall of his room, where the color spectrum was displayed. And when a part of this spectrum was passed through a second prism, no additional colors were created. From this, he concluded that glass in the prism did not create the colors, but that white light consists of a spectrum of colors, and the prism spread them out. Faced with these results, Newton decided that it was not possible to construct a telescope from lenses without color error. He concluded that the only way he could make a telescope without chromatic aberration was to use a mirror to focus starlight because all wavelengths obey the same law of reflection. The result was the Newtonian reflecting telescope. It wasn't until more than 90 years later that John Dollond devised a method for correcting chromatic aberration in glass lenses.

In this chapter, we will discuss the values and notation used to describe the colors of light. Then we will use the **OSsingletRev** to demonstrate chromatic aberration and determine the nature and size of the color error in an image. Finally, we will describe how this color error can be corrected with a doublet.

9.1 Refraction and Dispersion

Optical materials, such as glass, crystals, and plastics, possess many different material properties (e.g., density and hardness). For optical design, the two principal material properties are the refractive index and dispersion. The first of these determines the amount of ray bending across a surface (Fig. 1.4) according to Snell's law:

$$n' \sin i' = n \sin i, \qquad (1.2)$$

where i is the angle of incidence, and n is the refractive index on the one side of the interface; i' and n' are the same quantities on the other side. If the initial medium is air, then $n = 1$. But what refractive index should be used for the material on the other side of the interface? For example, in Section 3.3, an N-BK7 glass was entered in the LDE for the OSlens material (Fig. 3.6). When we checked the focal length of this lens in Section 3.5 using the lensmaker's equation [Eq. (1.4)], we determined that its refractive index was 1.5168. But this value is only correct at the wavelength of 587.56 nm, which was specified as part of the initial new lens setup (Fig. 3.1) in Section 3.1 System Data.

If we had used a different wavelength, we would have gotten a different index of refraction. The variation of the refractive index of a material with wavelength is known as the **dispersion** of the material. Figure 9.1 shows a plot of the index of refraction versus wavelength for N-BK7 over the visible spectrum (450–700 nm). The d, F, and C wavelengths are highlighted in the figure. These wavelengths are the same as the absorption lines of elements that were found in the solar spectrum by Joseph von Fraunhofer in 1814 and are used to compute lens performance across the visible region of the spectrum. The red line emitted by hydrogen ($\lambda_C = 656.3$ nm [Hα]) represents the long-wavelength end of the band, the yellow line in the helium spectrum ($\lambda_d = 587.56$ nm [He]) represents the center of the band, and the blue line of hydrogen ($\lambda_F = 486.13$ nm [Hβ]) represents the short-wavelength end of the band. (The symbols in brackets represent the elemental line for that wavelength.) This trio of lines is labeled from their wavelength subscripts as "d, F, C." It was the d-line (0.5876 μm) that was used for the initial design of the OSsinglet. Note that in the figure, the d-line is shown as a green line (not yellow) because the color yellow is difficult to "see" in figures.

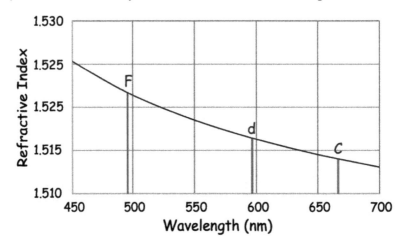

Figure 9.1 Refractive index versus wavelength for N-BK7 over the visible spectrum.

A measure of the dispersion of an optical material is given by its Abbe number or V-number,

$$V = \frac{n_d - 1}{n_F - n_C}, \tag{9.1}$$

where n_d is the refractive index of the yellow d-line, and n_F and n_C are the indices of the blue F-line and red C-line, respectively. Because the numerator $(n_d - 1)$ is a positive fraction less than one, the greatest contributor to the value of this number is the denominator. For glasses with high dispersion, the difference between the n_F and n_C refractive indices is large. And because this large difference is in the denominator of Eq. (9.1), their V-number will be small. These glasses are called **flints**. Those glasses with a small $n_F - n_C$ value have a low dispersion and are called **crown** glasses. Because the difference between n_F and n_C is small, the V-numbers of crown glasses are greater than those of the flints. When you plug $n_d = 1.516800$, $n_F = 1.522376$, and $n_C = 1.514322$ (the values for a common glass N-BK7) into Eq. (9.1), you find that the V-number is 64.2. Such a value indicates that this is a crown glass with modest dispersion.

The standard definition for V-number [Eq. (9.1)] is appropriate for optical systems operating in the visible portion of the electromagnetic spectrum. However, for optical systems designed to operate in other wavelength bands, there are corresponding V-numbers for those regions of the spectrum (see the text box, "Wavelength Bands").

Wavelength Bands

The visible wavelength band (400–700 nm) covers the range of the visual response and is the most common wavelength band for optical systems such as cameras, microscopes, and telescopes. However, optical systems can also be designed to operate over other wavelength bands stretching from the ultraviolet (10–400 nm) to the infrared [700 nm to about 15 μm (micrometers, or microns)]. Optical systems designed to work with infrared wavelengths can be used to detect heat loss in a structure, whereas the ultraviolet band can be used in forensic work to detect fingerprints and blood spatter. In some cases, an optical system, such as a terrestrial survey camera, uses several wavelength bands to record information about the ground over which a satellite passes.

Once a wavelength band is chosen, there is a need to establish the number of wavelengths, spread across the wavelength band, that enable a designer to analyze and correct a design so that its performance will be maintained across the entire wavelength band. For example, there are similar d, F, C triads for spectral bands outside the visible, and several of these are incorporated in OpticStudio's® preset spectrum selection (System Explorer > Wavelengths > Settings > Preset). These triads can be used for defining a general V-number V_{gen} for that wavelength band, using the long, mid, and short wavelengths in that band:

$$V_{gen} = \frac{n_{mid} - 1}{n_{short} - n_{long}}, \qquad (9.2)$$

so materials can be sorted into low- and high-dispersion materials for that wavelength band.

A plot of the central refractive index n_d as a function of the V-number for glasses available to the optical shop yields the Abbe diagram (Fig. 9.2), called the **glass map**. The map from Schott Glass, which is often used as a reference, locates a wide range of their glasses by the d-line index and dispersion. One notable feature of this plot is that the values on the V-number axis *decrease* from left to right. Crown glasses with low dispersion and therefore high V-numbers are on the left side of the diagram, whereas the flints are on the right. An interactive version of this map can be found at Libraries > Material Analyses > Glass Map.

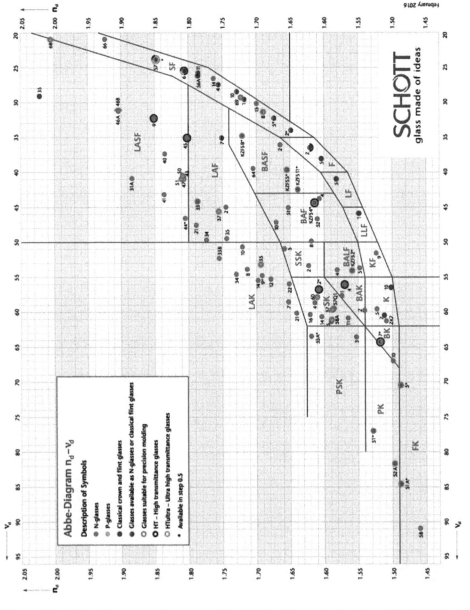

Figure 9.2 Abbe diagram or glass map. (Reprinted with permission from SCHOTT North America, Inc.)

Another way to specify a glass is to use the glass number, a six-digit number that contains both coordinates in the Abbe diagram. The first three digits are the three most significant digits of $n_d - 1$, and the last three digits are 10 times the glass's V-number. For N-BK7, the first three digits are $1.517 - 1 = 0.517$, and the last three digits are $64.2 \times 10 = 642$. Therefore, the glass number for N-BK7 is 517642. Recently, Schott has added an extension to the glass number with a decimal point followed by the three most significant figures of 100 times the density in g/cm^3; the glass number for N-BK7 becomes 517642.251. Although many glass vendors use common glass names, they are not universal, and a designer should be aware that other glass vendors use different glass names. For additional details, see the text box, "Name That Glass."

Name That Glass

There are a number of ways that glasses are named. Some glass companies include an "F" in the names of their flint glasses (e.g., F2), and the names of crown glasses contain a "K" (e.g., N-BK7; crown is spelled with a "K" in German). Other companies use numeric names. Usually, the companies provide a list of corresponding names between their products and those of other companies.

N-BK7 is a common borosilicate glass that is often used to begin a design. As the design is refined, other glasses will be added and, eventually, there may be a substitute for N-BK7. Originally, it was labeled as BK7. You will find this designation in older lens prescriptions. Because some old glasses contained "eco-unfriendly" chemicals such as lead and arsenic in their composition, they have been reformulated to remove those chemicals. Their replacements are designated by prefixing an "N" to their names. These new glasses (e.g., N-SF5) have nearly the same n_d and V-number as the originals (SF5), but because they differ somewhat, lenses made with the new glasses will have slightly different first-order values.

In the case of BK7 and N-BK7, there was no need to reformulate it, so these glass names can be used interchangeably. However, to avoid any confusion, N-BK7 will be used in all instances in this text.

9.2 Longitudinal Chromatic Aberration

In the previous chapters, we discussed both transverse and longitudinal measures of aberrations. This can also be done for errors caused by the dispersion of a lens. There is **longitudinal chromatic aberration**, usually called **axial color**, which causes the EFL of a lens to vary with wavelength. And then there is **transverse chromatic aberration**, usually called **lateral color**, which is a variation of the chief ray location as a function of wavelength. We will tackle longitudinal chromatic aberration first.

To demonstrate the longitudinal chromatic aberration in a lens, we start by looking at a simple N-BK7 singlet. Open the 50-mm EFL, $f/2.5$ **OSsingletRev** (Fig. 6.13) from Section 6.7 because its spherical aberration is less than the unreversed design. This allows us to illustrate the color error without being overwhelmed by the competing spherical aberration. To begin, we will use only a single axial field point (longitudinal chromatic aberration does not have a field dependence, so it is the same on and off axis). If the version of the lens you saved has off-axis fields, delete those fields in the FDE so we restrict the design to the axial field point, and delete the extra "Defocus" surface (S3) as it is not needed for the examples in this chapter. If you don't have the lens saved, you can type it in from the prescription given in Fig. 9.3.

	Surface Type	Comment	Radius	Thickness	Material	Clear Semi-Dia
0	OBJECT Standard ▼		Infinity	Infinity		0.000
1	STOP Standard ▼		32.464	5.000	N-BK7	10.000
2	Standard ▼		-120.000	47.376		9.669
3	IMAGE Standard ▼		Infinity	-		0.450

Title (OSsingletRev); EPD (20 mm); Field (0*); Wavelength (d-line)

Figure 9.3 LDE for the OSsingletRev with a single on-axis object point.

Next, the lens is changed from a monochromatic (single-wavelength) design to a polychromatic design that covers the visible spectrum. Although it is possible to input a triad of visible wavelengths directly, OpticStudio provides a series of preset wavelengths (System Explorer > Wavelengths > Settings > Preset) to be used as standard starting points for new designs. In the System Explorer, click on the Wavelengths disclosure arrow followed by Settings, to see the drop-down menu of Preset wavelengths for a wide variety of laser sources and common wavelength bands [Fig. 9.4(a)]. Choose the first one **F, d, C (Visible)** and then click on **Select Preset**. The trio of wavelengths within the visible spectrum, the F-line (0.486 μm, blue), the d-line (0.588 μm, yellow), and the C line (656 μm, red), will be listed below Select Preset [Fig. 9.4(b)]. Save this lens as **OSsingletRevBB** (BB stands for broadband) as we will use it later in this chapter.

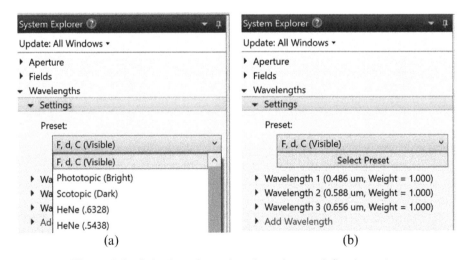

Figure 9.4 Selection of wavelengths using predefined spectra.

Chromatic Aberration

As shown in Fig. 9.4(b), the wavelengths are listed to three significant digits. However, the accuracy of these quantities within OpticStudio is much greater. The wavelengths (out to five significant digits) and the refractive indices for N-BK7 at these wavelengths (out to ten digits) can be displayed by opening a surface data report with Analyze > Reports > Surface Data and selecting surface 1. The (shortened) surface data listing for the refractive indices of N-BK7 is given in Table 9.1 (the remaining surface data in the report are not shown, for brevity).

Table 9.1 Indices for N-BK7 for the F-, d-, and C-lines to 10 places.

	Wavelength	Index
Index of Refraction:		
Glass: N-BK7		
F	0.48613	1.5223762897
d	0.58756	1.5168000345
C	0.65627	1.5143223473

Because the refractive index for each wavelength is different, the EFL of the lens (and therefore the focal position) will change with wavelength. To see this, use Ctrl-H to open the Prescription Data report (or go to Analyze > Reports > Prescription Data). If you scroll through this report, you'll see that it provides an enormous amount of valuable information about the current lens (including the index of refraction of all materials in the system, not just the one surface). This amount of information can be overwhelming to sort through. However, these reports can be easily tailored to a specific task by selecting **Settings** to open a complete list of data categories and picking only the data that are needed at that time. For this exercise, click the **Clear All** button and then check the boxes for **Index/TCE Data** and **Cardinal points**, as shown in Fig. 9.5.

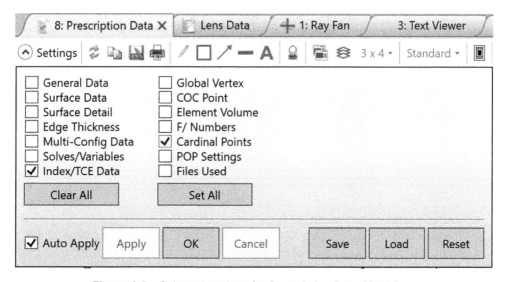

Figure 9.5 Selected settings for Prescription Data (Ctrl-H).

In the resulting report for the OSsingletRevBB is a full listing of the index of refraction (with the same values as those shown in Table 9.1) and thermal coefficient of expansion (TCE) data followed by a listing of the cardinal points for each wavelength that includes the focal length of the lens. Another compact list of the focal length output is provided in Table 9.2. The focal length (EFL) for the center (Primary) wavelength is 50 mm, as expected.

Table 9.2 EFLs for OSsingletRevBB for the F-, d-, and C-lines to 6 places.

```
Wavelength (µm)                    EFL (mm)
F          0.48613                 49.470223
d          0.58752   (Primary)     50.000000
C          0.65627                 50.239072
```

The EFL for the long wavelength, C, is 50.239 mm, while the EFL for the short wavelength, F, is 49.470 mm. The difference in focal length across this spectral band is therefore 50.239 – 49.470, or 0.769 mm. This error is known as **longitudinal chromatic aberration** and represents a focus shift along the optical axis with wavelength. Although the difference in the refractive index between the C and F lines is 0.008054 (relatively small), this variation in refractive index causes a large shift in focus and can limit the performance of a lens.

To graphically illustrate the longitudinal chromatic aberration for this trio of visible wavelengths in the OSsingletRevBB, open the Settings tab of a Layout window (Analyze > Cross-Section > Settings) and copy the settings shown in Fig. 9.6. The wavelength setting has been changed from 1 to All, and the rays are now colored by wavelength value (Wave #) instead of field. The result is a simple three-ray fan for the axial field point (Field 1) made up of the two marginal rays and an axial ray for each of the three colors (Fig. 9.7). To see the focal points for each of the colors, the cursor can be used to draw a rectangle and "zoom-in" to produce a magnified section of the focal area. As you can see, the focus position for each color shifts "longitudinally" along the optical axis. The blue (F) line is focused closer to the lens because of the larger refractive index.

Figure 9.6 Layout Settings to display longitudinal chromatic aberration in Fig. 9.7.

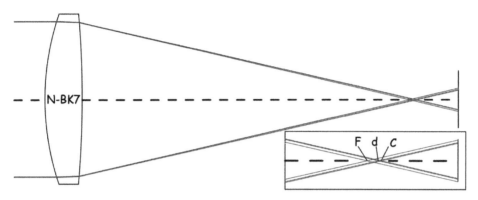

Figure 9.7 Longitudinal chromatic aberration in the OSsingletRevBB. The marginal focal points at the different wavelengths are labeled by their Fraunhofer letters. The vertical line to the right of the plot is the paraxial focal plane. The inset shows the region near the marginal ray focus.

Longitudinal chromatic aberration can also be observed in transverse ray plots (Analyze > Aberrations > Ray Aberration). If a lens has only longitudinal chromatic aberration, we would expect to see linear curves with a different slope for each wavelength (representing a defocus that changes with wavelength). However, as you will note in Fig. 9.8(a), the $f/2.5$ (20-mm EPD) OSsingletRevBB lens has a significant amount of spherical aberration that dominates the longitudinal chromatic aberration. To better illustrate the chromatic aberration, we can stop the lens down to EPD = 5 mm (System Explorer > Aperture > Aperture Value) so that the lens is working at $f/10$ and the spherical aberration is significantly reduced. This is shown in Fig. 9.8(b), where the vertical scale on the plot has been reduced by a factor of 10.

Figure 9.8(b) shows that the curves for the different wavelengths have different slope angles at the origin. The yellow wavelength curve (shown in green for clarity) is tangent to the x axis in Fig. 9.8, indicating that the evaluation plane is located at the paraxial focal point for d-line light (the primary wavelength). The slopes for the other wavelengths indicate the amount of chromatic EFL error that was illustrated in the inset of Fig. 9.7.

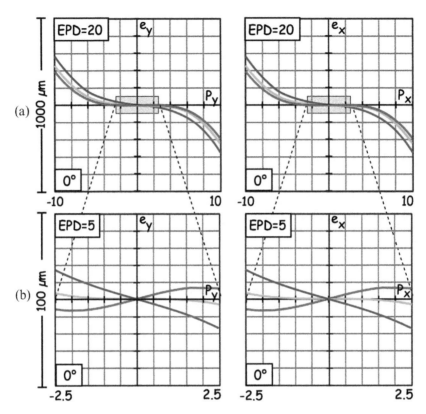

Figure 9.8 Transverse ray curves for the F, d, and C wavelengths for the OSsingletRevBB with N-BK7 glass (a) at $f/2.5$ and (b) stopped down to $f/10$.

Of course, N-BK7 is not the only available glass material for lenses (although, at times, it may seem so from its use in our examples). There are manufacturer catalogs of glasses from which an optical designer can choose lens materials. For example, the points on the Abbe diagram in Fig. 9.2 represent most of the available glasses from Schott for use in the visible spectrum. Using different materials in a design provides the designer with a means to reduce chromatic aberration, as we will show in Section 9.3.

Replace N-BK7 with F2, a flint glass, in the $f/10$ OSsingletRevBB by entering "F2" in the Material cell of surface S1 in the LDE. The **FIRST** macro shows that the EFL is 41.7329 mm. This is because F2 has a higher index of refraction (see Table 9.2) than N-BK7, resulting in a lens with a larger power and shorter focal length.

Table 9.2 Indices for F2 for the F, d, and C-lines to 10 places.

```
Index of Refraction:
Glass:   F2
     Wavelength               Index
  F    0.48613              1.6320814568
  d    0.58756              1.6200401372
  C    0.65627              1.6150316916
```

Chromatic Aberration

> **Exercise 9.1**
> Calculate the V-number for F2. How does this value compare to the V-number for N-BK7? Is F2 a crown or a flint glass? What is the glass number for F2?

To compare the aberrations of lenses with different glasses, it would be useful if we could switch glasses and still maintain the same focal length and $f/\#$. As was shown in Section 3.6, a marginal ray angle solve on the radius of curvature of the second surface of the lens can be used to maintain a 50-mm EFL. The angle needed for the solve is equal to the entrance pupil radius (2.5 mm) divided by the EFL (50 mm) or -0.05 rad. Click on the Solve box next to R2, select **Solve Type, Marginal Ray Angle,** and enter -0.05. This changes the R2 radius from -120 mm to -647.967 mm, and when you run FIRST, it shows that the OSingletRevBB with F2 glass is now a 50-mm EFL lens. Change the material back to N-BK7 and save this design as **OSsingletRevF10**. It is a 50-mm-EFL, $f/10$ singlet that can be used with any lens material and will be employed in future examples and exercises.

A comparison of the tangential (e_y) transverse ray curves (Analyze > Aberrations > Ray Aberration) between the 50-mm-EFL, $f/10$ N-BK7 lens [Fig. 9.9(a)] and the same lens with F2 glass [Fig. 9.9(b)] shows the increased separation between the C (red) and F (blue) focal points in this lens by the increased slope of curves for the F2 lens compared to those of the N-BK7 lens. Note: The sagittal transverse ray curves (e_x) are the same as the tangential ray curves for each lens and are omitted.

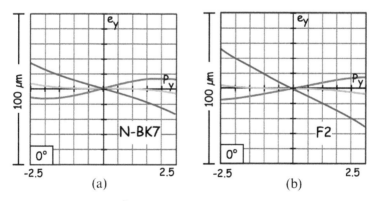

Figure 9.9 Comparison of tangential ray curves for the F, d, and C wavelengths at $f/10$ for the OSsingletRevF10 with (a) N-BK7 glass and (b) F2 glass.

Until this chapter, the aberrations we have discussed were for a single wavelength, 587.5618 nm. Now that the design has multiple wavelengths, the sizes of color errors can be displayed. These are listed in the previously neglected two columns, TAXC and TLAC, of the third-order aberration tables (**THIRD**). Table 9.3 lists the chromatic aberration coefficients of the $f/10$ OSsingletRevF10 for both the N-BK7 and F2 glasses. TAXC lists the axial color errors, and TLAC lists the lateral color errors. Lateral color is zero for a rotationally symmetric lens with only an on-axis field and will be discussed later. For now, note that the axial color is much greater for F2 than for N-BK7 due to its larger dispersion (see Exercise 9.1). Although the **THIRD** macro lists TAXC and TLAC with the third-order aberrations, they are actually first-order errors, showing differences in the paraxial marginal and paraxial chief ray locations with wavelength. These errors are included with the monochromatic third-order errors because they provide an important measure of what is wrong with a lens as a function of wavelength.

Table 9.3 Chromatic aberrations of the OSsingletRevF10 for two different materials.

N-BK7			F2		
	TAXC	TLAC		TAXC	TLAC
STO	-0.02045	0.00000	STO	-0.04052	0.00000
2	-0.01754	0.00000	2	-0.02656	0.00000
IMA	0.00000	0.00000	IMA	0.00000	0.00000
TOT	-0.03799	-0.00000	TOT	-0.06708	-0.00000

To further compare the chromatic aberration differences between an OSsingletRevF10 with F2 versus N-BK7, OpticStudio has a useful feature that plots the focal shift of a lens as a function of wavelength under Analyze > Aberrations > Chromatic Focal Shift. In Fig. 9.10, the chromatic focal shift curves are plotted for both N-BK7 and F2. For easy comparison, we have combined the plots for both glasses (see the text box "Overlaying Graphics" for more information on how to do this in OpticStudio). Because both singlets have angle solves to ensure a 50-mm EFL, the two curves intersect at the d-line. As expected, the focal shift is positive from yellow to red (the focal length is longer for red than it is for yellow) and negative from yellow to blue for both materials. The decreased plot slope (larger total focal shift) for the F2 flint indicates that it disperses the white light more strongly than the N-BK7 crown.

Not only will the dispersion of the glass differ from glass to glass, but the refractive index at the d-line will be different for each glass as well. Both the index of refraction of the material at the central wavelength in the spectrum and its dispersion play a part in any lens design. Consequently, to assist a designer in choosing a particular glass for a design, glasses in a catalog are ranked by their central-wavelength refractive index and their dispersion.

Overlaying Graphics

To combine two plots in OpticStudio, you first need to create two separate graphics windows. For example, for the OSsingletRevF10, create two chromatic focal shift plots, lock one of the plots with the lock icon at the top of the plot, and then change the lens material from N-BK7 to F2. You now should have two different plots, one for each material. Go to the plot that is not locked, right click on the plot, and choose "add/remove overlay series" from the menu listing. From the dialog box select "chromatic focal shift" and choose "primary" for the axis. The result is one graphics window with two different chromatic focal shift plots overlayed on it.

Chromatic Aberration

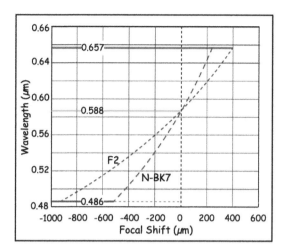

Figure 9.10 Plot of chromatic focal shift for the 50-mm-EFL singlet as a function of wavelength for two glasses, N-BK7 and F2.

Exercise 9.2
Change the material in the **OSsingletRevF10** to N-SK16. Calculate the V-number for N-SK16. Is N-SK16 a crown or a flint glass? What is the maximum focal shift range across the visible spectrum for this lens? How does this compare to the focal shift ranges in the earlier examples using N-BK7 and F2?

9.3 Correcting Longitudinal Chromatic Aberration

Nearly 100 years after Newton invented the reflective telescope, the art of instrument making had progressed to the point where a number of different glasses could be made, their refractive indices measured, and their dispersion curves determined. Researchers working in optics also invented new glasses by introducing various elements (e.g., lanthanum, barium) into silicon dioxide. Instrument makers also discovered that it was possible to reduce longitudinal chromatic aberration. The person who patented the technique and is credited with the invention was John Dolland. He (and others) figured out that by combining positive and negative lenses of different dispersions into a doublet, an optical system with reduced chromatic aberration could be made.

A simple relation between the powers of the two lenses and the dispersions of the glasses for a pair of thin lenses (in contact) can be derived with the help of the lensmaker's equation. If we use optical power instead of focal length, we get a compact set of equations that are easily manipulated. The idea is to combine two lenses, say, Lens A and Lens B, whose combined optical power will produce the required power at the central d-line. The power ϕ of a 50-mm-EFL lens is $1/f = 1/50 = 0.02$ mm^{-1}. The sum of the powers for this doublet (at the primary d wavelength) should be

$$\phi_A + \phi_B = \phi_d = 0.02 \, \text{mm}^{-1}. \tag{9.3}$$

The amount of chromatic aberration for each lens can be determined from the difference in optical powers between the two extreme wavelengths $\Delta\phi = \phi_F - \phi_C$ for each lens. Using the compact form of the lensmaker's equation given in Chapter 1,

$$\phi = (n-1)(c_1 - c_2) = (n-1)\beta, \tag{1.10}$$

we can express the power for the lens at the ends of the wavelength band using this form and then subtract the optical powers $\phi_F - \phi_C$; the difference in optical power across the band is

$$\Delta\phi = \phi_F - \phi_C = (n_F - 1)\beta - (n_C - 1)\beta = (n_F - n_C)\beta. \tag{9.4}$$

When $\Delta\phi$ is divided by the power ϕ_d for the central wavelength λ_d, we get the reciprocal of the dispersion, V:

$$\frac{\Delta\phi}{\phi_d} = \frac{\phi_F - \phi_C}{\phi_d} = \frac{(n_F - n_C)\beta}{(n_d - 1)\beta} = \frac{1}{V}. \tag{9.5}$$

For each of the lenses, their optical power difference can then be expressed as

$$\Delta\phi_A = \frac{\phi_A}{V_A} \text{ and } \Delta\phi_B = \frac{\phi_B}{V_B}.$$

If the chromatic aberration for the two lenses cancel each other (equal and opposite signs), the color error will be corrected. We can express this as

$$\Delta\phi_A + \Delta\phi_B = 0 \tag{9.6}$$

or

$$\frac{\phi_A}{V_A} + \frac{\phi_B}{V_B} = 0. \tag{9.7}$$

Because the V-numbers for standard optical materials are all positive, the only way that Eq. (9.7) can be satisfied is if one of the lenses is positive and the other is negative. However, to provide a lens that has the power specified in Eq. (9.3) (0.02 mm^{-1}), the power of the positive lens must be larger than the negative lens. Once the two glasses are chosen, the V-numbers are known, and the two equations [Eqs. (9.3) and (9.7)] can be solved to determine the powers (and focal lengths) of the two elements of the achromatic doublet. The solutions for the pair of equations are

$$\phi_A = \phi_d \frac{V_A}{V_A - V_B} \text{ and } \phi_B = \phi_d \frac{V_B}{V_B - V_A}. \tag{9.8}$$

Exercise 9.3
Find the powers (and focal lengths) of the two elements in a 100-mm-focal-length, achromatic thin lens doublet using N-FK5 and LASF35. Start by either calculating the V-number for each material using its index of refraction data or go to System Explorer > Material Catalogs > Schott and select the appropriate material from the list. The material's n_d and V_d are reported about half-way down the materials catalog output box. Having found the V_d for each glass, determine the power of each lens required to achromatize this combination.

9.4 An Example

Let's construct a 50-mm achromatic doublet using a N-BK7 crown, which has a V_A of 64.2, and an F2 flint with a V_B of 36.4. If we insert these values along with the power of the achromat $\phi_d = 0.02$ into Eq. (9.8), the powers of the two elements can be determined to be $\phi_A = 0.0462$ and $\phi_B = -0.0262$, corresponding to focal lengths of $f_A = 21.65$ mm and $f_B = -38.19$, respectively.

To create an achromatic doublet in OpticStudio, we will start with our $f/10$ broadband N-BK7 singlet **OSsingletRevF10** with only the axial field (F1). This is (as constructed above) a lens with a 5-mm EPD and an angle solve on the second surface radius to maintain a 50-mm EFL. Insert a surface after S1 (the stop surface) to create an interface between two lens elements and make its radius infinity. Then, specify F2 as the material for the second element (S2). The two glasses share the same surface curvature on S2, which we made flat. This type of lens is called a cemented doublet because the lens consists of two elements cemented together with a common radius of curvature. Note that in Chapter 7 we introduced a similar lens, the OSaplanat, but the two elements were separated by a small air space. In that case, the curvatures can be different, which introduces an additional degree of freedom when optimizing a design.

Once we have incorporated the F2 flint into the design, we need to determine the radii of curvature of the three surfaces and set some reasonable lens thicknesses. The values of the thicknesses are somewhat arbitrary and are usually chosen to make the elements "look good." For our doublet, we will make both lenses 2 mm thick.

Next, there are many different choices for the curvatures of each of the lenses. All that is required is that the difference of the curvatures of the two surfaces equals the shape factor β needed to satisfy Eq. 1.10, $\phi = (n - 1)\beta$. The simplest choice for the positive lens would be to equally split the curvature between the two surfaces. That is, make it a biconvex lens. The shape factor of Lens A is $\phi_A/(n-1) = 0.0462/(1.517 - 1) = 0.0893$. Dividing the shape factor by two (0.04465) and taking its reciprocal gives a radius of curvature for S1 of 22.4 mm. The second surface will have the same value but with a negative sign. The angle solve on S3 will change its radius to maintain a 50-mm-EFL lens at the primary d wavelength. At this point, the lens (Analyze > Cross-Section) and the LDE for your lens should look like Fig. 9.11. Save this lens as **OSachromat**.

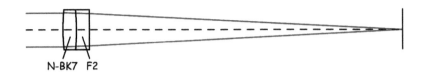

	Surface Type	Comment	Radius	Thickness	Material	Clear Semi-Di
0	OBJECT Standard ▼		Infinity	Infinity		0.000
1	STOP Standard ▼		22.400	2.000	N-BK7	2.500
2	Standard ▼		-22.400	2.000	F2	2.434
3	Standard ▼		-420.053 M	47.331 M		2.373
4	IMAGE Standard ▼		Infinity	-		1.284E-03

Title (OSachromat); EPD (5 mm); Field (0°); Wavelength (F d C lines)

Figure 9.11 LDE and cross-section of the OSachromat.

Using the Chromatic Focal Shift plot (Analyze > Aberrations > Chromatic Focal Shift) that showed the longitudinal chromatic aberration of the crown and flint singlets in Fig. 9.10, we have replotted those curves along with the chromatic focal shift curve for this achromatic doublet in Fig. 9.12. The doublet shows a dramatically smaller variation in focal shift with wavelength than the two singlets. A listing of the focal shift value in millimeters for each wavelength is given on the Text tab of the plot. The minimum shift of –0.00928 mm occurs at 537.18 nm, and the maximum shift of 0.03154 mm occurs at the C-line. The maximum focal shift range (listed on both the plot and the text tab) for the achromat is then only 40.81 μm. This compares to maximum focal shift ranges (between the C- and F-lines) of 755.25 μm and 1326.58 μm for the N-BK7 and F2 singlets, respectively.

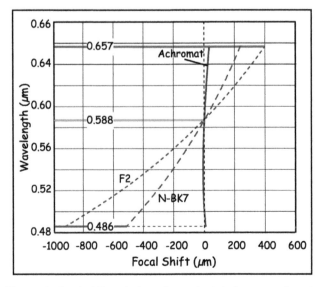

Figure 9.12 Chromatic focal shift variation of two singlets in comparison to an achromatic doublet (solid curve).

The equations [Eq. (9.8)] used to determine the powers of the two elements using a thin lens approximation provides an initial solution to reducing the color error for a 50-mm-focal-length *thin* lens. When real thickness is added to the elements, the chromatic aberration is no longer zero. It is possible to optimize the thick lens by slightly varying the radius of curvature of the surface between the two elements (slightly adjusting the powers of the two elements). In the initial solution with a biconvex crown lens, the intermediate radius of curvature is −22.4 mm. By manually entering different values for R2, we can see if the chromatic aberration can be further reduced by checking the chromatic focal shift plot and the maximum focal shift range on the plot.

For example, if you change R2 to −22 mm and check the Chromatic Focal Shift plot, you will see that the focal shifts at the ends of the curve are closer to being equal. By further changing R2 by small amounts, you can find the point where the curve is U-shaped (Fig. 9.13). This occurs with a radius for surface 2 of around −21.6 mm. The curve now crosses the zero focal shift line twice—once at the d-line (0.5876 μm, as it must because of the angle solve) and once at 527 nm. The focal shift difference between the C- and F-lines is now nearly zero (~0.0016 mm), and the total separation between the minimum (around 555 nm) and the C- and F-lines on the ends is only 28.6 μm. The chromatic focal shift of this lens has been plotted as the solid curve in Fig. 9.13 along with the initial thin lens achromatization (dashed line curve) with the biconvex crown element. Save this version of the achromat as **OSachromatOpt**.

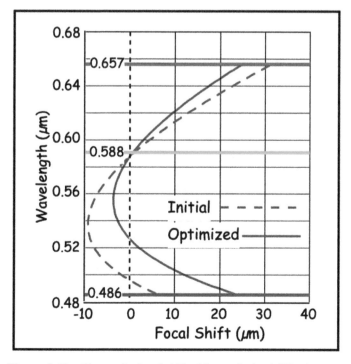

Figure 9.13 Chromatic focal shift of the optimized OSachromat.

The **THIRD** listing for color errors (TAXC and TLAC) for both the initial achromatization and the optimized version are shown in Table 9.4. The negative contributions to the axial color error from the two outer surfaces are balanced by the positive error from the N-BK7/F2 interface. And when optimized, the sum for TAXC is very close to zero.

Table 9.4 THIRD listing for TAXC and TLAC for the OSachromats.

	Achromat Initial			Achromat Optimized	
	TAXC	TLAC		TAXC	TLAC
STO	-0.02963	0.00000	STO	-0.02963	0.00000
2	0.05607	0.00000	2	0.05761	0.00000
3	-0.02771	0.00000	3	-0.02805	0.00000
IMA	0.00000	0.00000	IMA	0.00000	0.00000
TOT	-0.00127	-0.00000	TOT	-0.00007	-0.00000

This manual optimization shows that the variation of one lens parameter can improve the performance of a lens. The same approach is difficult to apply to more complex systems as the number of lens parameters increases and the number of lens errors to be reduced also grows. Instead, the variation of multiple parameters is carried out using an optimization routine within the lens design program that will be discussed in detail in Chapter 12.

9.5 Secondary Color and Spherochromatism

Even though we have corrected the longitudinal color in our OSachromatOpt, the lens still has other residual on-axis chromatic aberrations that need to be discussed. Figure 9.14 shows a transverse ray plot (Analyze > Aberrations > Ray Aberration) of the optimized achromat (OSachromatOpt). The slopes of the red and blue lines at the origin are the same, indicating that the C and F focal points are the same, but the d-line (shown in green), which is the primary wavelength, has a different slope and thus a different focal point. This tells us that this lens is not corrected for all three wavelengths at the same time. Although the total chromatic aberration of the doublet is much smaller than it would be for a single lens, there is still some residual chromatic aberration, an error referred to as a **secondary color**.

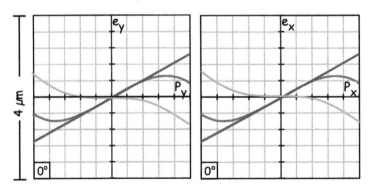

Figure 9.14 Transverse ray curves for the OSachromatOpt.

The second thing to notice in Fig. 9.14 is that while the blue curve is relatively straight, the green and red curves have a cubic shape to them, indicating that the spherical aberration of the lens is changing with wavelength. A **THIRD** command shows a total TSPH of −0.00079 at the primary wavelength (0.5876 μm). Switching the primary wavelength (System Explorer > Wavelengths > Settings) to 0.6563 μm (red) and executing **THIRD** again yields a similar TSPH value of −0.00100. The total TSPH for 0.4861 μm (blue) is nearly zero (−0.00026) as expected considering the straight blue transverse ray curve. This variation of SA with wavelength is referred to as **spherochromatism**. Unfortunately, compared to the reduction in axial color achieved by using a simple combination of two glasses with different dispersions, there is no simple strategy to reduce secondary color or spherochromatism. Glasses with special partial dispersions are needed (see Exercise 9.4).

Another way to display these two color errors is to show the longitudinal spherical aberration (LSA) as a function of normalized pupil radius (Analyze > Aberrations > Longitudinal Aberration). The longitudinal aberration plot for the OSachromatOpt is shown in Fig. 9.15.

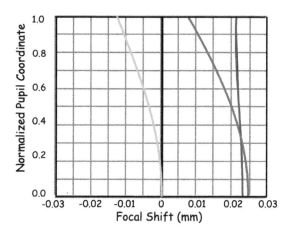

Figure 9.15 Longitudinal spherical aberration for the OSachromatOpt.

Note that in the LSA plot the C- and F-line curves intersect at the axis at nearly the same focus point (as expected), but the d-line curve doesn't fall on top of them. Rather, its focal point is about 0.025 mm in front of them. Another thing to notice in Fig. 9.15 is that the green and red curves show nearly the same curvature in the LSA plot, whereas the blue curve is nearly a straight line (indicating no spherical aberration for that wavelength).

At the beginning of Section 9.4, we stopped our singlet down to $f/10$ to be able to show chromatic aberration with very little spherical aberration. Now that we've corrected the chromatic aberration in the optimized doublet, open it up to $f/5$ by increasing the EPD to 10 mm and changing the angle solve on the last radius to −0.1. The resulting transverse ray aberration curves (Fig. 9.16) are more complicated than the third-order S-curves seen so far. Owing to the degree of correction provided by the additional element, much of the third-order spherical aberration seen in the singlet is corrected, and higher-order (fifth-order) spherical aberration can be seen at all wavelengths (e.g., the blue curve is no longer a straight line, and it starts to curve near the edges of the pupil).

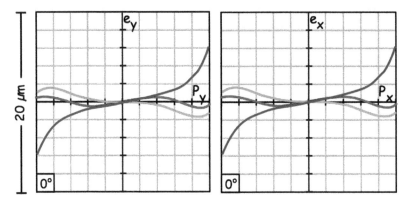

Figure 9.16 Opening the OSachromatOpt to $f/5$, its transverse ray curves show higher-order errors.

Exercise 9.4 Reduced Secondary Color

Reducing the secondary color of an achromat requires glasses with a special partial dispersion like N-FK51 (crown) and N-KZFS8 (flint). Design a thin-lens (zero thickness) 50-mm-EFL achromatic doublet with this material pair. The pertinent values for these glasses are N-FK51: $n_d = 1.4866$ and $V_d = 84.5$; N-KZFS8: $n_d = 1.7205$ and $V_d = 34.7$. Make the positive element bi-convex. Enter your values into the OSachromat lens and compare the performance of the lens over the same 5-mm EPD using the ray aberration, longitudinal aberration, and chromatic focal shift plots. Does this lens have more or less secondary color than the original OSachromat? How much spherochromatism is in the new doublet?

9.6 Lateral Color

The chromatic aberrations we've discussed so far (longitudinal color, secondary color, and spherochromatism) are color errors that do not change with field h and therefore have the same value for any field point. To see this, we can restore our $f/10$ broadband singlet **OSsingletRevF10** and set up three fields in the FDE at 0°, 3°, and 5°. If we plot the transverse ray aberration curves for this lens (Fig. 9.17), we see that the curves (especially their chromatic dependence) look very similar for each field point.

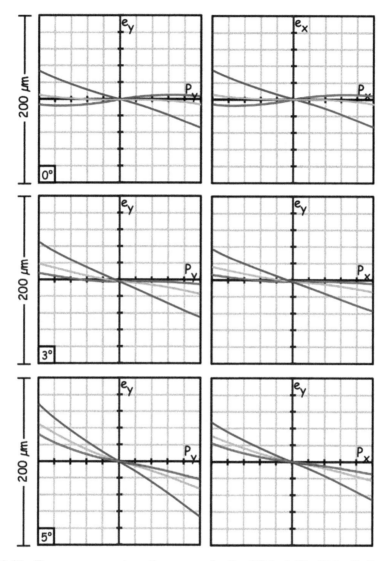

Figure 9.17 Transverse ray aberration curves for the OSsingletRevF10 with three fields; the stop is at the lens.

However, there is another chromatic aberration, **lateral color**, that does change with the field. Lateral color is highly dependent on the stop location (similar to coma and astigmatism). To see this aberration, we can insert a surface before the lens, shift the stop 10 mm in front of the lens, and replot the transverse ray plots (Fig. 9.18). The chief ray at the primary wavelength (587.56 nm) is at the origin of all transverse ray aberration curves. In Fig. 9.18, there is a clear shift of the tangential (y) ray curves around the chief ray (at the center of the pupil) as a function of wavelength, and the shift increases with field. The sagittal (x) ray curves do not shift since the object is defined only in y. This color error is also known as **transverse chromatic aberration**. It represents a change in image size with wavelength.

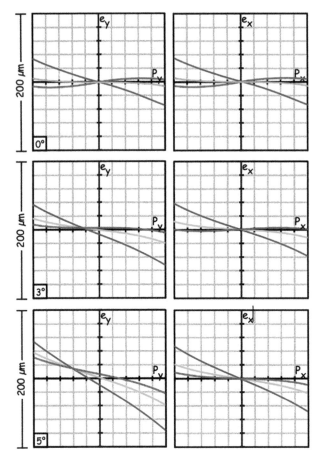

Figure 9.18 Transverse ray aberration curves for the OSsingletRevF10 with three fields; the stop is 10 mm in front of the lens.

We can quantify the lateral color by calculating how much the image size changes with wavelength by tracing a single chief ray (Analyze > Rays & Spots > Single Ray Trace) and finding its height in the image plane as a function of wavelength. Subtracting the values (4.36682 mm for red and 4.35126 mm for blue) gives a 0.01556-mm change in image height from red to blue. The amount of chief ray shift (in microns) from the primary wavelength can also be read directly off the text tab of the transverse ray curves for the largest field. The shift for blue is −10.77 μm, and the shift for red is 4.79 μm for a total shift from blue to red of 15.56 μm. We can also measure the amount of lateral color in a lens using the aberration coefficient TLAC from THIRD. Table 9.5 compares the TLAC of the lens with the stop at the lens (where the TOT is near zero) and the stop 10 mm in front of the lens (where the TOT is ~15.4 μm). Note that the axial color (TAXC) is unchanged with stop shift.

Table 9.5 THIRD listing for the OSsingletRevF10 with a 5° field.

	Stop at lens			Stop 10 mm in front of lens	
	TAXC	TLAC		TAXC	TLAC
STO	-0.02045	-0.02323	STO	0.00000	0.00000
2	-0.01754	0.02109	2	-0.02045	-0.03038
IMA	0.00000	0.00000	3	-0.01754	0.01495
TOT	-0.03799	-0.00214	IMA	0.00000	0.00000
			TOT	-0.03799	-0.01543

One of the easiest ways to correct lateral color is to make the optical system symmetric about the aperture stop (similar to the discussion for correcting distortion in Chapter 8). However, this means that broadband optical systems with external stops, such as eyepieces, are often limited by lateral color. To demonstrate this, we will use a Ramsden eyepiece. The system, designed by Jesse Ramsden in 1782, is simple, consisting of two plano-convex lenses separated by a distance less than their common focal length, as shown in Fig. 9.19. Its prescription is given in Fig. 9.20. The design for the eyepiece is inverted (with the eye located on the left and the intermediate image located on the right), as discussed in the text box, "Designing Eyepieces." Enter the lens with an EPD of 3 mm (the approximate diameter of the pupil of the eye), field angles of 0°, 7°, and 10°, the d, F, C wavelengths, then add a 10% CSD margin for the lenses. A Quick Focus for radial spot size was used to find the best image plane location (4.849 mm). Save the lens as **OSramsden**.

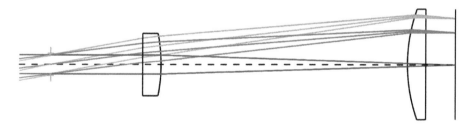

Figure 9.19 The Ramsden eyepiece.

	Surface Type	Comment	Radius	Thickness	Material	Clear Semi-Di
0	OBJECT Standard ▼		Infinity	Infinity		Infinity
1	STOP Standard ▼		Infinity	15.000		1.500
2	Standard ▼		Infinity	3.000	N-BK7	4.559
3	Standard ▼		-25.000	40.000		4.890
4	Standard ▼		25.000	3.000	N-BK7	8.657
5	Standard ▼		Infinity	4.849		8.545
6	IMAGE Standard ▼		Infinity	-		7.335

Title (OSramsden); EPD (3 mm); Fields (0° 7° 10°); Wavelengths (F d C lines)

Figure 9.20 LDE for the Ramsden eyepiece.

> **Designing Eyepieces**
>
> In Section 2.7, some optical systems with multiple elements were described. The Keplarian telescope shown in Fig. 2.14(a) has an objective lens and an eyepiece lens. The approach to the design of these two lenses is quite different. Whereas the objective lens is designed to provide a high-resolution image at the back focal plane, the eyepiece is designed to examine that image at a finite distance from the lens and produce an image at infinity. This is because the eye is most relaxed when viewing an image at infinity. Therefore, the intermediate image should be located at the front focal point of the eyepiece. Analyzing and optimizing an image at infinity can be difficult to do (and understand) in a design program, so these types of systems are typically flipped so that the "object" is located at infinity and the "image" is at the focal point of the eyepiece. This arrangement will be used for the Ramsden eyepiece.

Both lenses are made of N-BK7 glass, and there is no axial color correction, as a plot of the longitudinal spherical aberration for this lens shows (Fig. 9.21). Note that the three curves have the same shape, indicating that there is no spherochromatism in this design. However, the lens has a more significant lateral color problem (off-axis), as shown by the transverse ray aberration plots in Fig. 9.22(a) and the spot diagram (Analyze > Rays & Spots > Standard Spot Diagram) in Fig. 9.22(b). The spot diagrams for the 7° and 10° fields for the three wavelengths don't overlap, indicating a clear change in image location with wavelength.

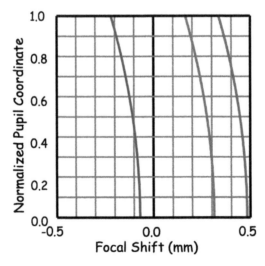

Figure 9.21 LSA for the Ramsden eyepiece.

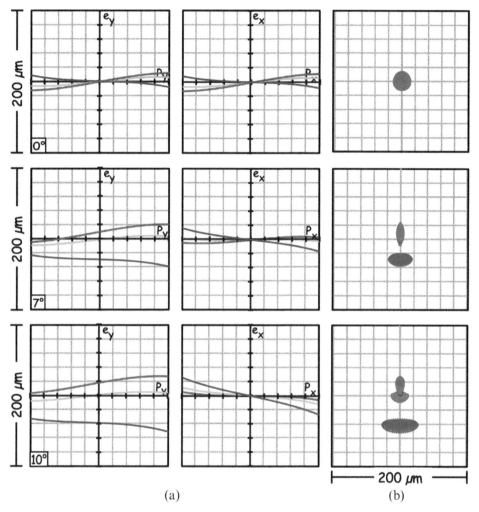

Figure 9.22 Plots for the Ramsden eyepiece showing lateral color: (a) transverse ray plots and (b) spot diagrams for the same design.

Perhaps the most evident examples of lateral color can be seen in images produced using simple lenses at large field angles. Usually, it is most evident in the corners of photographs and projector displays, particularly along the edges of areas where there is a large change in contrast. For example, Fig. 9.23(b) [an enlargement of an area of Fig. 9.23(a)] shows lateral color fringing (red) along one side of the white edge of the bill of the girl's cap and a blue fringe along the other edge.

Figure 9.23 Example of lateral color error.

Now that we have discussed the basic aberrations of a lens, we can begin to address them with the goal of improving the performance of the lens. We have already seen examples of this using lens bending (Sections 6.7 and 7.1.2). In the next chapter, this process can be carried out within OpticStudio using its optimization features where we demonstrate a number of strategies that can be employed to counteract certain aberrations.

Exercise Answers

Ex. 9.1
From Table 9.2:

Glass: F2
#	Wavelength	Index
1	0.48613	1.6320814568
2	0.58756	1.6200401372
3	0.65627	1.6150316916

$V_d = (1.620040 - 1)/(1.632081 - 1.615032) = 0.62004/0.017049 = 36.37$.

The F2 V-number (36.4) is ~2× smaller than the V-number for N-BK7, or 64.2. The small V-number makes it a flint. (Its name gives this away.) With $n = 1.620$ and $V = 36.4$, the glass number is $(1.620 - 1)$ with 10×36.4) or 620364.

Ex. 9.2
Change the material in the **OSsingletRevF10** to N-SK16. From the surface data report for N-SK16:

Glass: N-SK16

#	Wavelength	Index
1	0.48613	1.6275563488
2	0.58756	1.6204099651
3	0.65627	1.6172716599

$V_d = 0.620410 / (1.627556 - 1.617272) = 0.620410 / 0.010284 = 60.33$.

The high V-number indicates that it is a crown, but it is slightly more dispersive than N-BK7. The maximum focal shift range listed on the bottom of the chromatic focal shift plot is 803.6 μm which, as expected, is slightly more than that of N-BK7 (755.2 μm) and much less than that of F2 (1326.6 μm).

Ex. 9.3
N-FK5 $V_d = 70.4$
LASF35 $V_d = 29.1$

$\phi_d = 1/100$ mm $= 0.01$ mm^{-1}

Using Eq. (9.8),
$\phi_A = 0.01 \times 70.4 / (70.4 - 29.1) = 0.0170$ mm^{-1} (focal length of 58.8 mm);
$\phi_B = 0.01 \times 29.1 / (29.1 - 70.4) = -0.0070$ mm^{-1} (focal length of 142.9 mm).

Ex. 9.4
As in Ex. 9.3, the powers of the two lenses need to be determined, where
N-FK51 $n_d = 1.4866$ $V_d = 84.5$
N-KZFS8 $n_d = 1.7205$ $V_d = 34.7$

$\phi_d = 1/50$ mm $= 0.02$ mm^{-1}

Using Eq. (9.8),
$\phi_A = 0.02 \times 84.5 / (84.5 - 34.7) = 0.0339$ mm^{-1},
$\phi_B = 0.02 \times 34.7 / (34.7 - 84.5) = -0.0139$ mm^{-1}.

For the positive lens, split the power equally between the two surfaces such that $R1 = 2(n_A - 1) / \phi_A = 2 \times 0.4866 / 0.0339 = 28.708$ mm and $R2 = -28.708$ mm.

Open the **OSachromat** and resave it under another name so that you do not overwrite that lens. Replace the radii of curvature with the values calculated above, zero out the lens thicknesses, and update the lens materials. Plot the chromatic focal shift, ray aberration, and longitudinal aberration for this doublet and compare with the original OSachromat (with zero lens thicknesses). The new doublet has very little spherochromatism, about ½ of the secondary color, but more than 3× the spherical aberration.

	Surface Type		Comment	Radius	Thickness	Material	Clear Semi-Di
0	OBJECT	Standard ▼		Infinity	Infinity		Infinity
1	STOP	Standard ▼		28.708	0.000	N-FK51	2.500
2		Standard ▼		-28.708	0.000	N-KZFS8	2.634
3		Standard ▼		-64.332 M	50.000 M		2.633
4	IMAGE	Standard ▼		Infinity	-		0.878

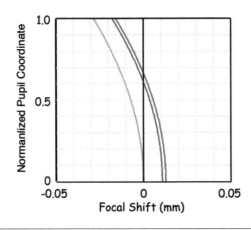

Chapter 10
Reducing Aberrations

Once you learn how to model an optical system and analyze its optical errors, you can determine if the performance of the existing design is good enough or if it needs to be improved. In some cases, it's acceptable to change the specifications to improve the performance of a design (e.g., stopping down the lens). In other cases, the design can be modified to improve performance, a process called **optimization**. For example, in Section 6.4, we used Optimize > Quick Focus to find the best image plane location. Later in Chapter 6, the OSsinglet was modified by "bending" the lens (by hand) to reduce its spherical aberration. If needed, lenses can be added or a completely different design form can be sought. Up to this point in the text, most of the optimization of lens performance has been done manually. But, optimization can be done automatically in OpticStudio® using the various features found on the Optimize tab.

In this chapter, we will focus on OpticStudio's local optimization (Optimize > Optimize!) to reduce the third-order aberrations that we identified in Chapters 6–8. This is not a comprehensive approach to optimizing lens performance. In practice, just reducing third-order aberrations is not advised because real lenses can have large apertures and/or large fields that produce higher-order aberrations. Targeting the third-order aberrations to zero prevents them from balancing with higher-order aberrations (typically resulting in worse performance). However, this approach provides an introduction to optimization and therefore is presented before we tackle more advanced optimization (including global optimization) of lens systems in Chapter 12.

10.1 The Merit Function

Optimization requires a starting design, a set of variable parameters (e.g., curvatures, thicknesses, and airspaces), and a **merit function** (a means of accounting for the change in performance during optimization). Optimization is carried out in a series of steps or cycles. During each cycle, the variables in the design are allowed to change in small increments. At the end of each cycle, the merit function value of the new design is computed and compared to that of the previous cycle. For local optimization, OpticStudio uses an actively damped least-squares algorithm to find the minimum of the specified merit function that is nearest to the starting point design. Then the designer must determine if this is an acceptable design ("good enough!") or whether more substantial changes to the starting point must be made and another round of optimization started.

But before we can optimize anything, we need to construct a merit function to monitor the changes in the optical system during optimization. In practice, the merit function has a single numerical value that is meant to represent how closely an optical system meets a set of specified goals. These goals can be performance related (e.g., RMS spot size), system constraint related (e.g., a 50-mm EFL), or manufacturing related (e.g., a 5-mm minimum edge thickness for a lens). In OpticStudio, these goals are called **operands**. Each operand has a target value (typically zero) and a weight. The merit function is then calculated by taking the square root of the weighted sum of the squares of the difference between the actual value and target value of each operand in the list. The merit function is defined this way so that the ideal value is zero. The optimization algorithm attempts to make the value of this function as small as possible.

In OpticStudio, the Merit Function Editor (MFE) is used to specify, modify, and review the operands involved in the construction of the merit function. Figure 10.1 shows the first 20 lines of an example merit function (we'll explain how to set this is up later). Each line of the MFE is an operand where the first column is the operand number and the four letters given in the second column (Type) determine the type of operand. The next eight data fields on the line are used to further define the operand. The first two fields are integer values (Int1 and Int2), and the next six are double precision values (Data1 through Data6). Note: Not all of the operands use all of the fields provided, so it is important to have your cursor in the operand line to activate the headers that describe the data field for that operand. Most of the performance operands are ray-based quantities and require the use of Data1 through Data4 to define the values of the normalized field (Hx and Hy) and pupil coordinates of the ray (Px and Py). For example, line 19 in Fig. 10.1 is a TRCY operand (transverse ray aberration in the y direction measured with respect to the centroid) for a ray at the edge of the field (Hy = 1) through the pupil at coordinates Px = 0.471 and Py = 0.816. The Int2 column (Wave) shows that the ray is being traced for wavelength 1.

	Type	Wave	Hx	Hy	Px	Py	Target	Weight	Value	% Contrib
1	EFFL	1					50.000	10.000	50.000	0.000
2	BLNK									
3	DMFS									
4	BLNK	Sequential merit function: RMS spot x+y centroid X Wgt = 1.0000 Y Wgt = 1.0000 GQ 3 rings 6 arms
5	BLNK	No air or glass constraints.
6	BLNK	Operands for field 1.
7	TRCX	1	0.000	0.000	0.336	0.000	0.000	0.436	-0.044	0.054
8	TRCY	1	0.000	0.000	0.336	0.000	0.000	0.436	0.000	0.000
9	TRCX	1	0.000	0.000	0.707	0.000	0.000	0.698	-0.436	8.637
10	TRCY	1	0.000	0.000	0.707	0.000	0.000	0.698	0.000	0.000
11	TRCX	1	0.000	0.000	0.942	0.000	0.000	0.436	-1.104	34.652
12	TRCY	1	0.000	0.000	0.942	0.000	0.000	0.436	0.000	0.000
13	BLNK	Operands for field 2.
14	TRCX	1	0.000	1.000	0.168	0.291	0.000	0.145	-0.015	2.035E-03
15	TRCY	1	0.000	1.000	0.168	0.291	0.000	0.145	-0.220	0.457
16	TRCX	1	0.000	1.000	0.354	0.612	0.000	0.233	-0.157	0.375
17	TRCY	1	0.000	1.000	0.354	0.612	0.000	0.233	-0.419	2.658
18	TRCX	1	0.000	1.000	0.471	0.816	0.000	0.145	-0.418	1.652
19	TRCY	1	0.000	1.000	0.471	0.816	0.000	0.145	-0.809	6.202
20	TRCX	1	0.000	1.000	0.336	0.000	0.000	0.145	-0.062	0.037

Merit Function: 0.307042179750782

Figure 10.1 Example merit function.

In the final columns of the MFE, each operand is given a target value (Target) and a weight (Weight), and the current value of the operand is listed under the Value column. The difference between the target and the current value of the operand is squared and summed (with weights) over all of the operands to yield the merit function value (shown at the top of the MFE). The last column (% Contrib) lists the relative contribution of each operand to the overall merit function as a percentage. This column is extremely useful for finding the biggest offenders in the merit function and/or setting an appropriate weight for an operand to make the optimizer "pay attention to" a particular constraint.

Like other OpticStudio editors, new operands can be added or deleted using the insert and delete keys. The merit function and the values of each operand can be updated by clicking the Update button ⟳ in the upper left-hand corner of the MFE. Note: When the weight of an operand is set to zero, the optimization algorithm calculates but ignores the operand. This is very useful for computing a result that does not have a specific target but might be used elsewhere in the merit function or if the value of the parameter is to be monitored during the design process.

When you first open the MFE, you will see that it is empty. Most merit functions consist of hundreds of operands, especially complex ray-based operands like the ones shown in Fig. 10.1. Instead of typing them in one by one (a daunting task, for sure), OpticStudio has an Optimization Wizard (Optimize > Optimization Wizard) that provides an easy way to create the majority of the operands that you will need for optimization. Figure 10.2 shows the Optimization Wizard's dialog box, where you will find a number of settings that define a default merit function (DMF). Once your settings have been selected and applied (Apply), the Wizard generates a list of operands below the DMFS (Default Merit Function Start) line in the MFE. It is important for proper operation of the DMF that all other user-entered operands be inserted and entered **above** this line, **never** below it. The placement of the DMFS line can be changed with the Start At setting in the wizard. Note: If new operands are inserted above the DMFS, the wizard automatically updates the Start At line number.

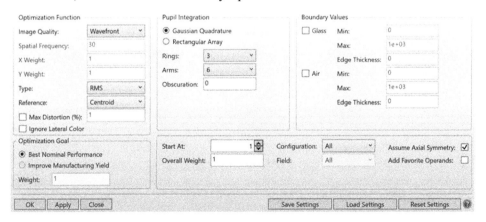

Figure 10.2 Starting Optimization Wizard settings.

The term "default" in the DMF may be confusing at first because it is not a default for all lenses in OpticStudio. Instead, the default wizard settings are those you have chosen for the optimization of the current lens. The default nomenclature is meant to separate the operands created by the wizard from any user-entered operands (that go above the DMFS). Although the DMF can be updated by changing the wizard settings, it is still referred to as the DMF even after the changes are applied.

For example, the starting **Image Quality** metric in the Optimization Wizard for any new lens is set to **Wavefront**. This setting generates optical path difference (OPD) operands and optimizes the wavefront of the lens as measured in the exit pupil. For some of the design examples in this text, we will use a DMF with the image quality set to **Spot** to generate centroid-based transverse ray aberration (TRC) operands to optimize the RMS spot size in the image plane. These operands and other wizard settings (and how they impact the DMF operands) will be discussed in detail as we walk through the optimization examples in Chapter 12. Note: An easy way to check your default wizard settings in the MFE is to look at the line below DMFS, where you will find a terse description of the DMF (e.g., **Sequential merit function: RMS spot x + y centroid X Wgt = 1.00 Y Wgt = 1.00 GQ 3 rings 6 arms** in Fig. 10.1).

10.2 Defocus

Although defocusing a lens does not reduce any of its aberrations, it serves as the simplest example of the automatic optimization of lens performance. In Section 6.4, we showed that the spot size for an axial ray bundle was smaller when the image plane was shifted (defocused) from the paraxial image plane. While the paraxial image plane is a very useful reference plane, unless the lens is perfect (no aberrations), it is not the best location for the image evaluation plane. For this reason, the image distance should be allowed to vary during the optimization of any lens. We will use Optimize > Optimize! to demonstrate this.

Restore the 50-mm-EFL, $f/2.5$ **OSsinglet** from Chapter 3 with only the on-axis field point in the FDE and a single wavelength (d-line at 0.5876 μm). In OpticStudio, a parameter is allowed to vary when the solve box menu is set to **Solve Type: Variable** (Ctrl-Z shortcut key). The box then displays a "V" to indicate its variable status. Change the thickness solve on surface 2 from marginal ray height (M) to variable (V), so that the resulting LDE looks like Fig. 10.3.

	Surface Type	Comment	Radius	Thickness	Material	Clear Semi-Dia
0	OBJECT Standard ▼		Infinity	Infinity		0.000
1	STOP Standard ▼		120.000	5.000	N-BK7	10.000
2	Standard ▼		-32.464	49.290 V		9.914
3	IMAGE Standard ▼		Infinity	-		1.350

Title (OSsinglet); EPD (20 mm); Field (0°); Wavelength (d-line)

Figure 10.3 LDE of the OSsinglet with a variable image distance before optimization.

If you were to try to optimize the lens now, nothing would happen. This is because a merit function hasn't been defined yet. Open the Optimization Wizard (Optimize > Optimization Wizard) and change the **Image Quality** from **Wavefront** to **Spot** and hit **OK**. In the first column of the resulting MFE (see Fig. 10.4), the DMF displays two transverse ray operands (TRCX and TRCY) for three ray positions in the pupil (Px: 0.336, 0.707, and 0.942) for a total of six operands. The starting merti function (MF) value equals 0.4602, where the largest contribution (79.9%) to the merit function value is the TRCX operand for the ray at the edge of the pupil (Px = 0.942). This is not surprising as the lens is currently at paraxial focus and the marginal rays at the edge of the pupil focus in front on the paraxial image plane due to spherical aberration.

	Type	Wave	Hx	Hy	Px	Py	Target	Weight	Value	% Contrib
1	DMFS ▾									
2	BLNK ▾	Sequential merit function: RMS spot x+y centroid X Wgt = 1.0000 Y Wgt = 1.0000 GQ 3 rings 6 arms								
3	BLNK ▾	No air or glass constraints.								
4	BLNK ▾	Operands for field 1.								
5	TRCX ▾		1	0.000	0.000	0.336	0.000	0.000	0.873	-0.044	0.126
6	TRCY ▾		1	0.000	0.000	0.336	0.000	0.000	0.873	0.000	0.000
7	TRCX ▾		1	0.000	0.000	0.707	0.000	0.000	1.396	-0.436	19.927
8	TRCY ▾		1	0.000	0.000	0.707	0.000	0.000	1.396	0.000	0.000
9	TRCX ▾		1	0.000	0.000	0.942	0.000	0.000	0.873	-1.104	79.948
10	TRCY ▾		1	0.000	0.000	0.942	0.000	0.000	0.873	0.000	0.000

Merit Function: 0.460209621002935

Figure 10.4 MFE for the OSsinglet before optimization.

Now when you optimize (Optimize > Optimize!), a Local Optimization window opens [Fig. 10.5(a)] and lists the active algorithm, number of targets (operands), number of variables, and the initial merit function (which is the same as the value listed at the top of the MFE). Click the Start button to start the optimization. OpticStudio then runs through its optimization routine, varying the image distance until a minimum MF is reached. The initial MF is reduced from 0.4602 to 0.1535, which is a factor of 3. To continue, you must close the Local Optimization window by clicking Exit.

The LDE [Fig. 10.5(b)] shows that the variable image distance is now 45.231 mm, which is 4.059 mm smaller than the distance before optimization. This is the same value that was computed using the Quick Focus feature in the Optimization tab at the end of Section 6.4. The spot diagrams at paraxial focus and best focus were shown in Fig. 6.7 and are reproduced here as Figs. 10.6(a) and (b). Not surprisingly, the spot size is approximately three times smaller at best focus, which agrees with the 3× improvement in the MF.

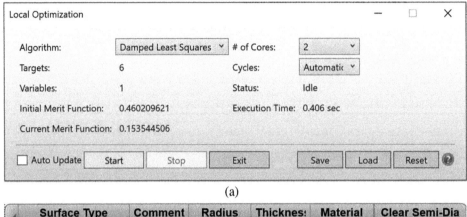

Figure 10.5 (a) Local Optimization window and (b) LDE for the OSsinglet after defocus optimization.

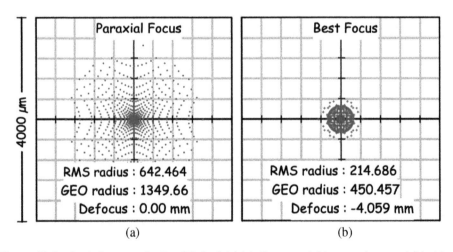

Figure 10.6 Spot diagrams for the OSsinglet (a) in the paraxial image plane and (b) at best focus.

10.3 Reducing Spherical Aberration

We described the manual optimization of the shape of the OSsinglet lens in Section 6.7 Lens Bending. We found the best lens shape that gave us minimum third-order spherical aberration by trial and error. Automatic optimization can be applied to the (focused) OSsinglet to perform the same operation. If you don't have your lens open from the last section, restore the **OSsinglet**, make the image distance variable, and apply a quick focus Optimize > Quick Focus. The LDE should have the same values as shown in Fig. 10.5(b).

The **THIRD** macro gives the third-order aberrations for the lens, listed in Table 10.1. Because we have only defined an axial field point, the spherical aberration coefficient (TSPH) is the only non-zero third-order aberration. Note that the TSPH for the first surface is –0.0032 and for the second surface is –1.1333; their sum is –1.1366. So, the strongly curved back surface is responsible for almost all of the third-order spherical aberration in this lens.

Table 10.1 Third-order aberration coefficients (THIRD) for the OSsinglet.

Surf	TSPH	TTCO	TAST	TPFC	TSFC	TTFC	TDIS
STO	-0.0032	-0.0000	-0.0000	-0.0000	-0.0000	-0.0000	-0.0000
2	-1.1333	0.0000	-0.0000	-0.0000	-0.0000	-0.0000	0.0000
IMA	-0.0000	-0.0000	-0.0000	-0.0000	-0.0000	-0.0000	-0.0000
TOT	-1.1366	0.0000	0.0000	0.0000	0.0000	0.0000	0.0000

Although you can vary almost every lens parameter (e.g., radii, thicknesses, and materials) in OpticStudio, for now we will simply optimize the shape of the OSsinglet by varying its two radii of curvature. In the LDE, change the **Solve Type** box next to each radius from **Fixed** to **Variable** (V). The resulting LDE is shown in Fig. 10.7. Note: The image distance is also still allowed to vary to find the best image plane location for the new lens shape.

	Surface Type	Comment	Radius	Thickness	Material	Clear Semi-Dia
0	OBJECT Standard		Infinity	Infinity		0.000
1	STOP Standard		120.000 V	5.000	N-BK7	10.000
2	Standard		-32.464 V	45.231 V		9.914
3	IMAGE Standard		Infinity	-		0.450

Figure 10.7 LDE with variables to optimize the lens shape of the OSsinglet.

For optimization, we can use the same RMS spot merit function that we used in Section 10.2 (see Fig. 10.4), where the Image Quality setting in the wizard was changed from Wavefront to Spot. However, now that both curvatures are allowed to vary, we need to introduce a user-entered operand (constraint) to maintain a 50-mm EFL during optimization. (Without this constraint, the EFL is free to vary and will quickly change.) This is done by entering an appropriate four-letter operand Type for the EFL to keep it near 50 mm. In this case, it is EFFL. There are many different optimization operands; the "User-Entered Operands" text box describes some useful help tools for quickly finding a desired operand in OpticStudio.

As discussed earlier, user-entered operands (like a focal length constraint) need to be added **before** the start line of the default merit function. Open the merit function editor (Optimize > Merit Function Editor) and insert three new operand lines above the DFMS in the MFE. Then, for operand #2, select the four-letter operand for effective focal length, **EFFL**, in the Type column and give it a target of 50 (mm) with a Weight of 1. The MFE after updating ⟳ is shown in Fig. 10.8, with a starting merit function value of 0.1426.

	Type	Wave					Target	Weight	Value	% Contrib
	▼ Wizards and Operands ◀ ▶						Merit Function:	0.142614451505521		
1	BLNK ▼									
2	EFFL ▼	1					50.000	1.000	50.000	0.000
3	BLNK ▼									
4	DMFS ▼									
5	BLNK ▼	Sequential merit function: RMS spot x+y centroid X Wgt = 1.0000 Y Wgt = 1.0000 GQ 3 rings 6 arms								
6	BLNK ▼	No air or glass constraints.								
7	BLNK ▼	Operands for field 1.								
8	TRCX ▼	1	0.000	0.000	0.336	0.000	0.000	0.873	0.232	31.594
9	TRCY ▼	1	0.000	0.000	0.336	0.000	0.000	0.873	0.000	0.000
10	TRCX ▼	1	0.000	0.000	0.707	0.000	0.000	1.396	0.166	26.096
11	TRCY ▼	1	0.000	0.000	0.707	0.000	0.000	1.396	0.000	0.000
12	TRCX ▼	1	0.000	0.000	0.942	0.000	0.000	0.873	-0.268	42.310
13	TRCY ▼	1	0.000	0.000	0.942	0.000	0.000	0.873	0.000	0.000

Figure 10.8 MFE for the OSsinglet with an EFL constraint (before optimization).

User-Entered Operands

There are hundreds of different operands in OpticStudio that can be entered by the user. Each operand is designated in the MFE by a four-letter Type. Since it would be difficult to memorize them all, OpticStudio provides some handy reference tables to quickly find the one that you are looking for. First, there is an Optimization Operands Summary Table (part of which is shown below), which categorizes the operands by general subject.

Category	Related Operands
First-order optical properties	AMAG, ENPP, EFFL, EFLX, EFLY, EPDI, EXPD, EXPP, ISFN, LINV, OBSN, PIMH, PMAG, POWF, POWP, POWR, SFNO, TFNO, WFNO
Aberrations	ABCD, ANAC, ANAR, ANAX, ANAY, ANCX, ANCY, ASTI, AXCL, BIOC, BIOD, BSER, COMA, DIMX, DISA, DISC, DISG, DIST, FCGS, FCGT, FCUR, LACL, LONA, OPDC, OPDM, OPDX, OSCD, PETC, PETZ, RSCE, RSCH, RSRE, RSRH, RWCE, RWCH, RWRE, RWRH, SMIA, SPCH, SPHA, TRAC, TRAD, TRAE, TRAI, TRAR, TRAX, TRAY, TRCX, TRCY, ZERN
MTF data	GMTA, GMTS, GMTT, MSWA, MSWS, MSWT, MTFA, MTFS, MTFT, MTHA, MTHS, MTHT

Then, for more detailed information on an individual operand, there is an alphabetical list of the MFE operands (the first part of which is shown below).

NAME	Description
ABCD	The ABCD values used by the grid distortion feature to compute generalized distortion. See "Grid Distortion" on page 184. The reference field number is defined by Ref Fld. The wavelength number is defined by Wave. Data is 0 for A, 1 for B, 2 for C, and 3 for D. See also "DISA" on page 496.
ABGT	Absolute value of operand greater than. This is used to make the absolute value of the operand defined by Op# greater than the target value.
ABLT	Absolute value of operand less than. This is used to make the absolute value of the operand defined by Op# less than the target value.
ABSO	Absolute value of the operand defined by Op#.
ACOS	Arccosine of the value of the operand defined by Op#. If Flag is 0, then the units are radians, otherwise, degrees.

Both of these reference tables can be found in the help documentation. The fastest way to locate them is to open the **Operand Properties** dialog box by clicking on the **Wizards and Operands** drop-down arrow in the MFE and then clicking the help question mark in the lower right-hand corner of the box.

After a local optimization (Optimize > Optimize!), the merit function is reduced to 0.0475 (~3× improvement). The MFE after optimization is shown in Fig. 10.9. A quick look at the value for the EFL operand (EFFL) shows that it is currently 50.001 and not quite the 50 mm we targeted. The % contribution to the merit function is 0.0003 (very small).

	Type	Wave					Target	Weight	Value	% Contrib
	Wizards and Operands						Merit Function:	0.047519344240443		
1	BLNK									
2	EFFL	1					50.000	1.000	50.001	3.051E-03
3	BLNK									
4	DMFS									
5	BLNK	Sequential merit function: RMS spot x+y centroid X Wgt = 1.0000 Y Wgt = 1.0000 GQ 3 rings 6 arms								
6	BLNK	No air or glass constraints.								
7	BLNK	Operands for field 1.								
8	TRCX	1	0.000	0.000	0.336	0.000	0.000	0.873	0.078	32.670
9	TRCY	1	0.000	0.000	0.336	0.000	0.000	0.873	0.000	0.000
10	TRCX	1	0.000	0.000	0.707	0.000	0.000	1.396	0.054	24.579
11	TRCY	1	0.000	0.000	0.707	0.000	0.000	1.396	0.000	0.000
12	TRCX	1	0.000	0.000	0.942	0.000	0.000	0.873	-0.090	42.747
13	TRCY	1	0.000	0.000	0.942	0.000	0.000	0.873	0.000	0.000

Figure 10.9 MFE for the OSsinglet with an EFL constraint (after optimization).

To get closer to 50 mm, the weight of the operand must be increased to increase the contribution to the merit function so the optimizer will pay more attention to the constraint. Increase the weight from 1 to 10 and update the merit function. The merit function value changes to 0.0318, but more importantly, the focal length operand % contribution to the merit function is larger, 0.031. After another round of local optimization, the change in the merit function is only in the fifth decimal place, but the focal length is now 50.000. This is an example of how to adjust the weight of user-entered operands during optimization.

The LDE for the optimized lens is shown in Fig. 10.10. As was the case when we flipped the lens in Section 6.7 Lens Bending, the singlet performance is significantly improved when the surface facing the image is much flatter (R2 = −171.3 mm) than the front surface (R1 = 30.1 mm). Figure 10.11(b) shows the spot diagram of the optimized singlet and compares it with the spot diagram for the OSsinglet at best focus [Fig. 10.11(a)]. The optimization has reduced the best RMS spot radius from 214.7 μm to 71.5 μm, a three-fold decrease.

	Surface Type		Comment	Radius	Thickness	Material	Clear Semi-Dia
0	OBJECT	Standard		Infinity	Infinity		0.000
1	STOP	Standard		30.127 V	5.000	N-BK7	10.000
2		Standard		-171.312 V	45.786 V		9.643
3	IMAGE	Standard		Infinity	-		0.148

Figure 10.10 LDE of the optimized OSsinglet for lens shape.

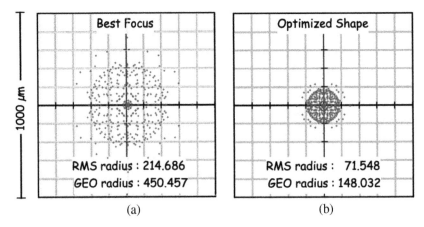

Figure 10.11 Spot diagrams of the OSsinglet (a) at best focus and (b) after lens shape optimization.

The third-order aberration errors after lens shape optimization are shown in Table 10.2. The TSPH for the optimized singlet is −0.3979. This value is quite close to the TSPH value of −0.3980 (Table 6.4) that was found after the manual optimization of the lens shape at the end of Section 6.7.

Table 10.2 Third-order aberration errors for the OSsinglet after lens shape optimization.

```
Surf    TSPH     TTCO     TAST     TPFC     TSFC     TTFC     TDIS
STO   -0.2054  -0.0000  -0.0000  -0.0000  -0.0000  -0.0000  -0.0000
  2   -0.1925   0.0000  -0.0000  -0.0000  -0.0000  -0.0000   0.0000
IMA   -0.0000  -0.0000  -0.0000  -0.0000  -0.0000  -0.0000  -0.0000
TOT   -0.3979   0.0000   0.0000   0.0000   0.0000   0.0000   0.0000
```

10.4 Reducing Coma

At the end of the Section 7.1, we showed that reversing the OSsinglet reduced the transverse coma (TTCO) six-fold. Because a lens and its reversed cousin provide two points on a curve of TTCO versus lens shape, it is safe to say that an optimization of the OSsinglet by lens bending would reduce its coma. If we add a constraint to our optimization, we can find the lens shape for a singlet that gives zero coma (see Exercise 10.1).

> **Exercise 10.1 Coma reduction in the OSsinglet by lens bending**
> Find the lens shape for the OSsinglet that results in zero coma. Start by restoring the OSsinglet and adding 0.7° and 1° off-axis field points. If paraxial ray aiming is off, turn it on. Make the two lens radii and the image distance variable. Set up a DMF spot size optimizer using the optimization wizard and add an operand to maintain the 50-mm EFL. Then find (see the text box on "User-Entered Operands") the operand needed to make coma zero, optimize, and check your result with **THIRD**.

10.4.1 Stop shifting

Lens bending is not the only way to change the coma in a lens. A second technique for reducing coma is called **stop shifting**. To demonstrate this effect, open a new lens (File > New) and enter the lens data for a new 50-mm EFL lens using the LDE values shown in Fig. 10.12(a). Note that the aperture stop is located on a dummy surface (S1) in front of, but separate from, the lens. Give the system an EPD of 4 mm, a d-line wavelength, and fields of 0°, 7°, and 10°. The 10° field will provide a field large enough to show the effects of stop shifting on the off-axis errors of the lens. If paraxial ray aiming is off, turn it on. This lens is shown in Fig. 10.12(b).

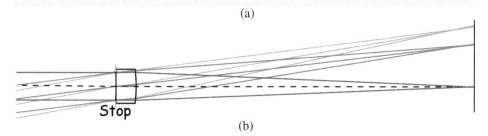

	Surface Type	Comment	Radius	Thickness	Material	Clear Semi-Dia
0	OBJECT Standard ▾		Infinity	Infinity		Infinity
1	STOP Standard ▾		Infinity	0.000		2.000
2	Standard ▾		-71.814	3.000	N-BK7	2.005
3	Standard ▾		-19.273	50.712		2.355
4	IMAGE Standard ▾		Infinity	-		8.985

Title (OSlandscape); EPD (4 mm); Field (0°, 7°, 10°); Wavelength (d-line)

(a)

(b)

Figure 10.12 (a) LDE of a 50-mm-EFL lens with the stop at the lens and (b) layout for the lens.

The third-order aberrations for this lens are listed in Table 10.3. The color aberrations, TAXC and TLAC, have been omitted because only the d-line (0.5876 μm) is being used for this example. For the present, our interest is in TTCO, the coma in the lens.

Table 10.3 Third-order aberration errors for the stop at the lens.

```
Surf    TSPH     TTCO     TAST     TPFC     TSFC     TTFC     TDIS
STO   0.0000   0.0000   0.0000   0.0000   0.0000   0.0000  -0.0000
  2   0.0001  -0.0023   0.0097   0.0074   0.0122   0.0220  -0.0775
  3  -0.0247   0.0761  -0.0520  -0.0275  -0.0535  -0.1055   0.0548
IMA  -0.0000  -0.0000  -0.0000  -0.0000  -0.0000  -0.0000  -0.0000
TOT  -0.0246   0.0738  -0.0423  -0.0201  -0.0412  -0.0835  -0.0227
```

When the stop is moved a distance of 7.6 mm before the lens [Fig. 10.13(a)] the off-axis rays strike the lens higher [Fig. 10.13(b)] than when the stop is at the lens. This significantly affects the values of some of the off-axis aberrations, including coma. Table 10.4 shows the third-order aberration errors for the lens after the stop is shifted. Moving the stop 7.6 mm in front the lens, coma (TTCO) is reduced from 0.0738 to 0.0243, about three-fold. Also, the tangential field curvature (TTFC) is reduced by more than four-fold. Note that spherical (TSPH) and Petzval (TPFC) are unchanged with stop shift.

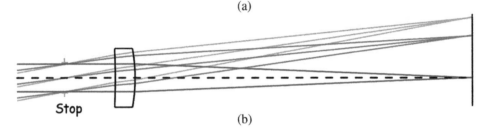

(a)

(b)

Figure 10.13 (a) LDE of a 50-mm-EFL lens with the stop in front of the lens and (b) layout for the lens.

Table 10.4 Third-order aberration errors for the stop at the front of the lens.

Surf	TSPH	TTCO	TAST	TPFC	TSFC	TTFC	TDIS
STO	0.0000	0.0000	0.0000	0.0000	0.0000	0.0000	-0.0000
2	0.0001	-0.0021	0.0078	0.0074	0.0113	0.0190	-0.0638
3	-0.0247	0.0263	-0.0062	-0.0275	-0.0306	-0.0368	0.0109
IMA	-0.0000	-0.0000	-0.0000	-0.0000	-0.0000	-0.0000	-0.0000
TOT	-0.0246	0.0243	0.0015	-0.0201	-0.0193	-0.0178	-0.0529

Very early camera lenses for photographing landscapes consisted of a simple singlet with a stop some distance in front of the lens because it resulted in much better correction over a wider field angle than a simple singlet whose aperture served as the stop. For this reason, a singlet with a shifted aperture stop is called a landscape lens. Save this lens (with a stop distance of 7.6 mm) as **OSlandscape**. It will be used later in the chapter.

It is possible to manually shift the stop and by trial and error to find the stop position that produces zero coma (TTCO = 0). However, it is much easier to use local optimization to automatically vary the distance between the stop and the first lens surface until coma equals zero. We only need one variable (the thickness of the stop surface) and one operand, COMA. Open the MFE (with only one operand, there is no need to use the Wizard on this one!) and enter the Operand Type as COMA with a target of zero and a weight of one. The operand COMA is equal to the Seidel aberration coefficient for coma (in Waves) and is proportional to the third-order error TTCO (see Table 10.4). Click the Update icon ↻ on the left. The starting value for COMA is –0.551 and since this is the only operand in the MFE (and it has a weight of one), the initial MF (Fig.10.14) value is also 0.551.

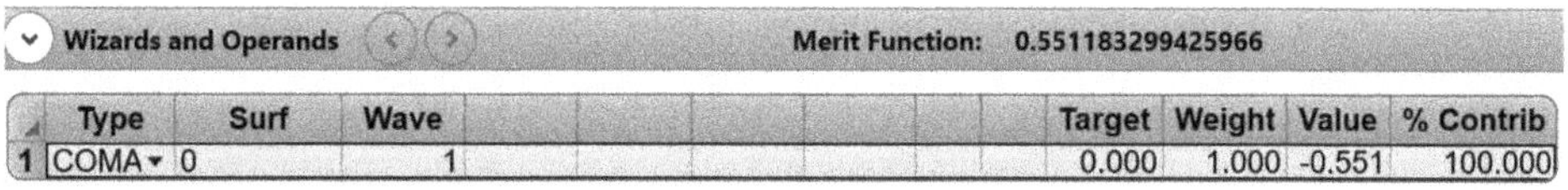

Type	Surf	Wave				Target	Weight	Value	% Contrib
1 COMA ▾	0	1				0.000	1.000	-0.551	100.000

Merit Function: 0.551183299425966

Figure 10.14 MFE for the OSlandscape to shift the stop for zero coma.

After a round of local optimization (Optimize > Optimize!), the initial MF is reduced to zero. An examination of the LDE shows that the stop has been shifted to a distance of 11.330 mm before the lens and the TTCO for this stop-shifted lens is zero (Table 10.5).

Table 10.5 Third-order aberration errors for the optimized OSlandscape lens.

Surf	TSPH	TTCO	TAST	TPFC	TSFC	TTFC	TDIS
STO	0.0000	0.0000	0.0000	0.0000	0.0000	0.0000	-0.0000
2	0.0001	-0.0019	0.0069	0.0074	0.0108	0.0177	-0.0577
3	-0.0247	0.0019	-0.0000	-0.0275	-0.0275	-0.0275	0.0007
IMA	-0.0000	-0.0000	-0.0000	-0.0000	-0.0000	-0.0000	-0.0000
TOT	-0.0246	-0.0000	0.0069	-0.0201	-0.0167	-0.0098	-0.0570

10.4.2 Flipping the lens

If this design is used for a camera lens, the fixed aperture stop could be replaced by an iris diaphragm to control the $f/\#$ of the lens and the image exposure. However, an external aperture stop is subject to dust and damage. If the stop were inside the optical system (i.e., between the lens and the image), it would be protected by the lens and its housing. So, there are two good design-form solutions for the landscape lens, one with the stop in the front and one with the stop in the back. To find a good starting point for this second solution, we can flip our stop-in-front landscape lens end for end (lens first, stop after). Although it is possible to manually flip the design by changing the signs of the radii of curvature and the sequence of the surfaces, this procedure is done often enough in lens design that it can be achieved with a single action, Reverse Elements ⇄, as we did at the beginning of Section 6.7.

To demonstrate this, reset the stop thickness to 7.6 mm in the current LDE or restore the **OSlandscape** that was saved **before** optimization in Section 10.4.1. Then select rows 1 through 3 and click the Reverse Elements icon⇄, which flips the surfaces between S1 and S3 about the y axis. Use quick focus (Optimize > Quick Focus) to adjust the image distance for best focus. The results (LDE and lens cross-section) are shown in Fig. 10.15.

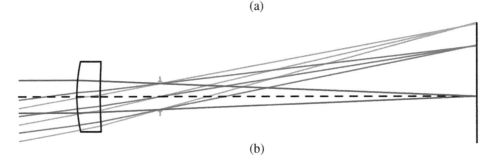

Surface Type	Comment	Radius	Thickness	Material	Clear Semi-Dia
0 OBJECT Standard ▼		Infinity	Infinity		Infinity
1 Standard ▼		19.273	3.000	N-BK7	4.093
2 Standard ▼		71.814	7.600		3.589
3 STOP Standard ▼		Infinity	38.944		1.590
4 IMAGE Standard ▼		Infinity	-		8.778

Title (OSlandscapeRearStop); EPD (4 mm); Field (0°, 7°, 10°); Wavelength (d-line)

(a)

(b)

Figure 10.15 The reversed OSlandscape lens before stop position optimization: (a) LDE and (b) lens cross-section.

An assessment of the third-order aberrations in Table 10.6 shows that the aberrations of the landscape lens have changed with the flip. This is because the conjugate distances were not flipped. That is, the entering parallel rays now encounter two convex surfaces instead of the two concave ones in the initial design (Fig. 10.13). The only aberration that stays the same is Petzval because neither the lens power nor the glass has changed. Save this lens as **OSlandscapeRearStop** for use in the next section.

Table 10.6 Third-order aberration errors for the OSlandscapeRearStop lens.

```
Surf    TSPH     TTCO     TAST     TPFC     TSFC     TTFC     TDIS
   1  -0.0063  -0.0118  -0.0049  -0.0275  -0.0299  -0.0348  -0.0187
   2  -0.0001   0.0032  -0.0311   0.0074  -0.0082  -0.0393   0.1189
 STO   0.0000  -0.0000   0.0000  -0.0000  -0.0000  -0.0000  -0.0000
 IMA  -0.0000  -0.0000  -0.0000  -0.0000  -0.0000  -0.0000  -0.0000
 TOT  -0.0063  -0.0086  -0.0360  -0.0201  -0.0381  -0.0741   0.1001
```

We can also shift the stop in this lens for zero coma by allowing the rear stop distance to vary and then running the optimization with the same COMA = 0 operand in the MFE that we used in the last section. The MF starts at a value of 0.1943 and drops to 0 after a local optimization. The stop moves to a point 10.581 mm beyond the lens (Fig. 10.16), yielding zero coma for the lens (Table 10.7). The lens has not been replotted because it differs little from Fig. 10.15 (the stop is farther from the lens, but we have not optimized the lens shape yet).

Reducing Aberrations 217

	Surface Type	Comment	Radius	Thickness	Material	Clear Semi-Dia
0	OBJECT Standard ▼		Infinity	Infinity		Infinity
1	Standard ▼		19.273	3.000	N-BK7	4.968
2	Standard ▼		71.814	10.581 V		4.453
3	STOP Standard ▼		Infinity	38.944		1.471
4	IMAGE Standard ▼		Infinity	-		9.652

Title (OSlandscapeRearStop); EPD (4 mm); Field (0°, 7°, 10°); Wavelength (d-line)

Figure 10.16 LDE of the OSlandscapeRearStop lens optimized for zero coma.

Table 10.7 Third-order aberrations for the rear-stop landscape lens with zero coma.

```
Surf    TSPH    TTCO    TAST    TPFC    TSFC    TTFC    TDIS
   1  -0.0063 -0.0033 -0.0004 -0.0275 -0.0277 -0.0281 -0.0049
   2  -0.0001  0.0033 -0.0330  0.0074 -0.0091 -0.0422  0.1372
 STO   0.0000 -0.0000  0.0000 -0.0000 -0.0000 -0.0000 -0.0000
 IMA  -0.0000 -0.0000 -0.0000 -0.0000 -0.0000 -0.0000 -0.0000
 TOT  -0.0063 -0.0000 -0.0334 -0.0201 -0.0368 -0.0702  0.1323
```

10.5 Reducing Distortion

Because distortion is, in effect, a chief ray error, the location of the aperture stop in an optical system can substantially affect this error. As a demonstration of this, let's compare the distortion of the two saved landscape lenses (front stop and rear stop) with the stop 7.6 mm from the lens. The third-order distortion coefficients (TDIS) from Tables 10.4 and 10.6 for both lenses are highlighted in Table 10.8. Flipping the lens changes the *sign* of the distortion, going from barrel distortion (–0.053) to pincushion (+0.100) when the stop is moved from the front to the back of the lens.

Table 10.8 Third-order aberration errors for the OSlandscape lens with a front stop and a rear stop.

```
Aperture stop 7.6 mm in front of first lens surface
Surf    TSPH    TTCO    TAST    TPFC    TSFC    TTFC    TDIS
TOT   -0.0246  0.0243  0.0015 -0.0201 -0.0193 -0.0178 -0.0529
-----------------------------------------------------------------
Aperture stop 7.6 mm behind the last lens surface
Surf    TSPH    TTCO    TAST    TPFC    TSFC    TTFC    TDIS
TOT   -0.0063 -0.0086 -0.0360 -0.0201 -0.0381 -0.0741  0.1001
```

Plots (Fig. 10.17) of the percent distortion (Analyze > Aberrations > Field Curvature and Distortion) show that the lens with the rear stop has twice the amount of distortion (1.15% versus –0.59%) as the same lens with the stop in front (at the same distance). Note: The relationship between the third-order aberration value (TDIS) and the Percent Distortion values shown on the distortion plots is discussed in Section 8.5.

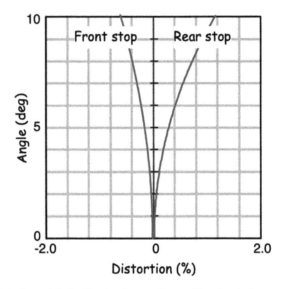

Figure 10.17 Distortion plots for the landscape lens with a front stop and a rear stop. Note the sign flip of distortion.

As described in detail in Section 8.5, distortion is the change in magnification across the image field. One of the biggest complaints regarding the landscape camera lens when used to photograph buildings was the noticeable distortion at the corners of the picture. A design geometry that guarantees no distortion is a symmetric arrangement of lenses about their aperture stop. A ray entering the lens at an angle to the optical axis will, by virtue of the symmetry, exit the system of lenses at exactly the same angle. This guarantees that there will be no change in the magnification of lens system across the entire field. Note that for exactly zero distortion, the symmetry requirement includes the object and image distances/conjugates.

To demonstrate stop symmetry and its effect on distortion, we will take the previous landscape lens (OSlandscapeRearStop) and create a symmetrical version of it. It is possible to do this by simply entering the correct radii of curvature, thicknesses, and glasses to achieve the desired symmetry. However, just as with the command for flipping a lens, for many designs there is a more useful way to assign the added parameters needed to complete the design. A **pickup** assigns the design parameter based on the value of another parameter in the system. The text box, "Pickups" describes pickup solves, and the next example illustrates their use.

Pickups

Some lens designs require a measure of symmetry to provide a certain performance metric (for example, low distortion). Other designs may use the same component more than once, thereby reducing the cost of manufacturing the system. During the optimization of these systems, we would like to be able to vary the parameters (radii, thickness, etc.) of the one component while maintaining the corresponding relationships with the other component(s). This is accomplished by using the pickup option within the solve box next to the quantities for radius, thickness, and material in the LDE.

The pickup with the simplest settings is that for Material. When you want to use the same material for two elements in a design, you can enter the glass for one of the elements and then use the solve box next to the material for the other surface that you want to have the same glass. From the drop-down menu for Solve Type, select Pickup. Then enter the surface number from which you want to pick up the glass ("2" in the example here) in the "From Surface" dialog box.

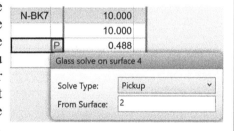

For a thickness solve, the pickup settings are somewhat more elaborate. Not only does the solve copy the thickness of another surface, but that value can be scaled and a constant value added to it (or subtracted from it) by way of a value in the Offset box. For the Radius pickup, the sign of the radius can be changed by inserting "–1" as the Scale Factor. This is particularly useful in creating double-pass systems or stop-symmetric systems.

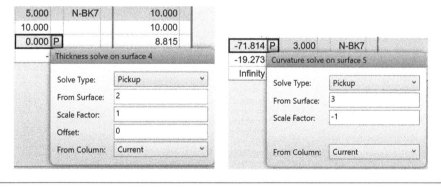

Start by restoring the **OSlandscapeRearStop** lens. Its LDE was shown in Fig. 10.15. Insert two surfaces after the stop. Set the thickness after the stop equal to the thickness before the stop by using a thickness pickup solve on T3 to pick up T2. Then make a copy of the first lens with the signs of the radii of curvature reversed to maintain the stop symmetry by following a sequence of steps (actions) listed in Actions 10.1. Use quick focus (Optimize > Quick Focus) to adjust the image distance for best focus. The LDE for this symmetrical lens is displayed in Fig. 10.18.

Actions 10.1 Setting up a symmetrical lens.

Enter a radius pickup on R4 with a –1 Scale Factor to copy –R2 onto R4.
Use a material pickup on G4 to copy G1 (N-BK7) onto G4.
Enter a thickness pickup on T4 to copy T1 onto T4.
Use a radius pickup on R5 with a –1 Scale Factor to copy –R1 onto R5.

	Surface Type	Comment	Radius	Thickness	Material	Clear Semi-Dia
0	OBJECT Standard ▼		Infinity	Infinity		Infinity
1	Standard ▼		19.273	3.000	N-BK7	4.093
2	Standard ▼		71.814	7.600 V		3.589
3	STOP Standard ▼		Infinity	7.600 P		1.590
4	Standard ▼		-71.814 P	3.000 P	N-BK7 P	2.963
5	Standard ▼		-19.273 P	18.913		3.332
6	IMAGE Standard ▼		Infinity	-		5.523

Figure 10.18 LDE of a symmetrical lens based on the landscape lens with a rear stop (EFL = 31.4 mm).

The **FIRST** macro for this symmetrical lens shows that the EFL is 31.4479 mm instead of the 50 mm that we have been using for our examples. We can use the Scale Lens feature (Setup > Scale Lens) to scale all dimensions of the lens in the LDE by a constant value to get a 50-mm EFL. Dividing 50 by 31.4479 gives a scale value of 1.58993. Enter this number into the Scale by Factor box in the Scale Lens window (Fig. 10.19). Change the First Surface from Object to 1 and press OK. Once you have done this, you have a 50-mm-EFL lens. Note: If you scale the lens from Object surface, the EPD will also be scaled, and you will need to change it back to 4.0 mm before continuing.

Figure 10.19 Scale Lens window to create a 50-mm-EFL lens.

The LDE and lens cross-section for the scaled 50-mm-EFL lens are shown in Figs. 10.20 and Fig. 10.21, respectively. The third-order aberrations (Table 10.9) show a distortion value of 0.03287, an approximately three-fold reduction in distortion when compared to the rear-stop landscape lens. Save this lens as **OSsymmetrical**. It will be used again in Section 10.6.2.

	Surface Type	Comment	Radius	Thickness	Material	Clear Semi-Dia
0	OBJEC Standard ▼		Infinity	Infinity		Infinity
1	Standard ▼		30.643	4.770	N-BK7	5.372
2	Standard ▼		114.179	12.083 V		4.596
3	STOP Standard ▼		Infinity	12.083 P		1.590
4	Standard ▼		-114.179 P	4.770 P	N-BK7 P	3.968
5	Standard ▼		-30.643 P	30.071		4.602
6	IMAGE Standard ▼		Infinity	-		8.764

Title (OSsymmetrical); EPD (4 mm); Field (0°, 7°, 10°); Wavelength (d-line)

Figure 10.20 LDE of the OSsymmetrical (EFL = 50 mm) lens.

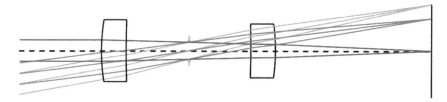

Figure 10.21 Layout for the OSsymmetrical lens.

Table 10.9 Third-order aberrations for the OSsymmetrical lens.

```
Surf    TSPH    TTCO    TAST    TPFC    TSFC    TTFC    TDIS
   1  -0.0016 -0.0047 -0.0031 -0.0173 -0.0188 -0.0219 -0.0187
   2  -0.0000  0.0013 -0.0196  0.0046 -0.0051 -0.0247  0.1189
 STO   0.0000 -0.0000  0.0000 -0.0000 -0.0000 -0.0000 -0.0000
   4   0.0004 -0.0058  0.0212  0.0046  0.0152  0.0364 -0.0829
   5  -0.0031  0.0075 -0.0041 -0.0173 -0.0193 -0.0234  0.0157
 IMA  -0.0000 -0.0000 -0.0000 -0.0000 -0.0000 -0.0000 -0.0000
 TOT  -0.0043 -0.0017 -0.0055 -0.0253 -0.0281 -0.0336  0.0329
```

Although the lens components are symmetrical, the distortion is not zero because the object is at infinity and the image is at finite distance (30.071 mm) from the lens. To achieve perfect symmetry about the stop (and therefore zero distortion), the object and image distances must be equal. This means that the object distance can no longer be infinity. For finite object distances, it is more appropriate to define the field in terms of object or image height than object angle. First, a finite distance needs to be established. Set the object distance to 100 mm. In the FDE (Fig. 10.22), change the Field Type to Object Height. The field values will change from degrees to 0, 7, and 10 mm. Put a marginal ray height solve on the last lens thickness to find the paraxial image plane. The resulting LDE is shown in Fig. 10.23.

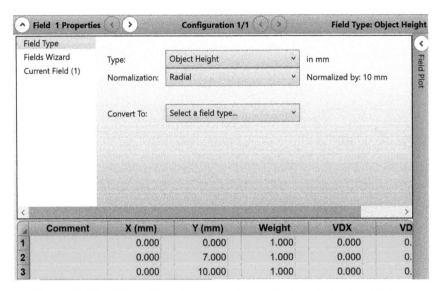

Figure 10.22 FDE settings to change from object angle to object height.

	Surface Type	Comment	Radius	Thickness	Material	Clear Semi-Dia
0	OBJECT Standard ▾		Infinity	100.000 V		10.000
1	Standard ▾		30.643	4.770	N-BK7	3.287
2	Standard ▾		114.179	12.083 V		2.912
3	STOP Standard ▾		Infinity	12.083 P		1.590
4	Standard ▾		-114.179 P	4.770 P	N-BK7 P	2.811
5	Standard ▾		-30.643 P	66.666 M		3.159
6	IMAGE Standard ▾		Infinity	-		7.225

Figure 10.23 LDE for the OSsymmetrical at a finite object distance of 100 mm.

For an object distance of 100 mm, the paraxial image distance is 66.666 mm. While we could keep changing the object distance manually until it equals the paraxial image distance, it is faster to use optimization to automatically find the ideal object thickness. When the object and image distances of a symmetrical lens system are equal, the system will have a paraxial magnification of unity. Open the MFE and change the COMA operand to PMAG (paraxial magnification), enter –1 for the Target (Fig. 10.24), and Update the MFE. The initial MF (after updating) is 0.279 (the difference between the current value and the target value for the magnification).

Wizards and Operands					Merit Function:	0.279248525089793	
Type	Wave				Target	Weight	Value	% Contrib
1 PMAG ▾	1				-1.000	1.000	-0.721	100.000

Figure 10.24 OSsymmetrical lens MF for unit magnification (M = –1) optimization.

Freeze the thickness of surface 2 (we no longer want it to vary) and vary the object thickness. After optimization (Optimize > Optimize!), the MF drops to zero (since the paraxial magnification equals –1) for an object distance of 80.63 mm (Fig. 10.25). The lens system cross-section is shown in Fig. 10.26 and the third-order aberrations are listed in Table 10.10.

	Surface Type	Comment	Radius	Thickness	Material	Clear Semi-Dia
0	OBJECT Standard ▾		Infinity	80.630 V		10.000
1	Standard ▾		30.643	4.770	N-BK7	3.535
2	Standard ▾		114.179	12.083		3.110
3	STOP Standard ▾		Infinity	12.083 P		1.590
4	Standard ▾		-114.179 P	4.770 P	N-BK7 P	3.106
5	Standard ▾		-30.643 P	80.626 M		3.531
6	IMAGE Standard ▾		Infinity	-		10.024

Title (OSsymmetrical (M=1)); EPD (4 mm); Field (0, 7, 10 mm); Wavelength (d-line)

Figure 10.25 LDE of a distortion-free lens system.

Figure 10.26 Distortion-free lens system.

Table 10.10 Third-order aberrations for the OSsymmetrical lens at $M = -1$.

```
Surf    TSPH     TTCO     TAST     TPFC     TSFC     TTFC     TDIS
  1   -0.0049   0.0075  -0.0025  -0.0111  -0.0124  -0.0149   0.0063
  2    0.0000  -0.0006   0.0031   0.0030   0.0046   0.0077  -0.0368
STO    0.0000  -0.0000   0.0000  -0.0000  -0.0000  -0.0000   0.0000
  4    0.0000   0.0006   0.0031   0.0030   0.0046   0.0077   0.0368
  5   -0.0049  -0.0075  -0.0025  -0.0111  -0.0124  -0.0149  -0.0063
IMA   -0.0000  -0.0000  -0.0000  -0.0000  -0.0000  -0.0000  -0.0000
TOT   -0.0098  -0.0000   0.0012  -0.0163  -0.0156  -0.0144  -0.0000
```

From Fig. 10.26, it is easy to see why such a system has no distortion. A chief ray from any point on the object will travel through the stop and by symmetry end up in the same location on the other side of the optical axis, so the mapping from object to image is exact for all points on the object. Table 10.10 shows how the distortion contribution from the first surface exactly cancels the contribution from the fifth surface and how the contribution from the second surface cancels the fourth surface, leaving a net sum of zero for distortion. Lens symmetry also has the added benefit of canceling coma in the same way: the coma from the lens in front of the stop exactly cancels the coma from the lens behind the stop.

The use of symmetry is another example of the reduction of a third-order aberration. This strategy works well on optical systems that require unit magnification with well-corrected distortion (e.g., lenses in copying machines), but most optical systems are not configured for unit magnification and typically have the object "at infinity." However, even for infinite conjugate systems, stop symmetry is an extremely powerful tool for reducing distortion (and coma and lateral color—all odd-order chief ray aberrations). Many camera lens designs use this principle.

10.6 Reducing Field Curvature

Section 10.3 showed that lens bending could minimize spherical aberration. Section 10.4 demonstrated that shifting the stop could eliminate coma. Both of these operations also changed the field curvature values of the lens (TTFC and TSFC) but did not change the fundamental Petzval curvature (TPFC). Section 8.3 described the TTFC and TSFC third-order coefficients as a combination of astigmatism and Petzval. In Fig. 8.6, the Petzval surface P was plotted together with the astigmatic field curves T and S to show their relationship. In effect, the Petzval surface can be considered the base curve for the field curvature plots, while the separation between the S curve and the T curve is a measure of the astigmatism in the lens. If astigmatism is eliminated, the S and T curves will overlap, and the only the remaining field curvature comes from the Petzval contribution. Therefore, reducing field curvature to produce a flat image surface requires both astigmatism and Petzval to be corrected. While both aberrations contribute to field curvature, the techniques used to correct them are significantly different.

10.6.1 Correcting astigmatism

When we shifted the stop of the OSlandscape lens to eliminate coma, the astigmatism (TAST) was also reduced while the Petzval (TPFC) was unchanged (see Tables 10.4 and 10.5). This means that the stop shift affects only the astigmatic portion of the field curvature. When the stop is at the lens, the field curves (Analyze > Aberrations > Field Curvature and Distortion) show a large undercorrected astigmatism where T is to the left of S [see Fig. 10.27(a)]. When the stop is 7.6 mm to the left of the lens, the astigmatism is nearly corrected (TTFC = TSFC = TPFC), and the T and S curves are nearly overlapping [see Fig. 10.27(b)]. When the stop is at the zero-coma position, the astigmatism is overcorrected, and T is to the right of S [see Fig. 10.27(c)]. Thus, the stop position has a large impact on the amount of astigmatism in the lens (similar to what we saw with coma).

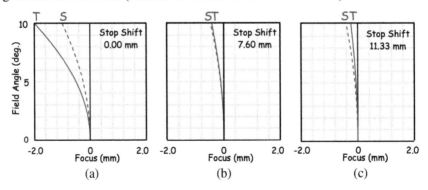

Figure 10.27 Field curves for the OSlandscape lens: (a) stop at the lens, (b) 7.6-mm stop shift, and (c) 11.33-mm stop shift (TTCO = 0).

It's interesting to note that overcorrecting the astigmatism reduces the overall field curvature (TTFC and TSFC) in the lens, as shown in Fig. 10.27 and Tables 10.4 and 10.5. The overcorrected astigmatism balances the undercorrected Petzval to reduce the overall field curvature. This is similar to using defocus to balance spherical aberration (Section 10.2). It can be shown that for a lens with some residual (uncorrectable) Petzval, the smallest field curvature values are obtained when the astigmatism is overcorrected such that the tangential field is flat (TTFC = 0). This is a design technique known as **artificially flattening the tangential field**.

To flatten the tangential field and keep coma at zero in the landscape lens, we need more variables than just a stop shift. In Section 10.3, when we bent the lens to minimize spherical, we also changed the astigmatic part of the field curvature (again, Petzval was unchanged). Therefore, astigmatism can be changed by both bending the lens and shifting the stop. Restore the **OSlandscape** lens [Fig. 10.13(a)]. Add a marginal ray height solve to the thickness of S3 to find the paraxial image plane. Vary both surface radii (R2 and R3) as well as the stop position (T1). The LDE before optimization is shown in Fig. 10.28.

	Surface Type	Comment	Radius	Thickness	Material	Clear Semi-Dia
0	OBJECT Standard ▼		Infinity	Infinity		Infinity
1	STOP Standard ▼		Infinity	7.600 V		2.000
2	Standard ▼		-71.814 V	3.000	N-BK7	3.326
3	Standard ▼		-19.273 V	50.712 M		3.684
4	IMAGE Standard ▼		Infinity	-		8.831

Title (OSlandscape); EPD (4 mm); Field (0°, 7°, 10°); Wavelength (d-line)

Figure 10.28 LDE for the OSlandscape with additional variable parameters.

The MFE for this optimization is shown in Fig. 10.29. Since we're varying the radii of the lens, the focal length needs to be held to 50 mm by the EFFL operand (row 2). As before, the COMA operand is added to the MFE with a zero target to minimize coma (row 3). Both operands have weights of 1. To flatten the tangential field, we need to target the TTFC to zero, but there is no built-in operand for this. However, we can create one using the built-in operands (ASTI and FCUR) and the relationship between the transverse aberration coefficients given in Table 6.2 (TTFC = 3TAST/2 + TPFC). Using, ASTI for astigmatism (TAST) and FCUR for Petzval (TPFC), and first multiplying ASTI by 1.5 and then adding that to FCUR, we can construct an operand in the MFE that is proportional to TTFC. Both of the built-in operands are first entered with zero weight (rows 5 and 6), so they do not contribute to the merit function, but their values are still accessible by the MFE. Then the operand PROB is entered on row 7 with a "5" indicating the row operand number (Op#1) and a factor of 1.500 in the Factor column. This multiplies the ASTI operand by 1.5 (1.5×-0.052) to give a Value of -0.079 in the PROB row. The weight for this operand is also set to zero. Finally, FCUR and PROB in rows 6 and 7 are summed by the operand SUMM (row 8) by setting the Values of

Op#1 = 5 and Op#2 = 6. This final operand, SUMM, is given a zero target and a weight of 1. When optimizing, if SUMM = 0, then the tangential field curvature will be zero (TTFC = 0).

	Type	Op#	Factor			Target	Weight	Value	% Contrib
1	BLNK ▼								
2	EFFL ▼		1			50.000	1.000	50.000	9.992E-08
3	COMA ▼	0	1			0.000	1.000	-0.551	45.293
4	BLNK ▼								
5	ASTI ▼	0	1			0.000	0.000	-0.052	0.000
6	FCUR ▼	0	1			0.000	0.000	0.684	0.000
7	PROB ▼	5		1.500		0.000	0.000	-0.079	0.000
8	SUMM ▼	6	7			0.000	1.000	0.606	54.707
9	BLNK ▼								

Figure 10.29 MFE to flatten the tangential field of the OSlandscape.

For optimization, we currently have three variables (two radii and one airspace) and three active operands in the MFE. The starting MF value is 0.4728. After a round of local optimization, it is reduced to zero and all constraints are met. The resulting lens shows a modest change in stop location with more substantial changes to the radii of curvature. This is evident in both the LDE [Fig. 10.30(a)] and the lens layout [Fig. 10.30(b)].

	Surface	Type	Comment	Radius	Thickness	Material	Clear Semi-Dia
0	OBJECT	Standard ▼		Infinity	Infinity		Infinity
1	STOP	Standard ▼		Infinity	7.550 V		2.000
2		Standard ▼		-28.294 V	3.000	N-BK7	3.297
3		Standard ▼		-13.994 V	51.806 M		3.717
4	IMAGE	Standard ▼		Infinity	-		8.822

(a)

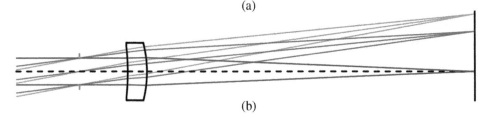

(b)

Figure 10.30 (a) LDE and (b) layout of the OSlandscape optimized for zero coma and zero tangential field curvature.

The **THIRD** listing (Table 10.11) shows that both the TTCO and TTFC are zero. In the field curve plots (Fig. 10.31), the tangential curve is a "flat" or a vertical line and the sagittal curve has the smallest field curvature of any of the previous lenses. One thing to note is that although the shape of this lens has improved the field curvature (and the performance at off-axis fields), the new lens shape has doubled the spherical aberration. This is one of the reasons this type of lens was primarily used for photographing landscape scenes where there was enough light that the lens could be stopped down to reduce the spherical aberration contribution. To use this form of lens indoors, a flash would be needed.

Table 10.11 Third-order aberrations of the OSlandscape optimized for zero coma and zero tangential field curvature.

```
Surf    TSPH     TTCO     TAST     TPFC     TSFC     TTFC     TDIS
STO    0.0000   0.0000   0.0000   0.0000   0.0000   0.0000  -0.0000
  2    0.0020  -0.0109   0.0133   0.0187   0.0254   0.0386  -0.0464
  3   -0.0512   0.0109  -0.0005  -0.0379  -0.0381  -0.0386   0.0027
IMA   -0.0000  -0.0000  -0.0000  -0.0000  -0.0000  -0.0000  -0.0000
TOT   -0.0492  -0.0000   0.0128  -0.0191  -0.0128   0.0000  -0.0437
```

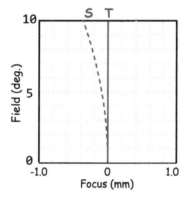

Figure 10.31 Field curves for the OSlandscape optimized for zero coma and zero tangential field curvature.

Exercise 10.2 TSFC = 0

What happens to the lens shape when the sagittal field curvature is targeted to zero (TSFC = 0) instead of the tangential field curvature? Is the total field curvature larger or smaller than when TTFC = 0? Restore the OSlandscape lens. Add a marginal ray height solve to the thickness of S3 to find the paraxial image plane. Vary both surface radii (R2 and R3) as well as the stop position (T1). Create a set of operands to optimize the sagittal field curvature and coma to zero while holding the 50-mm EFL. Compare the resulting field curves with Fig. 10.31.

10.6.2 Correcting Petzval curvature

Flattening the image surface requires that both astigmatism and Petzval curvature are corrected. In Section 10.6.1, we showed that we can use stop shifting and lens bending to overcorrect the astigmatism to artificially flatten the tangential field and reduce the overall field curvature. However, this technique did not change the fundamental Petzval curvature of the lens, and we could not completely flatten the image surface.

This section discusses another technique, the addition of an element, appropriately called a **field flattener,** which reduces the Petzval curvature and improves the performance of an optical design. As was noted in Section 8.3, the non-astigmatic part of field curvature, Petzval, depends only on the individual powers of the surfaces and index of refraction on each side of the surfaces. To reduce the Petzval curvature in a lens, we need to add an element, called a field lens, with the appropriate power (typically negative). This lens will be located close to the image plane so it has little effect on the spherical aberration in the lens.

Open the **OSsymmetrical** lens with the object plane at infinity (Fig. 10.21). Table 10.9 listed the third-order aberrations for this lens, where we can see that each (identical) lens contributes the same amount of TPFC to the Petzval curvature. The coma is nearly zero due to symmetry. The astigmatism (TAST) is well corrected by the separation of the stop from the lenses and is five times smaller than the Petzval (TPFC). A quick look at the field curves confirms that the lens has an inward-curving image plane where S and T are nearly overlapped. Because Petzval curvature limits the performance of this lens, it is an ideal candidate for a field flattener.

Add a field lens to the design by inserting two surfaces (S6 and S7) before the image. The field lens should be close to the image plane but not on top of it, so we will adjust the separation between the second lens and the field flattener (T5) to 28 mm. To provide a large power without a strong radius of curvature, the refractive index of a field flattener should be high. SF6 is a flint glass with a d-line index of 1.80512. Give the field lens a thickness of 1 mm (T6) and a material of SF6 (G6). Add a marginal ray solve (M) to the thickness before the image surface (T7) that will locate the image at the paraxial image plane. The result is shown in Fig. 10.32, and the LDE for this lens is given in Fig. 10.33. If you run **THIRD**, you will find that the third-order aberrations are essentially the same as those in Table 10.9 because at this point, we've only added a thin plate of glass to the lens with no power.

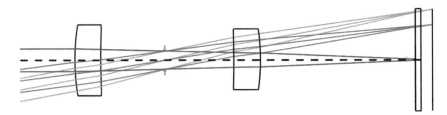

Figure 10.32 Layout for the OSsymmetrical with an initial field flattener plate.

	Surface Type	Comment	Radius	Thickness	Material	Clear Semi-Dia
0	OBJEC Standard ▼		Infinity	Infinity		Infinity
1	Standard ▼		30.643	4.770	N-BK7	5.372
2	Standard ▼		114.179	12.083 V		4.596
3	STOP Standard ▼		Infinity	12.083 P		1.590
4	Standard ▼		-114.179 P	4.770 P	N-BK7 P	3.968
5	Standard ▼		-30.643 P	28.000		4.602
6	Standard ▼	Flattener	Infinity	1.000	SF6	8.453
7	Standard ▼		Infinity	2.074 M		8.528
8	IMAGE Standard ▼		Infinity	-		8.883

Title (OSsymmetrical); EPD (4 mm); Field (0°, 7°, 10°); Wavelength (d-line)

Figure 10.33 LDE for the OSsymmetrical with an initial field flattener plate.

To show the effectiveness of the field flattener, the astigmatic field curves and spot diagrams are plotted (before optimization) in Figs. 10.34(a) and (b). Note that the S and T curves nearly overlap, indicating hardly any astigmatism in this lens. This is also evident in the spot diagram plot because the spot diagrams are all nearly circular. Thus, the remaining field curvature is due to Petzval.

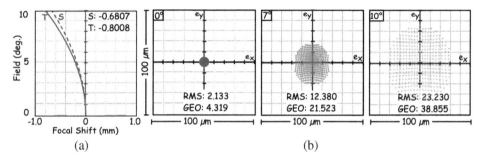

Figure 10.34 OSsymmetrical (a) field curves and (b) spot diagrams.

When you begin to work on a lens that has already been subject to any optimization, it is a good idea to freeze all parameters (Optimize > Remove All Variables) and then vary only the appropriate parameters for the new optimization. The strategy for this optimization is to use the field lens to introduce enough negative power to cancel the Petzval curvature in the symmetric doublet while maintaining its 50-mm EFL. This is accomplished by allowing the powers of the two identical lenses to increase by varying the shortest radius (R1). Its pickup (R5) will vary also. Because the power of only one surface in the field flattener is needed to provide the Petzval correction, the surface nearest the image surface (R7) can remain flat, while the curvature of the front surface of the flattener (R6) will be varied to flatten the field. In addition, the position of the field lens (T5) and the image distance (T7) will be allowed to vary. This results in the LDE shown in Fig. 10.35.

	Surface Type		Comment	Radius		Thickness		Material		Clear Semi-Dia
0	OBJECT	Standard ▾		Infinity		Infinity				Infinity
1		Standard ▾		30.643	V	4.770		N-BK7		5.372
2		Standard ▾		114.179		12.083				4.596
3	STOP	Standard ▾		Infinity		12.083	P			1.590
4		Standard ▾		-114.179	P	4.770	P	N-BK7	P	3.968
5		Standard ▾		-30.643	P	28.000	V			4.602
6		Standard ▾	Flattener	Infinity	V	1.000		SF6		8.453
7		Standard ▾		Infinity		2.074	V			8.528
8	IMAGE	Standard ▾		Infinity		-				8.883

Figure 10.35 LDE of the OSsymmetrical with a field lens prior to Petzval correction.

Now that we have more variables than constraints, we'll also want to add the RMS spot size to the MF while trying to minimize the Petzval. Clear the current MFE (the big red ✗ on the menu bar deletes all operands) and use the wizard to set up an RMS spot default merit function. This is the same one that we used in Section 10.2 (see Fig. 10.4), where the Image Quality setting in the wizard was changed from Wavefront to Spot. Insert five new operands above the DMFS. Add a focal length operand (EFFL) to hold the focal length to 50 mm with a weight of 10. The operand for Petzval curvature is PETC, which is given a weight of 100 and a target of zero. To prevent the field lens from ending up on the image plane during optimization, a >2 mm constraint on T7 is added with the operand, CTGT (Center Thickness Greater Than). The first 16 lines of the MFE before optimization are shown in Fig. 10.36. The starting MF value is 0.0152, where 97.3% of the MF is currently caused by the Petzval (PETC = 0) operand with its heavy weight of 100.

#	Type	Wave					Target	Weight	Value	% Contrib
1	BLNK									
2	EFFL	1					50.000	10.000	50.000	1.523E-06
3	PETC	1					0.000	100.000	-0.016	97.292
4	CTGT	7					2.000	1.000	2.000	0.000
5	BLNK									
6	DMFS									
7	BLNK	Sequential merit function: RMS spot x+y centroid X Wgt = 1.0000 Y Wgt = 1.0000 GQ 3 rings 6 arms								
8	BLNK	No air or glass constraints.								
9	BLNK	Operands for field 1.								
10	TRCX	1	0.000	0.000	0.336	0.000	0.000	0.291	-1.629E-04	2.839E-05
11	TRCY	1	0.000	0.000	0.336	0.000	0.000	0.291	0.000	0.000
12	TRCX	1	0.000	0.000	0.707	0.000	0.000	0.465	-1.525E-03	3.976E-03
13	TRCY	1	0.000	0.000	0.707	0.000	0.000	0.465	0.000	0.000
14	TRCX	1	0.000	0.000	0.942	0.000	0.000	0.291	-3.609E-03	0.014
15	TRCY	1	0.000	0.000	0.942	0.000	0.000	0.291	0.000	0.000
16	BLNK	Operands for field 2.								
17	TRCX	1	0.000	0.700	0.168	0.291	0.000	0.097	-2.378E-03	2.015E-03

Merit Function: 0.0152310603516444

Figure 10.36 MFE of the OSsymmetrical with field lens prior to Petzval correction.

After locally optimizing this lens (Optimize > Optimize!), the initial MF is reduced from 0.0152 to 0.000989, and the initial Petzval curvature, –0.016 in Fig. 10.36, is reduced to 0.0000284 mm (essentially flat) in Fig. 10.37. The radius of curvature of the front surface of the flattener lens is now –24.227 mm. The LDE for this lens is given in Fig. 10.38, and a cross-section of the lens is shown in Fig. 10.39. Save this lens as **OSfieldflattener** because it will be used again in the next section.

#	Type	Wave					Target	Weight	Value	% Contrib
1	BLNK									
2	EFFL	1					50.000	10.000	50.000	7.288E-07
3	PETC	1					0.000	100.000	2.838E-05	0.070
4	CTGT	7					2.000	1.000	2.000	1.427E-03
5	BLNK									
6	DMFS									
7	BLNK	Sequential merit function: RMS spot x+y centroid X Wgt = 1.0000 Y Wgt = 1.0000 GQ 3 rings 6 ar								
8	BLNK	No air or glass constraints.								
9	BLNK	Operands for field 1.								
10	TRCX	1	0.000	0.000	0.336	0.000	0.000	0.291	3.550E-03	3.195
11	TRCY	1	0.000	0.000	0.336	0.000	0.000	0.291	0.000	0.000
12	TRCX	1	0.000	0.000	0.707	0.000	0.000	0.465	6.134E-03	15.260
13	TRCY	1	0.000	0.000	0.707	0.000	0.000	0.465	0.000	0.000
14	TRCX	1	0.000	0.000	0.942	0.000	0.000	0.291	6.375E-03	10.302
15	TRCY	1	0.000	0.000	0.942	0.000	0.000	0.291	0.000	0.000

Merit Function: 0.000989131509042073

Figure 10.37 MFE of the OSsymmetrical with a field lens after Petzval correction.

	Surface Type	Comment	Radius		Thickness		Material	Clear Semi-Dia
0	OBJEC Standard ▼		Infinity		Infinity			Infinity
1	Standard ▼		27.985	V	4.770		N-BK7	5.476
2	Standard ▼		114.179		12.083			4.679
3	STOP Standard ▼		Infinity		12.083	P		1.541
4	Standard ▼		-114.179	P	4.770	P	N-BK7 P	3.968
5	Standard ▼		-27.985	P	23.118	V		4.609
6	Standard ▼	Flattener	-24.227	V	1.000		SF6	7.555
7	Standard ▼		Infinity		2.000	V		8.036
8	IMAGE Standard ▼		Infinity		-			8.906

Figure 10.38 LDE for the OSfieldflattener.

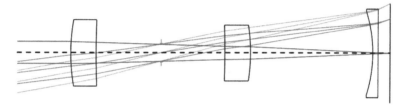

Figure 10.39 Plot of the OSfieldflattener.

So, how well did it work? The third-order aberration coefficients for this lens are listed in Table 10.12. The total TPFC coefficient is close to zero. If you add the TPFC contributions for S1–S5, you will find that they are canceled by the TPFC of the front surface of the field flattener S6. Another way to see the zero Petzval is from the astigmatic field curves [Fig. 10.40(a)]. Here, T7 has a Marginal Ray Height Solve (M), shifting the image plane to the paraxial image plane for these plots. If the Petzval contribution is zero, the tangential field curve will be three times larger than the sagittal field curve. This is also confirmed by the values for the TTFC (–0.02345) and TSFC (–0.00779) in Table 10.12. The totals for third-order coefficients for the OSsymmetrical (Table 10.9) are listed at the end of the table for comparison.

Table 10.12 Third-order coefficients for the OSfieldflattener with totals for the OSsymmetrical listed below them.

```
Surf     TSPH     TTCO     TAST     TPFC     TSFC     TTFC     TDIS
   1  -0.0020  -0.0043  -0.0020  -0.0189  -0.0199  -0.0220  -0.0141
   2  -0.0000   0.0021  -0.0244   0.0046  -0.0075  -0.0319   0.1298
 STO   0.0000  -0.0000   0.0000  -0.0000  -0.0000  -0.0000  -0.0000
   4   0.0004  -0.0066   0.0231   0.0046   0.0162   0.0393  -0.0852
   5  -0.0033   0.0060  -0.0024  -0.0189  -0.0201  -0.0225   0.0121
   6   0.0001   0.0009   0.0017   0.0286   0.0295   0.0311   0.0836
   7  -0.0001   0.0016  -0.0116   0.0000  -0.0058  -0.0174   0.0623
 IMA  -0.0000  -0.0000  -0.0000  -0.0000  -0.0000  -0.0000  -0.0000
 TOT  -0.0050  -0.0002  -0.0157   0.0000  -0.0078  -0.0235   0.1885

OSsymmetrical:
 TOT  -0.0043  -0.0017  -0.0055  -0.0253  -0.0281  -0.0336   0.0329
```

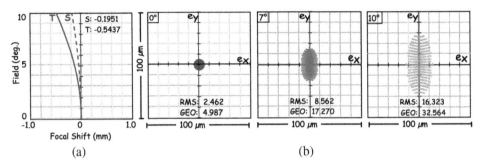

Figure 10.40 The OSfieldflattener at paraxial focus: (a) field curves and (b) spot diagrams.

10.6.3 Finishing the Job

As expected, the field flattener was able to correct the Petzval and in doing so introduced very little spherical aberration (see Table 10.12). However, when comparing Figs. 10.34(b) and 10.40(b), we see that the spot sizes for the lens before and after optimization are essentially the same—we have just swapped Petzval for astigmatism! To correct the remaining astigmatism in the lens, we need to add the lens shape (lens bending) and the lens distance to the stop (stop shifting) as variables in the optimization.

Restore the **OSfieldflattener** (Fig. 10.38) and add the R2 radius and the T2 stop distance as variables (Fig. 10.41). With the pickups on the second lens, all of the lens surface radii (except the back of the field lens) and all of the air spaces are variable. To control astigmatism, the operand ASTI is added to the MFE with a target of 0 and a weight of 10. The starting MFE is shown in Fig. 10.42 with an initial MF value of 0.1494.

	Surface Type	Comment	Radius		Thickness		Material		Clear Semi-Dia
0	OBJECT Standard ▼		Infinity		Infinity				Infinity
1	Standard ▼		27.985	V	4.770		N-BK7		5.476
2	Standard ▼		114.179	V	12.083	V			4.679
3	STOP Standard ▼		Infinity		12.083	P			1.541
4	Standard ▼		-114.179	P	4.770	P	N-BK7	P	3.968
5	Standard ▼		-27.985	P	23.118	V			4.609
6	Standard ▼	Flattener	-24.227	V	1.000		SF6		7.555
7	Standard ▼		Infinity		2.000	V			8.036
8	IMAGE Standard ▼		Infinity		-				8.906

Figure 10.41 LDE for the OSfieldflattener with added variables.

Reducing Aberrations 233

	Type	Surf	Wave				Target	Weight	Value	% Contrib
1	BLNK ▾									
2	EFFL ▾		1				50.000	10.000	50.000	2.942E-11
3	PETC ▾		1				0.000	100.000	2.838E-05	2.833E-06
4	ASTI ▾	0	1				0.000	10.000	0.533	99.996
5	CTGT ▾	7					2.000	1.000	2.000	5.759E-08
6	BLNK ▾									
7	DMFS ▾									
8	BLNK ▾	Sequential merit function: RMS spot x+y centroid X Wgt = 1.0000 Y Wgt = 1.0000 GQ 3 rings 6 arms								
9	BLNK ▾	No air or glass constraints.								
10	BLNK ▾	Operands for field 1.								
11	TRCX ▾		1	0.000	0.000	0.336	0.000	0.000	0.291	3.550E-03	1.290E-04
12	TRCY ▾		1	0.000	0.000	0.336	0.000	0.000	0.291	0.000	0.000
13	TRCX ▾		1	0.000	0.000	0.707	0.000	0.000	0.465	6.134E-03	6.159E-04
14	TRCY ▾		1	0.000	0.000	0.707	0.000	0.000	0.465	0.000	0.000

Merit Function: 0.149449272647238

Figure 10.42 MFE for reducing astigmatism in the OSfieldflattener.

After optimization, the MF drops to 0.00033 (Fig. 10.43) and the astigmatism and the Petzval curvature are essentially zero, indicating that the optimization succeeded. A comparison of the previous LDE (Fig. 10.41) to that for the new design (Fig. 10.44) shows that the radii of curvature for the symmetrical lenses are a bit shorter, as is the distance between the lenses and the stop.

	Type	Wave					Target	Weight	Value	% Contrib
1	BLNK ▾									
2	EFFL ▾	1					50.000	10.000	50.000	2.251E-07
3	PETC ▾	1					0.000	100.000	-4.628E-08	1.556E-06
4	ASTI ▾ 0	1					0.000	10.000	1.610E-06	1.882E-04
5	CTGT ▾ 7						2.000	1.000	2.000	9.426E-05
6	BLNK ▾									
7	DMFS ▾									
8	BLNK ▾	Sequential merit function: RMS spot x+y centroid X Wgt = 1.0000 Y Wgt = 1.0000 GQ 3 rings 6 arms								
9	BLNK ▾	No air or glass constraints.								
10	BLNK ▾	Operands for field 1.								
11	TRCX ▾	1	0.000	0.000	0.336	0.000	0.000	0.291	1.651E-03	5.760
12	TRCY ▾	1	0.000	0.000	0.336	0.000	0.000	0.291	0.000	0.000
13	TRCX ▾	1	0.000	0.000	0.707	0.000	0.000	0.465	1.106E-03	4.132
14	TRCY ▾	1	0.000	0.000	0.707	0.000	0.000	0.465	0.000	0.000

Merit Function: 0.000328868789832305

Figure 10.43 MFE for the OSfieldflattener after optimization for zero Petzval and astigmatism.

	Surface Type	Comment	Radius		Thickness		Material		Clear Semi-Dia
0	OBJECT Standard ▾		Infinity		Infinity				Infinity
1	Standard ▾		21.299	V	4.770		N-BK7		5.335
2	Standard ▾		47.716	V	11.350	V			4.458
3	STOP Standard ▾		Infinity		11.350	P			1.524
4	Standard ▾		-47.716	P	4.770	P	N-BK7	P	3.813
5	Standard ▾		-21.299	P	24.006	V			4.530
6	Standard ▾	Flattener	-25.182	V	1.000		SF6		7.626
7	Standard ▾		Infinity		2.000	V			8.093
8	IMAGE Standard ▾		Infinity		-				8.916

Figure 10.44 LDE for the OSfieldflattener after an optimization with an increased number of variables.

Even though the resulting lens does not look very different from the previous version, the improvement in field curvature with the additional variables is remarkable (Fig. 10.45). Table 10.13 displays zeroes for all third-order coefficients related to field curvature. The field flattener has done its work.

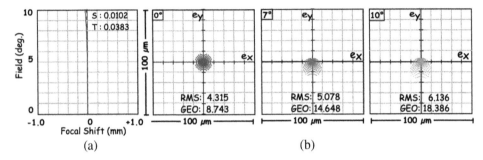

Figure 10.45 The OSfieldflattener (at paraxial focus) after an optimization with an increased number of variables: (a) field curves and (b) spot diagrams.

Table 10.13 Third-order aberrations for the OSfieldflattener after an optimization with an increased number of variables.

Surf	TSPH	TTCO	TAST	TPFC	TSFC	TTFC	TDIS
1	-0.0046	-0.0022	-0.0002	-0.0249	-0.0250	-0.0252	-0.0039
2	-0.0000	-0.0009	-0.0107	0.0111	0.0058	-0.0049	0.0997
STO	0.0000	-0.0000	0.0000	-0.0000	-0.0000	-0.0000	-0.0000
4	0.0009	-0.0093	0.0203	0.0111	0.0213	0.0416	-0.0699
5	-0.0050	0.0023	-0.0002	-0.0249	-0.0250	-0.0252	0.0038
6	0.0001	0.0008	0.0014	0.0275	0.0282	0.0296	0.0760
7	-0.0000	0.0015	-0.0106	0.0000	-0.0053	-0.0159	0.0558
IMA	-0.0000	-0.0000	-0.0000	-0.0000	-0.0000	-0.0000	-0.0000
TOT	-0.0087	-0.0078	-0.0000	-0.0000	-0.0000	-0.0000	0.1614

10.6.4 Final Comment

This chapter has introduced some of the concepts that are used to define, perform, and analyze the optimization of lenses. We limited the discussion to individual third-order aberrations that were reduced or eliminated using a number of strategies, including lens bending, stop shifting, symmetry, and the introduction of a field flattener. In the next chapter, the objective will be to propose a real design problem and describe the performance that must be met to solve this problem using optimization. In doing so, the optimization process will be described in more detail, as will the criteria that determine the lens performance.

Explorations

Up to this point in the text, we have used ray tracing to present the basic concepts needed to analyze an optical system and to measure its performance with its third-order aberrations. All changes to a system or its surroundings were entered manually. In Chapter 6, for example, the lens was bent by entering different values of the radius curvature of the first surface. The exercises in the first ten chapters were devised to help you to understand the design concepts and methods that are used to improve a system. In some cases, our exercises were variations on the examples in the text, while others were extensions of the demonstrations.

Now that optimization has been introduced, a design can be investigated and improved with greater insight and efficiency. Optimization can be used to explore a design in a search for alternatives for optimized performance. One consequence of this freedom to explore is that there is no single correct answer. Instead, optimizing a design can provide a series of trade-offs to be evaluated as to which gives the best results. In effect, they are open-ended exercises.

So, for this chapter, along with the two Exercises, a number of "Explorations" have been devised. These are intended to get you to explore optical design space using OpticStudio. There are no right answers, but your efforts and results can be used as points of discussion with others. Here are three explorations that incorporate the material we have just finished discussing. The answer to the first Exploration is given after the Exercise Answers.

Exploration 10.1 Location of the stop for zero distortion in the landscape lens

Where should the stop for the landscape lens be located to produce an undistorted image? You could vary the stop location manually, run THIRD, and look for zero distortion. However, automatic optimization is now at your disposal, so what can you do? First, restore the **OSlandcape** lens (this is the one with the stop in front of the lens). Make the stop distance (T1) variable and optimize for zero distortion using the DIST operand. Where does the stop go? Does this make sense? Did the overall lens performance get better or worse? The answer to this Exploration is given after the Exercise Answers.

Exploration 10.2 Reducing astigmatism

As discussed in Section 10.6.1, shifting the stop of the **OSlandscape** lens has a significant impact on the amount of astigmatism in a lens. What happens if you target the third-order astigmatism (ASTI operand) to zero during optimization? Look at the resulting field curves—are T and S overlapped as expected? Is the performance better or worse than that of the lens with zero coma? Is the performance better or worse than the performance when the tangential (or sagittal) field curve is artificially flattened? Is the performance better or worse than a lens optimized with no third-order aberration constraints at all?

Exploration 10.3 Achromatization

Starting with the **OSachromat** in Section 9.4, vary the surface curvatures while maintaining the 50-mm EFL, and use optimization to reduce the longitudinal chromatic aberration (operand AXCL) to a minimum. How does this result compare with the manual optimization at the end of Section 9.4? Is the performance better or worse than that of a lens optimized with no chromatic constraints at all?

Exercise Answers

Exercise 10.1 Coma reduction in the OSsinglet by lens bending

Find the lens shape for the OSsinglet that results in zero coma. Start by restoring the OSsinglet and add 0.7° and 1° off-axis field points. If paraxial ray aiming is off, turn it on. Make the two lens radii and the image distance variable. Set up a DMF spot size optimizer using the optimization wizard and add an operand to maintain the 50-mm EFL. Then find (see text box on "User-Entered Operands") the operand to make coma zero, optimize, and check your result with THIRD.

The operand to control coma is COMA. After optimization, the LDE is:

	Surface Type	Comment	Radius	Thickness	Material	Clear Semi-Dia
0	OBJECT Standard ▾		Infinity	Infinity		Infinity
1	STOP Standard ▾		28.839 V	5.000	N-BK7	10.000
2	Standard ▾		-233.839 V	45.643 V		9.660
3	IMAGE Standard ▾		Infinity	-		1.004

From THIRD:
```
Surf   TSPH     TTCO     TAST     TPFC     TSFC     TTFC     TDIS
STO   -0.2341  -0.0354  -0.0012  -0.0009  -0.0015  -0.0027  -0.0001
  2   -0.1658   0.0354  -0.0017  -0.0001  -0.0009  -0.0026   0.0001
IMA   -0.0000  -0.0000  -0.0000  -0.0000  -0.0000  -0.0000  -0.0000
TOT   -0.3999   0.0000  -0.0029  -0.0010  -0.0024  -0.0053  -0.0000
```

Exercise 10.2 TSFC = 0

What happens to the lens shape when the sagittal field curvature is targeted to zero (TSFC = 0) instead of the tangential field curvature? Is the total field curvature larger or smaller than when TTFC = 0? Restore the **OSlandscape** lens. Add a marginal ray height solve to the thickness of S3 to find the paraxial image plane. Vary both surface radii (R2 and R3) as well as the stop position (T1). Create a set of operands to optimize the sagittal field curvature and coma to zero while holding the 50-mm EFL. Compare the resulting field curves with Fig. 10.31.

The MFE needed to target TSFC to zero is:

	Type	Op#	Factor			Target	Weigh	Value	% Contrib
	Wizards and Operands					Merit Function:		0.495656062259165	
1	BLNK								
2	EFFL		1			50.000	1.000	50.000	9.093E-08
3	COMA	0	1			0.000	1.000	-0.551	41.220
4	BLNK								
5	ASTI	0	1			0.000	0.000	-0.052	0.000
6	FCUR	0	1			0.000	0.000	0.684	0.000
7	PROD	5	0.500			0.000	0.000	-0.026	0.000
8	SUMM	6	7			0.000	1.000	0.658	58.780
9	BLNK								

The only difference between this and Fig. 10.29 is that the astigmatism coefficient is multiplied by 0.5 instead of 1.5. After optimization, the third-order coefficients are:

```
Surf   TSPH     TTCO     TAST     TPFC     TSFC     TTFC     TDIS
STO    0.0000   0.0000   0.0000   0.0000   0.0000   0.0000  -0.0000
  2    0.0357  -0.0750   0.0351   0.0490   0.0666   0.1017  -0.0467
  3   -0.2044   0.0750  -0.0061  -0.0635  -0.0666  -0.0727   0.0081
IMA   -0.0000  -0.0000  -0.0000  -0.0000  -0.0000  -0.0000  -0.0000
TOT   -0.1688   0.0000   0.0290  -0.0145  -0.0000   0.0290  -0.0386
```

Note the TTFC coefficient when TSFC = 0 is 0.0290, which is more than twice the value of TSFC, −0.0128 mm (Table 10.11) when TTFC is zero. The result is a much larger field curvature, so it is better to flatten the tangential field.

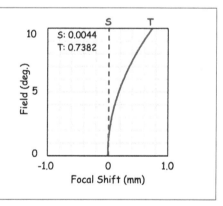

Exploration 10.1
Restore the **OSLandscape** lens. Make the stop distance (T1) variable and optimize for zero distortion using the DIST operand. The resulting LDE is shown below:

	Surface Type	Comment	Radius	Thickness	Material	Clear Semi-Dia
0	OBJECT Standard		Infinity	Infinity		Infinity
1	STOP Standard		Infinity	-2.517 V		2.000
2	Standard		-71.814	3.000	N-BK7	2.451
3	Standard		-19.273	50.712		2.150
4	IMAGE Standard		Infinity	-		9.068

The stop location is now negative! This indicates that the stop needs to be on the other side of the lens for zero distortion. Move the stop surface to the other side of the lens and reoptimize for zero distortion. The stop position for zero distortion is 0.595 mm behind the lens:

	Surface Type		Comment	Radius	Thickness	Material	Clear Semi-Dia
0	OBJECT	Standard ▼		Infinity	Infinity		Infinity
1		Standard ▼		-71.814	3.000	N-BK7	2.464
2		Standard ▼		-19.273	0.595 V		2.163
3	STOP	Standard ▼		Infinity	50.117 M		2.005
4	IMAGE	Standard ▼		Infinity	-		9.071

Chapter 11
Analyzing the Performance of a Lens

In the previous chapters, we've used a series of lenses to illustrate the properties and limitations of optical systems. What has not been addressed, until now, is what a lens will be used for and how good it needs to be. That is, if we want to design a lens to accomplish something in the real world, how do we go about it? While lenses can be used to focus a laser or to direct the output of a light source to illuminate a surface, the lens designs discussed in this text are targeted at imaging a real-world scene onto an appropriate sensor with the necessary resolution.

But what does it mean to "resolve" some detail of an image? In crime dramas on television or in the movies, there have been scenes in which a detective (or technical wizard) examines an image on a computer. The detective magnifies an area, clearly revealing a license plate number or a fuzzy portrait of someone near the scene of the crime. Sometimes screenwriters invoke the use of highly sophisticated software to "improve" the resolution from a small area of a single frame of a camera video. But how realistic is this? In this chapter, we will discuss different ways to measure the resolution of an optical system. We will show that it is a function of both the sensor and the lens, and that it can be limited by aberrations, diffraction, and/or the sensor itself.

11.1 Sensors

In an optical system, the image of an object is typically captured by a sensor. Ideally, there is a one-to-one mapping between each point in the object and each point in the image. The sensor stores the information as the difference (contrast) in light distribution between adjacent areas on the sensor. In your eye, for example, images are collected by your retina using photosensitive structures called rods and cones (~ 2 µm in size). Depending on the lens that the sensor is coupled to, the size of these light-gathering structures can impact (and ultimately limit) the resolution of the optical system.

For most of the 20th century, the medium for the permanent recording of images was photographic film—more precisely, silver halide on a transparent medium. The size of the silver grains dictated the number of stored elements in the image. Although film was manufactured in a number of formats, most photographs were taken on a sprocketed strip of film that was 35 mm wide and was referred to as 35-mm film. This film was used to record images on black-and-white or color negatives that were enlarged to provide positive photographic prints or to create color slides for projecting images.

In the last few decades, the world has "gone digital." Digital sensors (e.g., CCDs and CMOS sensors) have become the preferred media for capturing images (essentially making film obsolete), and digital displays (e.g., computer monitors, digital televisions, and mobile phone screens) are used to view the captured images. Both digital sensors and digital displays are patterned into a "checkerboard," and each "square" of the checkerboard records/displays one picture element, or one **pixel**, in the image. Currently, these digital devices can collect/display thousands to millions of pixels at once. Note that each "pixel" is usually composed of several primary color pixels (red/green/blue) to display/record the pixel's color.

Both digital displays and digital sensors are characterized by their pixel size, number of pixels, and aspect ratio (the ratio of the length to the width). For example, most high-definition televisions (4K UHD) have a height of 2160 pixels and a width of 3840 pixels (broadcast standards set the aspect ratio to 16:9). While the sizes of digital displays vary greatly (from watch faces to large screens in stadiums), digital sensors are typically quite small, allowing them to fit inside handheld optical instruments, cameras, or mobile phones. Another difference between digital displays and sensors is the total number of pixels in the display/sensor. For displays, there is little benefit to adding more pixels than those that can be resolved by the human eye. Beyond this, no information can be perceived. Therefore, most home displays have as many as 3840 pixels in the horizontal direction. In comparison, digital sensors can have two to five times as many pixels because we want to acquire as much detail in an image as possible.

Figure 11.1 shows the relative sizes of a series of common digital sensors; the dimensions of these sensors are listed in Table 11.1. As digital technology started to replace film, it was common to compare the size of the digital sensor to film. The standard size of a 35-mm film image was 36 mm wide by 24 mm high, and this image size had a 3:2 aspect ratio. Digital camera sensors of the same size are called "full frame" sensors. The table shows that the larger digital sensors are typically camera format sensors, which have a 3:2 aspect ratio, whereas the smaller digital sensors have moved away from film to an aspect ratio of ∼4:3.

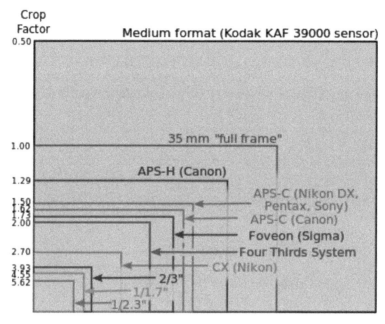

Figure 11.1 Comparative dimensions of sensor sizes. (Image reprinted from Wikipedia courtesy of Creative Commons.)

Table 11.1 Dimensions of image sensors.

Sensor	Width (mm)	Height (mm)	Aspect Ratio
1/3"	4.80	3.60	4:3
1/2.3"	6.17	4.55	4:3
1/1.7"	7.60	5.70	4:3
2/3"	8.80	6.60	4:3
CX (Nikon)	13.20	8.80	3:2
4/3 system	17.30	13.00	4:3
Foveon (Sigma)	20.70	13.80	3:2
APS-C (Canon)	22.30	14.90	3:2
APS-C (Nikon, et al.)	23.60	15.60	3:2
35-mm full frame	36	24	3:2
KAF-39000 (Kodak)	49	36.8	4:3

Another feature of a digital sensor that can be compared to film is sensitivity. The response of film to light is dependent on the size of halide crystals in the film. The larger the crystals the more rapidly an image can be recorded. (This response is referred to as the "speed" of the film. Its counterpart, the speed of a lens, is a measure of the amount of light that can be delivered to an image plane, which will also affect the speed of image recording.) But a large crystal will reduce the image resolution, so there is a trade-off between resolution and sensitivity in films. A similar relationship exists between pixel size and light response in digital sensors.

11.2 Spot Diagrams

There are a number of ways to analyze the performance of an optical system. One method that we've used in previous chapters is to represent the object with three point sources (three field points) and examine the size (and shape) of the image of each of these point sources using spot diagrams. These are ray-based calculations that ignore the effects of diffraction (see the Second Hiatus). Many optical systems are not diffraction limited. This may be because the application doesn't require a diffraction-limited system, or it would be either too costly or too large for the project at hand. So long as the aberrations of the optical system are much larger than the diffraction limit, the spot diagram for each field should provide a good measure of the performance of the lens. In this section, we'll start by taking a more detailed look at spot diagrams and their connection to sensor pixel size. Later in the chapter, we'll introduce a new diffraction-based measure of the image of a point source, called a point spread function, that takes into account the wave nature of light for analyzing diffraction-limited systems.

We will begin with the OStriplet (Fig. 11.2) that we used in Chapters 3 and 8. You can either restore the **OStriplet** from a saved file or enter it using the values in the LDE (Fig. 11.3). Open the FDE and add three fields of 0°, 7°, and 10°, if they are not present. The lens should have an EPD of 10 mm. Be sure to check that Paraxial Ray Aiming is turned on. Update the wavelength settings to the preset trio of visible wavelengths (F, d, C) in the System Explorer. Save this lens as **OStripletMod**.

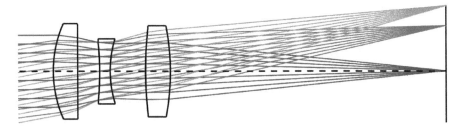

Figure 11.2 The OStripletMod.

	Surface Type	Comment	Radius	Thickness	Material	Clear Semi-
0	OBJEC Standard ▼		Infinity	Infinity		Infinity
1	Standard ▼		15.000	3.500	N-SK16	5.887
2	Standard ▼		Infinity	3.100		5.268
3	STOP Standard ▼		-37.530	1.500	N-F2	3.912
4	Standard ▼		14.080	5.000		4.008
5	Standard ▼		50.260	3.500	N-SK16	5.289
6	Standard ▼		-31.380	38.515 M		5.617
7	IMAGE Standard ▼		Infinity	-		8.834

Title (OStripletMod); EPD (10mm); Fields (0°, 7°, 10°); Wavelengths (F d C)

Figure 11.3 LDE for the OStripletMod.

To gain some feel for the size of spots of the point source images that will fall on the sensor, we will use a spot diagram (Analyze > Rays & Spots > Standard Spot Diagram). The plots shown in Fig. 11.4(a) are with the sensor in the paraxial image plane. An Airy Disk circle was overlayed on the spot diagram for each field by checking the "Show Airy Disk" box on the Spot Diagram Settings drop-down menu. Note: We have also added a box (outside of OpticStudio®) that represents a 20 μm×20 μm detector pixel centered on the chief ray. Because the grid in the diagram is a 100-μm square, a 2×2 square box in the grid corresponds to the area of a 20-μm pixel. Run a Quick Focus (Optimize > Quick Focus), as was done in Section 6.4. This deletes the marginal ray solve and shifts the image plane to 38.421 mm, or a defocus of −0.094 mm from the paraxial image plane. After this refocus, the point source images nearly fit within each pixel box, as shown in Fig. 11.4(b).

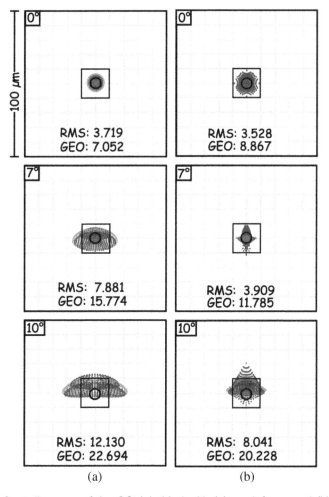

Figure 11.4 Spot diagrams of the OStripletMod with (a) no defocus and (b) −0.094-mm defocus. The squares represent a 20 μm×20 μm detector pixel centered on the chief ray, and the circles show the size of the Airy disk.

How well would this lens work with a sensor that has a standard format? Suppose we assume that our pixel size is on the order of our current spot diameter (~20 μm) and we use a Nikon CX formatted sensor with a height of 13.2 mm (13200 μm) and 8.8 mm (8800 μm). Then the sensor with 20-μm pixels would only contain an array of 660 × 440 pixels, which is very coarse resolution these days. For comparison, the main camera of an iPhone 14 Pro has a sensor that can produce 48-MP pictures (8000 × 6000 pixels) with its 1.22-μm pixels (RAW format) or 12-MP pictures with an effective pixel size of 2.44 μm (this is the default setting for the camera and helps to capture more light per pixel in low-light situations by binning the pixels).

It is clear then that modern digital camera lenses must be designed to provide very high image quality with much smaller spots than our current lens. But how small should the spot be? It turns out that the answer to this question is a bit more complicated than simply specifying a geometric spot diameter that fits inside a pixel. First, the geometric spot diameter completely ignores the effects of diffraction. Second, the sensor must capture the irradiance variations within the finest image detail across the sensor face. If changes in the image occur **within** a single pixel of a sensor, they will be averaged out and not captured by the image sensor. As a result, we need more than 1 pixel (and in many cases more than 4 pixels) across any spot diameter to match the sensor capability to the lens performance. These issues will be examined in much greater detail in the next two sections.

11.3 Point Spread Function

A point source at infinity produces plane waves at the entrance pupil of a lens. The lens then shapes these plane wavefronts into converging wavefronts that focus on the image plane. If the lens is perfect, the wavefronts will be a series of converging spherical surfaces. When analyzed with ray optics, the image of the point source object is a perfect (infinitesimally small) point. However, if diffraction and the wave nature of light are taken into account, the light distribution will be a 3D Airy function (a 2D Airy function rotated about the chief ray; see the Second Hiatus). It is the broadening of the central peak due to diffraction that is principally responsible for the limitation on lens resolution in both a perfect lens and a **diffraction-limited lens** (one whose aberrations have been sufficiently suppressed through design).

As discussed in the previous chapters, real lenses are not perfect. Wavefronts exiting a lens with aberrations will not be converging spherical surfaces. Instead, the aberrated wavefront is a distorted concave surface that results in a light distribution that will not be as narrow or as smooth as an Airy function. The aberrations reduce the light in the central peak and distribute it to the rings surrounding it. Because spot diagrams are purely ray-based calculations and do not account for diffraction, a spot diagram does not provide a complete evaluation of the lens if the aberrations of an optical system are comparable to the diffraction limit of the lens. What is needed is a reliable analysis that incorporates the effects of both aberration and diffraction. This is called a **point spread function (PSF)**, or a plot of the response of the lens to a point source object at a given field point. It is a little misleading to call this a "function" in the sense of an Airy function, which can be written as a compact mathematical function. The PSF would be better called a point spread distribution because it is the distribution of light from a point source focused onto a plane.

We can use OpticStudio to see what the PSF of an optical system looks like. Let's start with a lens without any aberrations—in a sense, a "perfect" lens. Open a new lens (File > New), set the EPD to 20 mm, and change the Surface Type of surface 1 (the stop surface) from Standard to Paraxial. (To quickly navigate to Paraxial after opening the drop-down menu in the Surface Type column, press the "P" key.) The focal length of a paraxial surface defaults to 100 mm. This can be changed if needed by entering a new value in the focal length column for the surface. For now, we'll stick with the 100-mm EFL and put the image plane at the paraxial focus of the perfect lens surface by entering 100 mm for the image distance (T1). The resulting LDE for the perfect lens is shown in Fig 11.5 and the lens in Fig. 11.6.

	Surface Type	Comment	Radius	Thickness	Semi-Diameter	Focal Length	OPD Mode
0	OBJECT Standard ▼		Infinity	Infinity	0.000		
1	STOP Paraxial ▼			100.000	10.000	100.000	1
2	IMAGE Standard ▼		Infinity	-	0.000		

Figure 11.5 LDE for a perfect lens.

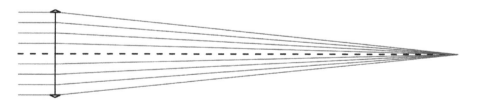

Figure 11.6 A perfect lens! All rays meet at the paraxial image plane!

There are several different ways to calculate and display the PSF in OpticStudio. These options are found on the Analysis tab (Analyze > PSF). Click on the drop-down menu for PSF and choose Huygens-PSF. When this is first selected, the plot that is displayed is not as revealing as it should be. The default pupil and image sampling is set very low (32×32), resulting in a large loss in resolution in the plot. In the Settings dialog box for the PSF plot (Fig. 11.7), increase the Pupil Sampling to 128×128 and the Image Sampling to 256×256 and replot the PSF.

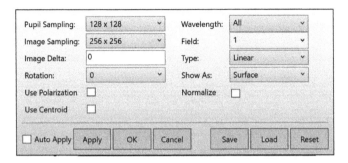

Figure 11.7 Settings dialog box for the Huygens-PSF plot.

Figure 11.8 shows two PSF plots of the diffraction-limited 3D Airy pattern of a perfect (aberration-free) lens. The initial PSF plot [Fig. 11.8(a)] shows the central peak surrounded by diffraction rings that are weak and therefore difficult to see. A logarithmic plot of the same pattern [Fig. 11.8(b)] reveals the rings. To get the second plot, in the Settings dialog box change the "Type" from "Linear" to "Log −5."

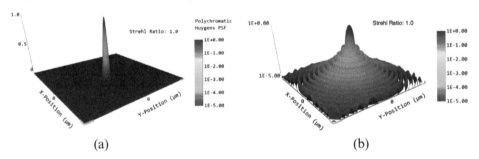

Figure 11.8 3D Airy pattern of a perfect lens: (a) linear plot (diffraction rings not visible) and (b) logarithmic plot showing diffraction rings.

Real lenses have aberrations that reduce the light in the central peak and distribute it to the rings surrounding it. So, what does the PSF of a real lens look like? Open the modified triplet lens, **OStripletMod**, saved in Section 11.2. Run a Quick Focus (Optimize > Quick Focus) to shift the image plane from the paraxial image plane to a best focus position (T6 = 38.421 mm). Plot the Huygens PSF (Analyze > PSF > Huygens-PSF) for the axial field (Field 1) using the same settings as the perfect lens above. The result is shown in Fig. 11.9(a).

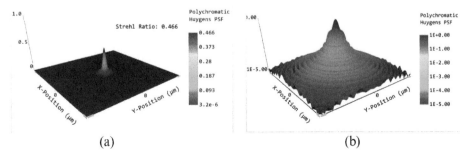

Figure 11.9 Point spread function for the on-axis field point of the OStripletMod: (a) linear plot and (b) logarithmic plot.

The on-axis PSF [Fig 11.9(a)] looks similar to the Airy function for the perfect lens [Fig. 11.8(a)]. However, the peak of the plot is only about half as high, and it appears that the missing energy has been distributed out to the sides of the peak. To see this more clearly, in Settings change the "Type" from "Linear" to "Log −5" and replot the PSF. The result is shown in Fig. 11.9(b). Comparing this plot with Fig. 11.8(b), it is clear that there is more energy in the first ring for the aberrated lens than for the perfect lens.

While there is not as much information about the dimensions of the blur spot in these PSF graphics as there is in a spot diagram (Fig. 11.4), each PSF plot lists a Strehl ratio in the title bar at the bottom of the plot. The **Strehl ratio** is the ratio of the peak of the PSF to that of a diffraction-limited lens with the same EPD. The Strehl ratio of a perfect lens is 1. For real lenses, it is a performance measure that represents how far the lens is from perfect. In Fig. 11.9, the Strehl ratio for the on-axis field (field 1) is 0.466. This means that the peak of the PSF is only 46.6% of the diffraction-limited Airy function.

To see the PSF plots for the other field points, you will need to change the field value in Settings and replot. The PSF plot for the (highly aberrated) 10° field point is shown in Fig. 11.10(a). The Strehl ratio is only 0.072, and almost all of the energy in the peak has been distributed out to the rings to the point where even on a logarithmic scale [Fig. 11.10(b)], the basic structure of the Airy function is no longer visible. This field point is far from diffraction limited, and the Strehl ratio starts to become an irrelevant performance measure.

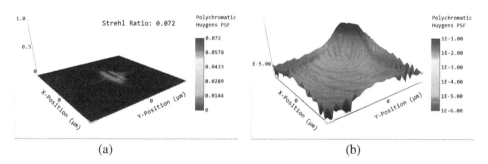

Figure 11.10 Point spread function for the 10° field point of the OStripletMod: (a) linear plot and (b) logarithmic plot.

To get a better measure of the performance of the lens, the "Show as" drop-down menu in the Settings dialog box permits the user to replot the 3D PSF as a few different 2D plots (changing the view or color scheme). Figure 11.11 shows 2D "False Color" PSFs for the on-axis and the full-field points. These plots give the designer an idea as to the spot size of each field point when diffraction is included in the analysis. They can be compared to the ray-based spot diagrams for the axis and 10° fields shown in Fig. 11.4(b).

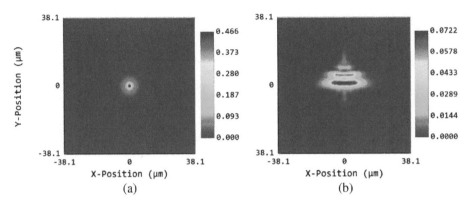

Figure 11.11 2D plot of the PSF (a) on axis and (b) at the 10° field for the defocused OStripletMod. Compare these to the spot diagrams shown in Fig. 11.4(b).

11.4 Measuring Resolution

Shortly after World War II, some standardized tests were devised to determine the performance of an optical system. For example, the United States Air Force (USAF) 1951 chart consists of an array of sets of black-and-white rectangular bars of decreasingly smaller size (see Fig. 11.12). The pattern of bars is laid out in groups of six elements, where each numbered element is half the size of the element with the same number in the next largest group. Each of the elements within a group is the sixth root of one-half ($\sqrt[6]{1/2} = 0.8909$) smaller than the next larger one. The group number is located above the set of bars, while each element is labeled to the right or left of the bars. A full description of the chart can be found on the following Wikipedia page: https://en.wikipedia.org/wiki/1951_USAF_resolution_test_chart

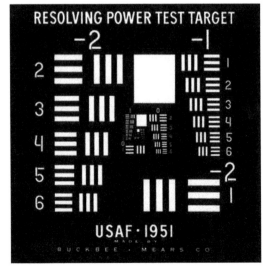

Figure 11.12 USAF 1951 bar chart.

Originally, these charts were printed on high-quality white cardboard and posted on the wall to test the image quality of an optical system. In practice, the largest element observed *without* distinct image **contrast** (the ratio of the separation between the peaks and valleys to the maximum value of the peak) indicates the approximate resolution limit of the optical system. As the bars get smaller and closer together, both diffraction and aberrations will reduce the contrast of the observed image. Modern high-resolution test setups use charts photographically printed on small glass slides to measure the spatial resolution of an imaging system.

While the width of the smallest line pair that can be "seen" provides an indication of lens resolution, a more useful and intuitive measure of performance uses the reciprocal of this quantity, its **spatial frequency**. Just as musical notes are characterized by their audio frequencies [middle C is 262 cycles/second (c/s) or 262 Hz], periodic bar patterns such as the elements of the USAF 1951 chart can be labeled by their spatial frequencies. This way, the resolution of the lens is expressed as the highest spatial frequency [in line pairs per millimeter (lp/mm)] of the smallest resolved element. For example, a 40-μm line pair has a spatial frequency of 25 lp/mm, where

$$\frac{1 \text{ line pair}}{40 \text{ }\mu\text{m}} \times \frac{1000 \text{ }\mu\text{m}}{1 \text{ mm}} = 25 \text{ lp/mm}.$$

Most objects are more elaborate than the three simple point sources we've used so far in our optical designs. It would be nice if we could take a complex scene (e.g., a landscape or a USAF 1951 test target) and simulate the performance of our lens with that specific object. OpticStudio has several different geometric and diffraction-based options (Analyze > Extended Scene Analysis) that show you how the image of an object scene will look. The object is defined as a bitmapped image file (.bmp) and is scaled to a particular field size. For the diffraction-based options, if you have taken courses on linear systems theory, you will recall that an image is simulated by convolving the PSFs of the lens with the object scene. An easy way to visualize this is to imagine that the PSF is like the size of a paintbrush used to paint the image. The smaller the paintbrush the more details can be drawn in the image. Because the PSF changes over the field of view, it is like using a different-sized paintbrush for different regions of the image.

We could take the simulated image of a USAF 1951 chart from one of these image simulation options and digitally measure the resolution of our design as long as the scaling of the elements is known. However, this approach can be quite tedious to implement if you want to find the resolution at many points across the field. Obviously, there needs to be a better approach to evaluate the performance of the lens at a number of field angles and ranges of spatial frequencies. This is discussed in the next section.

11.5 Modulation Transfer Function

Another approach to evaluate the performance of a lens is to compute the response of the lens to a broad range of spatial frequencies. Like the audio response by a stereo system when reproducing music from a variety of musical sources, the response of an optical system is measured by the **optical transfer function (OTF)**. The OTF is the Fourier transform of the point spread function (the image of a point source). The **modulation transfer function (MTF)** is then the modulus of the complex OTF. Similar to measurements of a USAF 1951 chart, the MTF measures the image contrast as a function of spatial frequency. However, the analysis does not require a bar chart with element-by-element analysis. Instead, the wavefront emerging from the exit pupil of a lens from a single field point can be constructed in optical design software from the optical paths of traced rays and used to compute an MTF via a Fourier transform.

Figure 11.13 shows a basic MTF plot for one field point. The modulation (or contrast) is plotted for a series of increasing spatial frequencies for both sagittal (vertical) lines and tangential (horizontal) lines. For an optical system with no aberrations and no diffraction, the MTF would be unity at all frequencies. The object shown in the top right of the figure represents a sinusoidal wave of 30 cycles/mm. Note: A cycle is the equivalent of a line pair on the USAF target. The MTF curves on the left and the profiles on the right represent the responses of the lens. If the lens were a perfect lens having no aberrations, its MTF would only be due to diffraction, given by the solid black line. In a real lens, the contrast is degraded by both diffraction and aberrations. By default, the MTF is plotted to the optical cut-off frequency, which is the spatial frequency beyond which an optical system cannot transmit information and for incoherent light equals $2(NA)/\lambda$ (or $1/\lambda f/\#$). Here, the MTF is plotted out to 90 cycles/mm.

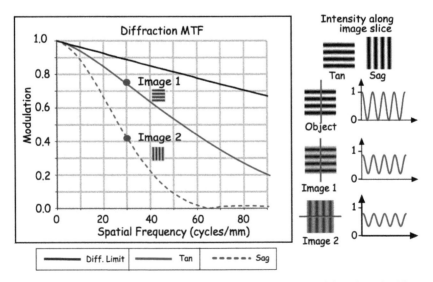

Figure 11.13 MTF chart for one field point showing both tangential and sagittal bars and the diffraction limit.

Restore the OStripletMod (Fig. 11.3) and change the image distance to 38.421 mm (for best focus). Like the PSF, there are a number of ways to produce an MTF plot in OpticStudio. For example, a fast Fourier transform (FFT) MTF plot can be generated by clicking on the MTF icon in the Analysis tab. Within the drop-down menu that appears, choose FFT MTF (Analyze > MTF > FFT MTF). At first, this feature creates an MTF plot out to the cutoff frequency of 410 cycles/mm, a frequency range that far exceeds the ability of this lens to reproduce such fine features. To display the contrast of this lens at a more appropriate frequency range, open Settings (Fig. 11.14) and change the Maximum Frequency to 100. Increase the Sampling to 128×128 and add a diffraction-limited (solid black) line to the plot by checking the Show Diffraction Limit box.

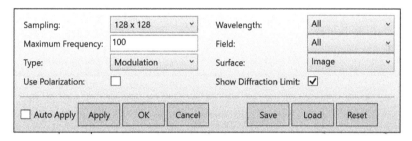

Figure 11.14 Settings for FFT MTF.

Click OK. The results are shown in Fig. 11.15. The program generates a plot of the MTF versus spatial frequency for the 0°, 7°, and 10° fields. For the axial field, the sagittal and tangential curves are the same and overlap so only one curve is shown. For the other two fields, the tangential responses are plotted as solid lines, while the sagittal responses are dotted lines. The MTF plot also shows the curve for a diffraction-limited lens (Diff. Limit) with the same $f/\#$ and operating wavelengths. It's quite clear that OStripletMod is not diffraction limited as the contrast for all fields is far below the diffraction limit. This agrees with our PSF analysis in Section 11.3.

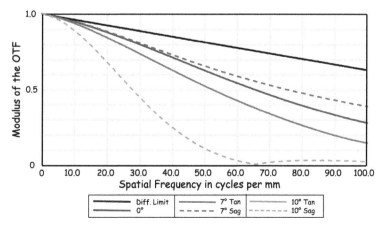

Figure 11.15 MTF plot of the OStripletMod lens.

An MTF plot provides a quantitative measure of how much of the fine detail in the object the lens can resolve. But what are the requirements for a digital sensor to record this image? That is, how many pixels are required to capture the image and what should be their separation? It would be foolish to design a high-resolution lens if the array of pixels could not record the details in the image, or to use a costly multi-megapixel array with a lens that only has modest performance.

For example, suppose that the OStripletMod lens images onto a CMOS array made up of 20-μm pixels, similar to the one used at the beginning of Section 11.2. We can use the Nyquist theorem (see the text box "Nyquist Theorem") to calculate the highest spatial frequency that a digital sensor can resolve (one over twice the pixel pitch). The maximum spatial frequency that can be recorded (the **Nyquist frequency**) is $1 / (2 \times 20 \ \mu m) = 25$ cycles/mm. In the MTF Settings, change the Maximum Spatial Frequency to 25. After you click OK, an MTF plot that reflects the sensor cut-off frequency will be created, as shown in Fig. 11.16.

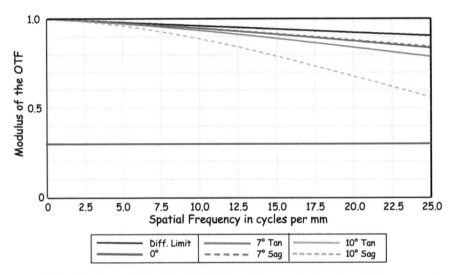

Figure 11.16 MTF plot of the OStripletMod plotted out to the Nyquist frequency for a sensor with 20-μm pixels. The (added) blue line indicates the minimum required 30% contrast.

Nyquist Theorem

Suppose we have a sensor that is constructed from an array of square pixels. It might seem like the best approach is to match the spot diameter (or Airy diameter) of a point image to the pixel dimension of the sensor. In the following figure (a), the image of a series of bars falls on an array of square pixels. If we were to follow the logic that the pixel size should match the image spot size, then the white bar would fall on a single row of pixels and produce the maximum signal V_{max} for that row of pixels, and the black bar would fall on the next row, producing a minimum signal V_{min} there. Very neat. But there's a problem: suppose that the pixel rows relative to the bars are aligned so that the pattern is shifted one-half a pixel width. Then the maxima and minima of the patterns fall halfway between the pixel rows, as illustrated in (b). That is, the centers of the bars fall between the pixels. In that case, the pixels would detect exactly the same levels of illumination and produce a signal $V_{50\%}$, and no bar pattern could be recorded. That is hardly the result we're after. We need more pixels.

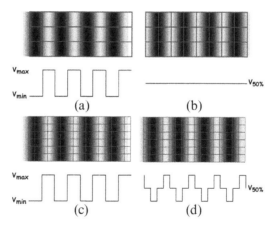

To demonstrate this, the number of pixels is doubled in the array in (c), and the periodic signal is detected with the same contrast as (a). Now when the bar pattern is shifted one-half a pixel width, so that the centers of the white and black bars are aligned with alternate rows of pixels, the periodic pattern is still recorded (d). Obviously, as the number of pixels is increased, the array will more precisely reproduce the bar pattern for any shifted position of the pattern.

So, the spatial period of the digital array must be less than half that of the smallest image period. To put this in terms of spatial frequencies, the spatial frequency of the digital array (determined from the pixel spacing) should be twice the maximum spatial frequency delivered by the lens. This is an example of the Nyquist theorem, which comes from signal theory. It states that for a periodic signal of a specific spatial frequency, the digital receiver should sample the signal at twice the spatial frequency of the signal. This leads to the calculation of a **Nyquist frequency**, which equals one over twice the pixel pitch (pp) or [1/(2pp)] and represents the highest spatial frequency that a digital sensor can resolve.

A typical specification for the performance of this lens could be written as >30% contrast (MTF = 0.3) at 25 cycles/mm for all fields. A quick look at Fig. 11.16 shows that this lens easily meets this specification, where a blue line was added to the plot to show the 30% target specification. The lowest contrast at 25 cycles/mm is ~55% for the sagittal curve at the last field. While the MTF plot is a great visual aid when quickly looking at lens performance, you might wonder if we can get the actual numbers for the plot. The answer is "Yes." The Text tab contains the data used to make the plot. If you click on the Text tab and scroll through the output, you will find output tables for the three fields and the diffraction limit. Table 11.2 shows the values for 25 cycles/mm.

Table 11.2 MTF data of the modified OStripletMod.

Field	Tangential	Sagittal
Diff. Limit	0.905780	0.905780
0°	0.844815	0.844815
7°	0.836583	0.845091
10°	0.788011	0.562604

Although spot sizes and simulated USAF charts can provide the designer with some measure of performance, it is really an MTF analysis of the lens across its several fields that provides the best evaluation of whether the lens can meet their needs. In the next chapter, we will start with a design with a single lens element, and a number of approaches for improving the design will be applied. The end performance of the lens will be evaluated against a specification using the MTF.

Explorations

Exploration 11.1 Image simulation
Explore the use of image simulation (Analyze > Extended Scene Analysis > Image Simulation) with some of our lenses using the various built-in image input files. Try changing the field height and pupil and image sampling. Compare geometric simulations to diffraction simulations (under "Aberrations"). How much better is the **OStripletMod** than the **OSsinglet**? How does changing the focus position affect the simulation?

Exploration 11.2
Restore the **OSsinglet** and determine the maximum frequency that it would provide with at least 10% contrast. Limit the maximum spatial frequency to that and then replot the MTF.

Exploration 11.3
Examine the MTF of some of the lenses that are discussed in the text: **OSachromat** (Section 9.4), **OSaplanat** (Section 7.2), **OSfieldFlattener** (Section 10.5.2), **OSlandscape** (Section 10.3.1), **OSlandscapeRearStop** (Section 10.3.2), and **OSsymmetrical** (Section 10.4). How do they compare to each other?

Exploration 11.4
The PSF incorporates all wavelengths specified for the design. How much does this affect the MTF of a lens specified for the C-, d-, and F-lines when compared to the monochromatic performance of the same design at the d-line? Use some of the designs listed in Exploration 11.3 and don't forget to focus the lens for best focus when changing wavelengths.

Exploration 11.5
Restore the **OSprotar** from Exercise 3.7. Use three fields of 0°, 7°, and 10° and change the wavelength to the visible spectrum. Plot the MTF of the lens at its best focus position. What is the largest spatial frequency that this lens can resolve if the target contrast is >50% for all fields? What pixel size would this correspond to?

Chapter 12
Designing a Lens

Thus far, we have shown how to enter a lens design into OpticStudio®, describe its faults, and then improve its performance. Our objective in this chapter is to design a lens to meet the requirements for a specific application. Starting with a simple lens system, it will be evaluated to see if it satisfies our requirements. If not, the design will be modified and re-evaluated through a series of steps or iterations until we arrive at a design that can perform to the specified conditions.

12.1 Defining the Problem

Now let's use the concepts that have been demonstrated in previous chapters to design a lens for a simple security camera. Modern security cameras can be quite complex, operating in both the visible and near-infrared wavelength bands with very large fields of view (FoVs) and mounted on a motorized stage to scan and zoom the scene. For now, we will focus on a much simpler security camera, whose purpose is to record objects and activity within a moderately sized (fixed) field and provide surveillance in the visible band only. In most instances, such cameras are mounted overhead or on ceilings (Fig. 12.1). Therefore, the objects of interest will be located far enough away from the lens that we may assume that the object is at infinity.

Figure 12.1 Ceiling-mounted security camera.

This type of optical system consists of an imaging lens and a CMOS sensor. But which lens and which sensor should we use? Do we need a custom design for the lens? A custom sensor? In the real world, the cost of materials, manufacturing, or assembly for an optical system must be considered before arriving at a final design solution. Although custom lenses are fairly common, the choice of a sensor is usually limited to one of the existing standard formats (see

Table 11.1) because the cost of the design and fabrication of a custom sensor would be prohibitive. For this reason, a CMOS sensor is usually chosen first in many optical design projects, and then a custom lens is designed to match the sensor to the requirements of the application.

12.2 Specifying the System

Figure 12.2 illustrates a simple first-order thin lens layout of the system. The object is located at infinity, the aperture stop is located at the lens, and a CMOS sensor is placed in the paraxial image plane. Because the image size is limited by the size of the CMOS sensor, the sensor is the field stop of the system. Although an overall size for the compact security camera has not been specified, it is clear by the intended application that both the sensor and the lens must be small. Therefore, a standard 1/3" CMOS sensor, one of the smallest formats, will be used. This sensor, which was listed in Table 11.1, has a 4:3 format with a 6-mm diagonal dimension, leading to a 4.8 mm (H)×3.6 mm (V) frame size.

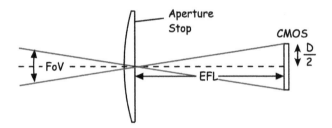

Figure 12.2 Field of view of the security camera.

The FoV of the camera is determined by the angle that the full field chief ray makes with the axis (see Section 5.6.1). The distance from the lens to the sensor is the EFL, and the FoV is then given by

$$\tan(\text{FoV}/2) = (D/2)/\text{EFL},$$

where D is the CMOS sensor diagonal. Solving for the EFL,

$$\text{EFL} = D/[2\tan(\text{FoV}/2)]. \tag{12.1}$$

To choose a suitable FoV (and focal length) for your security camera, you must first estimate both the size of the scene you want to capture and the approximate distance the scene is from the camera. These two values give you the required angular FoV. For example, to capture a 5' 10" person 4' from the camera, we would need a camera with a 72° FoV. Although this FoV would provide broad surveillance for most home or small-office security systems, to simplify our design example, we are going to design a camera with a smaller, 50° full FoV (this means that the person would need to be more than 6' from the camera to capture their full height). After entering this value and the 6-mm CMOS sensor diagonal into Eq. (12.1), the EFL for our lens is 6.43 mm.

Now that we have the focal length and the FoV, we must also specify the wavelengths and $f/\#$ of the camera. Some security cameras are designed for low-light, nighttime applications and need to be very fast ($< f/3$) infrared cameras. In contrast, our camera is planned for daytime use in a well-lit area. Accordingly, the C-, d-, and F-lines of the visible spectrum will be used. And to simplify the design example, we will limit its $f/\#$ to $f/10$ (which will also keep it small).

Finally, we need a performance specification. If we choose a sensor with a video graphics array (VGA) format (640×480 pixels), then the individual pixel size will be 7.50 μm. Based on the Nyquist criterion (see Chapter 11), the smallest line pair (lp) that can be detected by this sensor would be two pixels wide, or 15 μm/lp. Therefore, the minimum spatial frequency that can be resolved by the camera would be the reciprocal of 15 μm/lp, or 66.67 cycles/mm. As will be shown in Chapter 13, the nominal design performance does not account for the manufacturing errors that will further reduce the lens performance when it gets built. So, for the nominal design, we will require a lens with greater than 50% MTF at 50 cycles/mm (75% of the Nyquist frequency). Table 12.1 lists the specifications of our security camera.

Table 12.1 Specifications for a fixed-field security camera.

Object location	Infinity
Aperture	f/10
Field	50 deg full field of view
Wavelengths	Visible (656.3 587.6 486.1 nm)
Focal length	6.43 mm
Performance	> 50% MTF @50 cycles/mm

12.3 Step 0: The Initial Assessment

We will begin with a 0.5-mm-thick plano-convex lens made of N-BK7 glass. For a 6.43-mm-EFL thin lens, the radius of curvature of the first surface should be 3.33 mm [see Eq. (1.9)]. Obviously, this is, in comparison to the other lenses we have investigated, a very tiny lens. Open a new lens file (File > New) and follow the steps listed in Actions 12.1 to both enter this lens and update the system settings to match those in Table 12.1. Note: We've included an offset surface with a 1-mm thickness to plot the incoming rays and added a 10% margin to the clear semi-diameters to make your lens cross-sections look better. Figure 12.3 shows the resulting LDE. Don't forget to check your ray aiming settings; they should be set to paraxial ray aiming on.

The YZ cross-section of the lens (Analyze > Cross-Section) is shown in Fig. 12.4. It is obvious from the figure that there is a great deal of field curvature. This is to be expected because the lens system consists of a single positive element with the stop very close to the lens. The relative sizes of the third-order aberration coefficients (**THIRD**) for the lens (Table 12.2) also show that field curvature (TTFC and TSFC) dominates. The small amounts of spherical aberration and coma are due to the low speed of this $f/10$ lens.

Actions 12.1 Actions Setting up the OSsecureCam0.

In the System Explorer:
  Change the aperture type to Image Space F/#.
  Enter Aperture Value = 10.
  Enter Clear Semi Diameter Margin% = 10.
  In the FDE, enter three fields: 0°, 17°, and 25°.
  Select the F, d, C (Visible) preset in the wavelength data editor.
  Turn paraxial ray aiming on.
  Title this lens "OSsecureCam0."
In the LDE:
  Insert 3 surfaces in front of the Stop surface.
  Enter T1 = 1, R2 = 3.33, T2 = 0.5, G2 = N-BK7, and T3 = 0.1.
  Use a marginal height solve to locate the image plane at paraxial focus.

	Surface Type	Commen	Radius	Thickness	Material	Clear Semi-Dia
0	OBJECT Standard ▼		Infinity	Infinity		Infinity
1	Standard ▼		Infinity	1.000		1.108
2	Standard ▼		3.330	0.500	N-BK7	0.574
3	Standard ▼		Infinity	0.100		0.395
4	STOP Standard ▼		Infinity	6.014 M		0.301
5	IMAGE Standard ▼		Infinity	-		3.232

Title (OSsecureCam0); f/# (10); Fields (0°, 17°, 25°); Wavelength (F d C)

Figure 12.3 The LDE for the preliminary design OSsecureCam0.

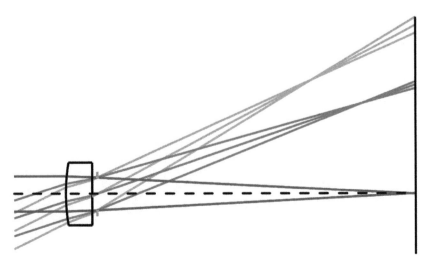

Figure 12.4 Ray trace for the OSsecureCam0 (note the obvious field curvature).

Table 12.2 Third-order aberration coefficients for the OSsecureCam0.

```
Surf    TSPH     TTCO     TAST     TPFC     TSFC     TTFC     TDIS
   1  0.0000   0.0000   0.0000   0.0000   0.0000   0.0000  -0.0000
   2 -0.0007  -0.0082  -0.0226  -0.0231  -0.0344  -0.0570  -0.1429
   3 -0.0002   0.0065  -0.0431   0.0000  -0.0216  -0.0647   0.2155
 STO  0.0000  -0.0000   0.0000   0.0000   0.0000   0.0000  -0.0000
 IMA -0.0000  -0.0000  -0.0000  -0.0000  -0.0000  -0.0000  -0.0000
 TOT -0.0009  -0.0017  -0.0658  -0.0231  -0.0560  -0.1217   0.0727
```

Figure 12.5 shows a plot of the ray aberrations (Analyze > Rays & Spots > Ray Aberration) and a spot diagram (Analyze > Rays & Spots > Standard Spot Diagram) for this lens. As expected, the linear shape of the curves in the transverse ray plots that increase in slope with field [Fig. 12.5(a)] indicate that field curvature (both Petzval and astigmatism) is present. Because the slopes of the tangential and sagittal ray fans are different and the spot diagrams show elliptical spot patterns [Fig. 12.5(b)], the astigmatism is larger than the Petzval curvature, which agrees with the third-order coefficient values.

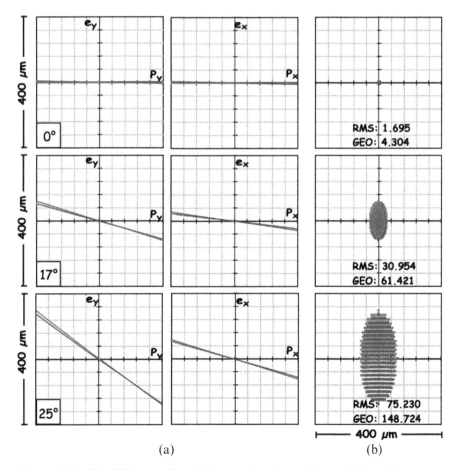

(a) (b)

Figure 12.5 The OSsecureCam0: (a) ray aberration plots and (b) spot diagrams.

As was noted in Section 11.5 on MTF, there are a number of ways to produce an MTF in OpticStudio. The FFT option (Analyze > MFT > FFT MTF) will result in the plot shown in Fig. 12.6. A diffraction-limited (solid black) line has been added by checking the Show Diffraction Limit box in Settings. By default, the maximum value for the x-axis is set to 250 cycles/mm (past the lens cut-off frequency of 205.8 cycles/mm). However, for a simple lens operating with such high field angles (and this much field curvature), the plot has many oscillations due to the severe aberrations, and for the larger fields in the plot much of the detail is lost. It does, however, show that the on-axis curve closely follows the Diff. Limit curve, indicating that the lens is nearly diffraction limited on axis. This is because the lens is slow ($f/10$) and oriented with its flat side facing the focal plane (Section 6.7), resulting in a very small amount of spherical aberration.

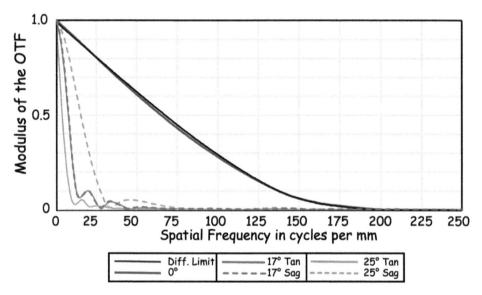

Figure 12.6 The MTF for the OSsecureCam0 plotted out to a maximum frequency of 250 cycles/mm.

To get a better assessment of lens performance, we can limit the MTF curve to the region we care about (remember that our specification is at 50 cycles/mm) by changing the maximum plotted frequency to 100 cycles/mm in the MTF (Maximum Frequency) settings. The MTF curve for the lens is shown in Fig. 12.7. The large differences between the tangential and sagittal curves for fields 2 and 3 indicate that astigmatism is present in the design. This is in accord with the spot diagrams and transverse ray curves presented earlier. The only place that the MTF is acceptable is on axis, where the curve closely tracks the curve for diffraction-limited performance. Save this lens as **OSsecureCam0**.

Designing a Lens

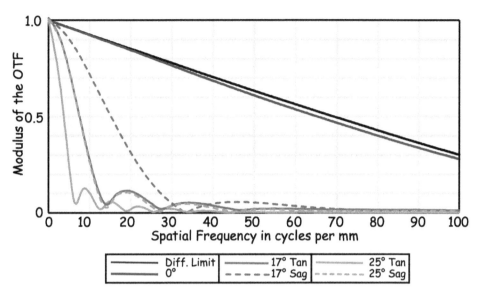

Figure 12.7 The MTF for the OSsecureCam0 plotted out to a maximum frequency of 100 cycles/mm.

12.4 Step 1: Bend the Lens

We have specified our design, constructed a first-order starting point, and analyzed its aberrations so that they can be compared with improvements in later designs. Now, we will start our search for a better design by optimizing the shape of the lens and the location of the sensor (defocus) while keeping the stop position fixed. At each step in the design process, the newest version of the lens will be saved with the integer at the end of the filename incremented. The title of the lens should also be accordingly updated at each step (System Explorer > Title/Notes).

The image plane for the OSsecureCam0 was placed at the paraxial image location using a marginal ray solve. While this is a good location for the on-axis image quality, the off-axis image quality can be greatly improved by defocusing the lens. It is often good practice to find the best image plane location for all fields using a quick focus (Optimize > Quick Focus) before starting a more complex lens optimization. This deletes the marginal ray height solve on the last thickness and changes the image distance from 6.014 mm (the paraxial image plane location) to 5.107 mm. Make the image distance variable as we will continue to let it change during optimization. To optimize the lens shape, the lens radii of curvature also need to be changed from fixed to variable. The resulting LDE is shown in Fig. 12.8.

	Surface Type	Comment	Radius	Thickness	Material	Clear Semi-Dia
0	OBJECT Standard ▼		Infinity	Infinity		Infinity
1	Standard ▼		Infinity	1.000		1.108
2	Standard ▼		3.330 V	0.500	N-BK7	0.574
3	Standard ▼		Infinity V	0.100		0.396
4	STOP Standard ▼		Infinity	5.107 V		0.301
5	IMAGE Standard ▼		Infinity	-		2.700

Figure 12.8 LDE of the OSsecureCam1 before optimization.

As discussed in Chapter 10, use the optimization wizard (Optimize > Optimization Wizard) to set up a merit function for minimizing the spot radius by following the steps outlined in Actions 12.2. To enable us to compare the optimized lens to its predecessor, the EFL must be held constant during optimization. The top 10 lines of the resulting merit function editor (before optimization) are shown in Fig. 12.9.

Actions 12.2 Setting up the merit function for the OSsecureCam1.

In the Optimization Wizard:
   Change the Image Quality from Wavefront to Spot.
   Check that the pupil integration is set to GQ with 3 rings/6 arms.
   Make sure that no boundary values are checked.
   Click "OK."
In the MFE:
   Insert 2 lines above the DFMS.
   Enter an EFFL operand in the first row.
   Set the Target for the EFFL operand to 6.43.
   Set the Weight for the EFFL operand to 10.
   Update the merit function.

Merit Function: 0.0164695794606411

	Type	Wave					Target	Weight	Value	% Contrib
1	EFFL ▼	2					6.430	10.000	6.443	41.251
2	BLNK ▼									
3	DMFS ▼									
4	BLNK ▼	Sequential merit function: RMS spot x+y centroid X Wgt = 1.0000 Y Wgt = 1.0000 GQ 3 rings 6 arms								
5	BLNK ▼	No air or glass constraints.								
6	BLNK ▼	Operands for field 1.								
7	TRCX ▼	1	0.000	0.000	0.336	0.000	0.000	0.097	0.014	0.444
8	TRCY ▼	1	0.000	0.000	0.336	0.000	0.000	0.097	0.000	0.000
9	TRCX ▼	1	0.000	0.000	0.707	0.000	0.000	0.155	0.030	3.107
10	TRCY ▼	1	0.000	0.000	0.707	0.000	0.000	0.155	0.000	0.000

Figure 12.9 First 10 lines of the OSsecureCam1 MFE before optimization.

Designing a Lens 265

Now that the variables are assigned and the merit function is set up, perform a local optimization with Optimize > Optimize!. At the beginning of the optimization, the merit function (MF) is 0.0165. After one local optimization, the MF is reduced to a fourth (0.00393) of the initial MF. Figure 12.10 shows the LDE for the optimized lens. The lens no longer has a planar back surface, and both radii of curvature are positive, as shown in Fig. 12.11. It is evident from the change in the focus of the off-axis bundles in Fig. 12.11 that the field curvature is considerably improved compared to those shown in Fig. 12.4.

	Surface Type	Comment	Radius	Thickness	Material	Clear Semi-Dia
0	OBJECT Standard ▼		Infinity	Infinity		Infinity
1	Standard ▼		Infinity	1.000		1.136
2	Standard ▼		1.077 V	0.500	N-BK7	0.558
3	Standard ▼		1.341 V	0.100		0.338
4	STOP Standard ▼		Infinity	4.980 V		0.266
5	IMAGE Standard ▼		Infinity	-		2.967

Figure 12.10 The OSsecureCam1 LDE after optimization of the lens.

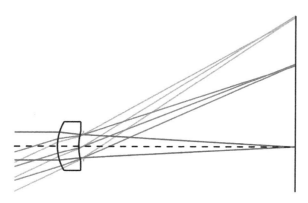

Figure 12.11 The OSsecureCam1 after optimization.

A comparison of the third-order coefficients in Table 12.2 for the initial design with those of its optimized version given in Table 12.3 shows that the TSPH and TTCO have gotten worse (due to the change in shape of the lens), the field curvature (TTFC and TSFC) has improved substantially, and the Petzval curvature (TPFC) is a little more than half of the earlier value.

Table 12.3 Third-order aberration coefficients for the optimized OSsecureCam1.

```
Surf    TSPH     TTCO     TAST     TPFC     TSFC     TTFC     TDIS
   1   0.0000   0.0000   0.0000   0.0000   0.0000   0.0000  -0.0000
   2  -0.0192  -0.0453  -0.0237  -0.0711  -0.0830  -0.1067  -0.0652
   3   0.0011   0.0110   0.0252   0.0571   0.0697   0.0949   0.2398
 STO   0.0000  -0.0000   0.0000   0.0000   0.0000   0.0000  -0.0000
 IMA  -0.0000  -0.0000  -0.0000  -0.0000  -0.0000  -0.0000  -0.0000
 TOT  -0.0181  -0.0343   0.0015  -0.0140  -0.0133  -0.0118   0.1746
```

The plot of the ray aberration curves, shown in Fig. 12.12(a), and the spot diagrams, shown in Fig. 12.12(b), demonstrate the dramatic reduction in astigmatism. Notice that the scale of the plots is much smaller than before. (The ray aberration curve scale was reduced from 400 μm to 100 μm.) The U-shaped aberration curves and the comatic spot diagrams for the off-axis fields indicate that the majority of the remaining aberration is now coma.

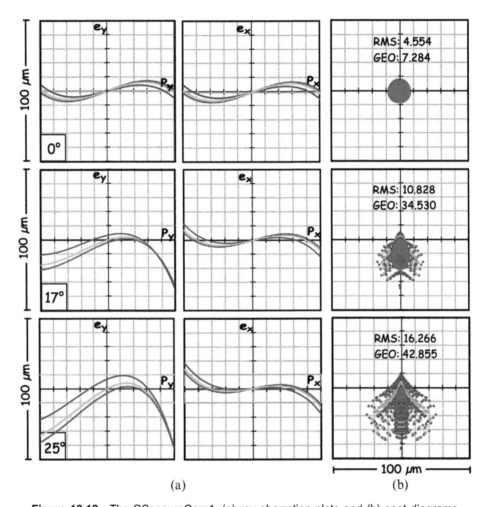

Figure 12.12 The OSsecureCam1: (a) ray aberration plots and (b) spot diagrams.

Finally, you can see that, compared to the curves in Fig. 12.7, the MTF shown in Fig. 12.13 has been significantly improved and is now better balanced across the fields (although the performance of the on-axis field is worse). Save this lens as **OSsecureCam1**.

Designing a Lens

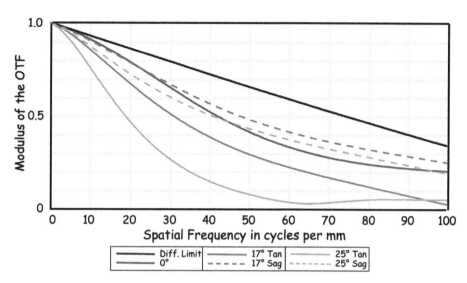

Figure 12.13 The MTF for the optimized OSsecureCam1.

12.5 Step 2: Shift the Stop

In an earlier discussion (Section 10.3), it was shown that coma could be reduced by varying the stop position. As part of the initial design, the stop was placed next to the back surface of the singlet on a surface with no power. To make the stop position a variable, the distance from the second lens surface to the stop (T3) should be varied in the LDE. Change the lens title to OSsecureCam2 and then optimize with the same merit function and EFL operand from Section 12.4.

The design begins with MF = 0.00393. After local optimization (Optimize > Optimize!), this has been reduced to 0.00252 and the stop has moved 0.467 mm beyond the lens, as indicated in the LDE (Fig. 12.14). The lens remains a meniscus lens, but the radii of curvatures are longer. Because the location of the stop is 0.467 mm beyond the last surface of the lens, the lens apertures (Fig. 12.15) had to increase substantially to permit rays from the 25° field to get through the lens to the image plane.

	Surface Type	Comment	Radius	Thickness	Material	Clear Semi-Dia
0	OBJECT Standard ▼		Infinity	Infinity		Infinity
1	Standard ▼		Infinity	1.000		1.407
2	Standard ▼		1.265 V	0.500	N-BK7	0.782
3	Standard ▼		1.768 V	0.467 V		0.571
4	STOP Standard ▼		Infinity	4.851 V		0.255
5	IMAGE Standard ▼		Infinity	-		3.031

Figure 12.14 The OSsecureCam2 LDE after stop shift optimization.

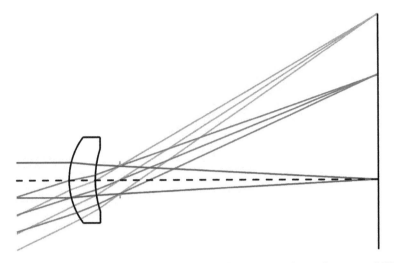

Figure 12.15 Ray trace of OSsecureCam2 after optimization using stop shifting.

If you compare the optimized third-order aberration coefficients listed in Table 12.4 to those of the lens before the optimization, you will see that the coma (TTCO) has been reduced by more than 3× with only a slight increase in Petzval curvature (TPFC).

Table 12.4 Third-order aberration coefficients for the optimized OSsecureCam2.

Surf	TSPH	TTCO	TAST	TPFC	TSFC	TTFC	TDIS
1	0.0000	0.0000	0.0000	0.0000	0.0000	0.0000	-0.0000
2	-0.0118	-0.0105	-0.0021	-0.0605	-0.0615	-0.0636	-0.0181
3	0.0002	0.0027	0.0074	0.0433	0.0470	0.0544	0.1895
STO	0.0000	-0.0000	0.0000	0.0000	0.0000	0.0000	-0.0000
IMA	-0.0000	-0.0000	-0.0000	-0.0000	-0.0000	-0.0000	-0.0000
TOT	-0.0116	-0.0077	0.0053	-0.0172	-0.0146	-0.0092	0.1714

A comparison between the ray aberration curves before [Fig. 12.12(a)] and after [Fig. 12.16(a)], both plotted on the same scale, shows the great reduction in coma due to shifting the stop. It can be seen from the separations between the aberration curves of different wavelengths that the design is now limited by lateral color at the edge of the field. This is also quite evident from the spot diagram for the 25° field [Fig. 12.16(b)]. Finally, a comparison of the MTF for this design state (Fig. 12.17) with that before the stop shift shows that the spatial frequency response has improved. Save this lens as **OSsecureCam2**.

Designing a Lens

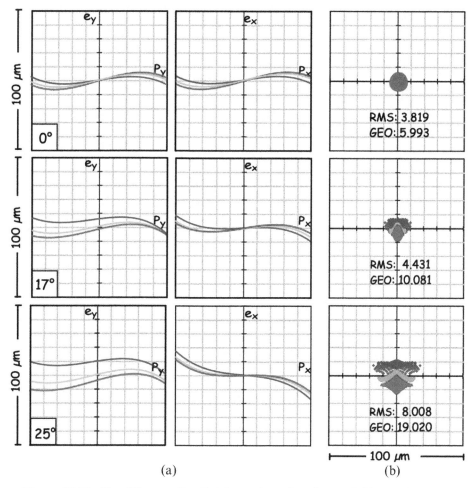

Figure 12.16 The OSsecureCam2: (a) ray aberration plots and (b) spot diagrams.

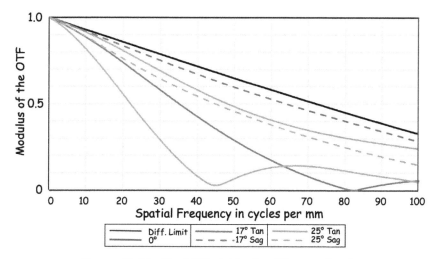

Figure 12.17 The MTF plot for the OSsecureCam2.

12.6 Step 3: Turn a Singlet into a Doublet

We know from Chapter 9 that the primary way to reduce chromatic aberration is to combine positive and negative lenses of different dispersions in an achromatic doublet. For example, given the power of the achromat ϕ_d and two glasses, a crown and a flint, the power of each element can be calculated using the relationships in Eq. (9.8). However, for this exercise, we will simply add a negative N-SF4 (flint) lens to the positive N-BK7 (crown) in the OSsecureCam2 and then let OpticStudio determine the best powers and radii for the two elements of the doublet.

Using Actions 12.3, insert an interior surface between the existing lens surfaces to create a cemented doublet. Note: When the new surface is first inserted, OpticStudio copies the 1.768-mm radius from S4 to S3. This (and the change in material for the back half of the doublet) results in a significant change in focal length from 6.43 mm to 5.52 mm. Since it is usually best practice to start a lens optimization at or very close to the target focal length, we adjusted the S3 radius to 2.235 mm to yield a 6.43-mm EFL. After a Quick Focus (Optimize > Quick Focus), the LDE of the lens before optimization should look like Fig. 12.18, and the lens should look like Fig. 12.19.

Actions 12.3 Turn a singlet into a cemented doublet with a 6.43-mm EFL.

Insert a surface between S2 and S3.
Set its thickness T3 = 0.2 mm.
Set its material G3 = N-SF4.
Change the radius R3 to 2.235 (to recover the 6.43-mm EFL).
Vary the new R3 radius (R2 and R4 should still be varying).
Change the title of the lens to OSsecureCam3.

	Surface Type	Comment	Radius	Thickness	Material	Clear Semi-Dia
0	OBJECT Standard ▼		Infinity	Infinity		Infinity
1	Standard ▼		Infinity	1.000		1.539
2	Standard ▼		1.265 V	0.500	N-BK7	0.880
3	Standard ▼		2.235 V	0.200	N-SF4	0.687
4	Standard ▼		1.768 V	0.467 V		0.576
5 STOP	Standard ▼		Infinity	4.534 V		0.236
6 IMAGE	Standard ▼		Infinity	-		3.126

Figure 12.18 The LDE for the OSsecureCam3 before optimization.

Designing a Lens 271

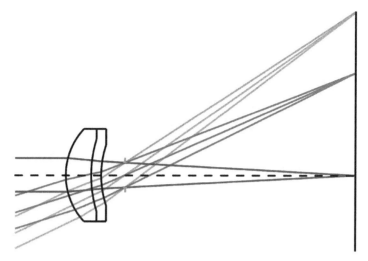

Figure 12.19 The OSsecureCam3 with an added lens before optimization.

In the MFE, update the existing merit function. Because the lens was (somewhat arbitrarily) split in two and a second glass added, the initial merit function has increased from 0.00252 to 0.00291. After local optimization (Optimize > Optimize!), this is reduced to 0.00202 (almost a 50% improvement from the optimized singlet). The resulting doublet is shown in the LDE (Fig. 12.20) and plotted in Fig. 12.21.

	Surface Type	Comment	Radius	Thickness	Material	Clear Semi-Dia
0	OBJECT Standard ▾		Infinity	Infinity		Infinity
1	Standard ▾		Infinity	1.000		1.742
2	Standard ▾		1.534 V	0.500	N-BK7	1.056
3	Standard ▾		9.701 V	0.200	N-SF4	0.931
4	Standard ▾		3.151 V	0.749 V		0.797
5	STOP Standard ▾		Infinity	4.521 V		0.235
6	IMAGE Standard ▾		Infinity	-		3.102

Figure 12.20 LDE for the OSsecureCam3 doublet after optimization.

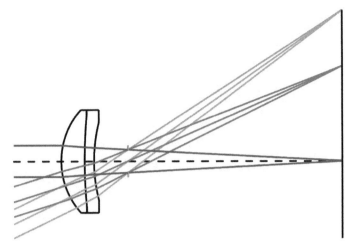

Figure 12.21 The OSsecureCam3 doublet after optimization.

The monochromatic aberration coefficients (Table 12.5) for spherical aberration (TSPH) and coma (TTCO) have decreased, while the Petzval curvature (TPFC) has slightly increased. The chromatic aberration values are not shown, but it is obvious when comparing the ray aberration plot for the stop-shifted singlet [Fig. 12.16(a)] with the ray aberration plot for this doublet [Fig. 12.22(a)] that the chromatic aberrations are more than 2× smaller. The color error reduction is also obvious in the spot diagram [Fig. 12.22(b)], particularly for the 25° field. Finally, the MTF (Fig. 12.23) for the tangential full field has dramatically improved. The pronounced dip in the curve at 45 cycles/mm has been eliminated. Save this lens as **OSsecureCam3**.

Table 12.5 Third-order aberration coefficients for the optimized OSsecureCam3.

Surf	TSPH	TTCO	TAST	TPFC	TSFC	TTFC	TDIS
1	0.0000	0.0000	0.0000	0.0000	0.0000	0.0000	-0.0000
2	-0.0066	0.0056	-0.0011	-0.0499	-0.0505	-0.0515	0.0143
3	0.0001	-0.0034	0.0233	-0.0021	0.0096	0.0328	-0.0977
4	-0.0000	-0.0018	-0.0162	0.0307	0.0226	0.0064	0.3007
STO	0.0000	-0.0000	0.0000	0.0000	0.0000	0.0000	-0.0000
IMA	-0.0000	-0.0000	-0.0000	-0.0000	-0.0000	-0.0000	-0.0000
TOT	-0.0066	0.0004	0.0060	-0.0213	-0.0183	-0.0123	0.2173

Designing a Lens

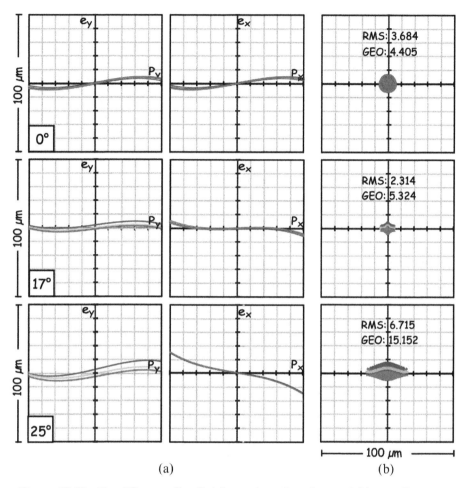

Figure 12.22 The OSsecureCam3: (a) ray aberration plots and (b) spot diagrams.

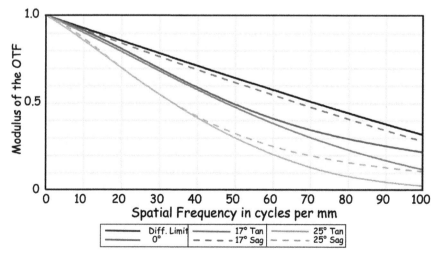

Figure 12.23 The MTF plot for the OSsecureCam3.

12.7 Step 4: Add a Field Flattener

Now that the color error is under control, how else can the design be improved? The MTF plot of the doublet, shown in Fig. 12.23, offers a clue. If there were no astigmatism, the tangential and sagittal curves for each field would lie above each other. The third-order aberration coefficients indicate that there is also some residual Petzval curvature in the lens. The field curvature plot showing only the d-line curve, Fig. 12.24(a), confirms that there is a large separation between the tangential and sagittal field curves. The tangential curve is also to the right of the sagittal curve, indicating that the lens has overcorrected astigmatism that is balancing the undercorrected Petzval curvature. Section 10.5 demonstrated that introducing a lens (termed a field flattener) close to the image surface can be a useful tool for reducing this field curvature (astigmatism + Petzval).

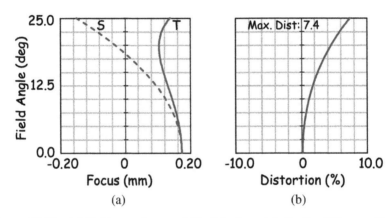

Figure 12.24 (a) Field curvature and (b) distortion plots for the OSsecureCam3.

Add a field flattener to our lens system following Actions 12.4. After a Quick Focus (**Optimize > Quick Focus**), the image distance is set to 2.245 mm. The LDE of the lens before optimization should look like Fig. 12.25. A cross-section of this design (Fig. 12.26) shows the OSsecureCam3 doublet with a slab of glass behind it. Because the added element has no optical power, the EFL (6.43 mm) has not changed from that of the earlier design.

Actions 12.4 Add a field flattener to the OSsecureCam3.

> Insert two surfaces, S6 and S7, after the Stop.
> Locate the field flattener 2 mm behind the Stop, T5 = 2.
> Set the field flattener thickness T6 = 0.5 mm.
> Set its material G6 = N-SK16 (a high-index crown glass).
> Vary the radii of the field flattener, R6 and R7.
> Vary the distance to the image plane, T7.
> Change the title of the lens to OSsecureCam4.

Designing a Lens

	Surface Type	Comment	Radius	Thickness	Material	Clear Semi-Dia
0	OBJECT Standard ▾		Infinity	Infinity		Infinity
1	Standard ▾		Infinity	1.000		1.742
2	Standard ▾		1.534 V	0.500	N-BK7	1.056
3	Standard ▾		9.701 V	0.200	N-SF4	0.931
4	Standard ▾		3.151 V	0.749 V		0.797
5 STOP	Standard ▾		Infinity	2.000 V		0.235
6	Standard ▾		Infinity V	0.500	N-SK16	1.654
7	Standard ▾		Infinity V	2.245 V		1.846
8 IMAGE	Standard ▾		Infinity	-		3.102

Figure 12.25 The LDE for OSsecureCam4 with a field flattener before optimization.

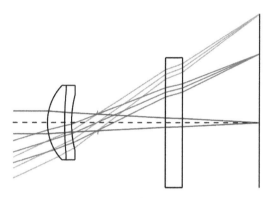

Figure 12.26 The lens cross-section for the OSsecureCam4 before optimization.

Perform a local optimization (Optimize > Optimize!). The initial merit function is 0.00204. After optimization, this is reduced to 0.0005 (a 4× improvement!). However, a quick look at the lens cross-section (Fig. 12.27) reveals some significant issues with the resulting optical design. First, the field lens is currently resting on the image plane surface and, second, the first element in the doublet has such a small edge thickness that it would be very difficult to fabricate. If the edge of a lens is too thin, the chance of chipping the lens when making or mounting it is high.

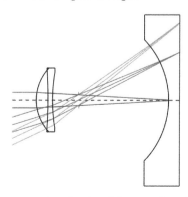

Figure 12.27 The lens cross-section for the OSsecureCam4 after optimization (with no boundary value operands).

As we just discovered, OpticStudio does not automatically constrain the values of lens thicknesses, edge thicknesses, or airspaces during optimization. This means that when varying element thicknesses, it is up to the designer to keep the lenses from getting too thick or too thin, or else the optimization of a lens can lead to some very strange-looking lenses—ones that would be hard to make, test, or mount. It is even possible to end up with lenses (or airspaces) with negative thicknesses! If an optimized lens looks very odd, the designer needs to intervene. To that end, constraints should be added to the merit function, and it is usually best to add these constraints before the optimization gets out of hand.

Luckily, these issues can be fixed by operands that monitor the value of a lens parameter (e.g., lens thickness) and increase the merit function when the lens parameter gets too large or too small. For example, the optimization wizard has some easy-to-define boundary operands (see the text box "Adding Boundary Value Operands to the Merit Function") for element thicknesses and airspaces.

Adding Boundary Value Operands to the Merit Function

The optimization wizard has a convenient set of boundary operands (Boundary Values) for both the glass (Glass) and the airspaces (Air) that are easy to use. See the highlighted right-hand side of the starting optimization wizard settings below.

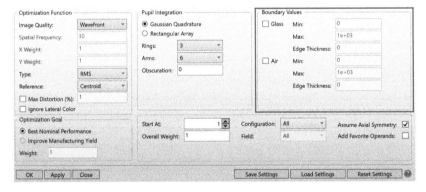

As shown above, the initial settings for the Wizard are such that all boundary values are inactive (unchecked). Min is the minimum center thickness, Max is the maximum center thickness, and Edge Thickness is the minimum edge thickness. Once the boundary values are checked, operands to control center and edge thicknesses of both glass (designated by a G) and air (designated by an A) surfaces are added to the merit function for every surface in the system (e.g., MNCG, MXCG). These surfaces are listed at the top of the DMF, and only those that are active (outside of the boundary) contribute to the merit function value. When surfaces are added to the lens system, operands for the new surfaces are automatically updated in the merit function.

Note: It is important when updating an existing merit function with the optimization wizard that the "Start At" value at the bottom of the wizard matches the current line number of the DMFS when you click "Apply" or your existing operands (e.g., EFFL) may be overwritten.

At this point, it is best to undo (shortcut key F3) this last (wayward) optimization and go back to the lens LDE shown in Fig. 12.25 before continuing with our design example. Actions 12.5 summarize a series of changes that we want to make to the lens and the merit function for an improved optimization of our OSsecureCam4. For example, after the last optimization, both the edge thickness of element 1 and the image distance were way too small. Before adding constraints to increase these thicknesses, we should vary the thickness of element 1 and the other two elements to give the optimizer more variables to satisfy the extra constraints without sacrificing any performance. Then we can use the optimization wizard to add appropriate boundary value operands to the merit function.

As shown in Fig. 12.28, we want to add boundary value operands with a minimum value of 0.1 mm and a maximum value of 0.7 mm for the thickness of the lens elements ("Glass"). These constraints put a heavy penalty on any optimization step that that would result in a lens thickness smaller than 0.1 mm or greater than 0.7 mm. While these values might seem too small, remember that this is also a very small-diameter lens system. We can also add a constraint to prevent the edge thickness of any lens element from being smaller than 0.25 mm. The remaining boundary values for "Air" keep the airspaces from becoming too small (>0.1 mm), and a minimum value of 0.25 mm is set for the air edge thickness.

The distortion plot in Fig. 12.24 showed that our lens has significant pincushion distortion (>7%) that is too much for our security camera. It's important to note that the optimizer does not automatically control distortion unless a distortion operand (DIMX) is specifically added to the merit function (all spot sizes are measured with respect to the real chief ray not the paraxial ray). To add a 5% maximum distortion operand, check the Max Distortion box in the optimization wizard (see Fig. 12.28) and enter a value of 5. This specifies an upper bound for the absolute value of the distortion across the entire field as the maximum distortion does not always occur at the maximum field coordinate.

Actions 12.5 Updating the lens and MF for the OSsecureCam4.

In the LDE:
   Vary all lens thicknesses (T2, T3, and T6).
In the Optimization Wizard:
   Set the Glass and Air boundary values to those shown in Fig. 12.28.
   Check the Max Distortion box and enter 5%.

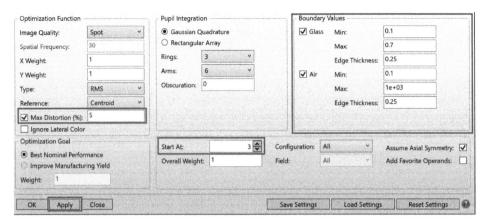

Figure 12.28 Optimization Wizard settings to control glass, airspaces, and maximum distortion.

Having made these changes to optimization wizard settings, check that the "Start At" box is set for line 3 (the start of the DMFS) and click "Apply" at the bottom left of the frame. Figure 12.29 shows the top 15 lines of the new merit function editor with the added distortion constraint and boundary value operands applied.

	Type	Field	Wave	Absolute			Target	Weight	Value	% Contrib
1	EFFL ▼		2				6.430	10.000	6.430	3.693E-10
2	BLNK ▼									
3	DMFS ▼									
4	BLNK ▼	Sequential merit function: RMS spot x+y centroid X Wgt = 1.0000 Y Wgt = 1.0000 GQ 3 rings 6 arms								
5	BLNK ▼	Default individual air and glass thickness boundary constraints.								
6	DIMX ▼	0	2	0			5.000	1.000	6.544	99.327
7	MNCA ▼	1	1				0.100	1.000	0.100	0.000
8	MXCA ▼	1	1				1000.000	1.000	1000.000	0.000
9	MNEA ▼	1	1	0.000	0		0.250	1.000	0.250	0.000
10	MNCG ▼	1	1				0.100	1.000	0.100	0.000
11	MXCG ▼	1	1				0.700	1.000	0.700	0.000
12	MNEG ▼	1	1	0.000	0		0.250	1.000	0.250	0.000
13	MNCA ▼	2	2				0.100	1.000	0.100	0.000
14	MXCA ▼	2	2				1000.000	1.000	1000.000	0.000
15	MNEA ▼	2	2	0.000	0		0.250	1.000	0.250	0.000

Merit Function: 0.201157895256589

Figure 12.29 The top 15 lines of the MFE after setting up boundary operands.

To keep the final lens at least 2 mm away from the image plane (often called the image clearance), we need to scroll down the merit function editor and change the boundary value operands for both the minimum air center thickness (MNCA operand #43) and the minimum air edge thickness (MNEA operand #45) for surface S7 to a target value of 2. Figure 12.30 shows the updated operands.

Designing a Lens 279

	Type	Surf1	Surf2					Target	Weight	Value	% Contrib
41	MXCG ▼	6	6					0.700	1.000	0.700	0.000
42	MNEG ▼	6	6	0.000	0			0.250	1.000	0.250	0.000
43	MNCA ▼	7	7					2.000	1.000	2.000	0.000
44	MXCA ▼	7	7					1000.000	1.000	1000.000	0.000
45	MNEA ▼	7	7	0.000	0			2.000	1.000	2.000	0.000
46	MNCG ▼	7	7					0.100	1.000	0.100	0.000
47	MXCG ▼	7	7					0.700	1.000	0.700	0.000
48	MNEG ▼	7	7	0.000	0			0.250	1.000	0.250	0.000
49	BLNK ▼	Operands for field 1.									
50	TRCX ▼		1	0.000	0.000	0.336	0.000	0.000	0.097	2.300E...	2.138E-05

Merit Function: 0.201157895256589

Figure 12.30 Updated image clearance operands for S7 in the MFE.

We're finally ready to optimize (Optimize > Optimize!). With the added (out of specification) distortion constraint, the initial merit function value jumped up to 0.20116. After a quick local optimization, the merit function is reduced to 0.00051, again a considerable improvement in performance. However, this time, the added boundary value operands keep the lenses manufacturable and keep the final lens clear of the image plane. The optimized design is listed in the LDE (Fig. 12.31). and the cross-section is plotted as Fig. 12.32. The increase in performance is evident in both the aberration curves and the spot diagrams (Fig. 12.33). Note that the scale of both plots has been reduced and the RMS spot radii are below 3 μm. Save this lens as **OSsecureCam4**.

Note: Now that we have a significant number of variables, it is likely that you will not get exactly the same values for radii, lens thicknesses, and airspaces as those shown in Fig. 12.31. There are many equivalent solutions with nearly the same level of performance. Your optimization result depends on the specific version of OpticStudio, the computer platform, the number of cycles that are run, etc. However, your optimized lens should "look" similar to the one shown in Fig. 12.32, and your lens performance should "match" Fig. 12.33.

	Surface Type	Comment	Radius		Thickness		Material	Clear Semi-Dia
0	OBJECT Standard ▼		Infinity		Infinity			Infinity
1	Standard ▼		Infinity		1.000			1.636
2	Standard ▼		1.569	V	0.536	V	N-BK7	0.981
3	Standard ▼		5.515	V	0.644	V	N-SF4	0.799
4	Standard ▼		2.318	V	0.298	V		0.468
5	STOP Standard ▼		Infinity		1.955	V		0.229
6	Standard ▼		4.012	V	0.644	V	N-SK16	2.110
7	Standard ▼		5.969	V	2.402	V		2.154
8	IMAGE Standard ▼		Infinity		-			2.800

Figure 12.31 The LDE for the OSsecureCam4 after optimization with boundary value operands.

280 Chapter 12

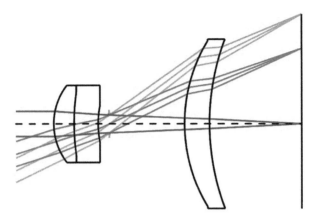

Figure 12.32 The OSsecureCam4 with boundary value operands and added distortion constraints after optimization.

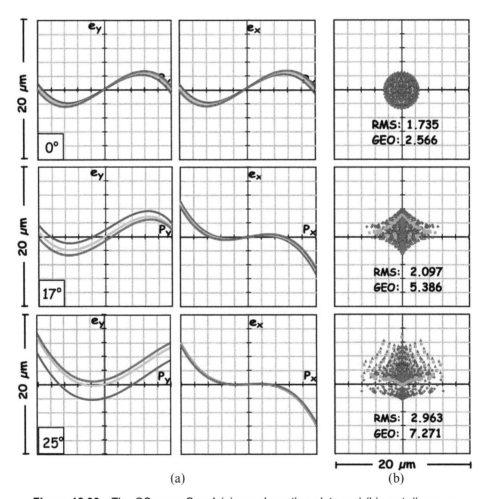

Figure 12.33 The OSsecureCam4 (a) ray aberration plots and (b) spot diagrams.

The field curvature and distortion plots for the OSsecureCam4 are shown in Fig. 12.34 plotted on the same scales as Fig. 12.24. There is a significant reduction in field curvature, and the distortion at full field has dropped from 8% (pincushion) to −5% (barrel).

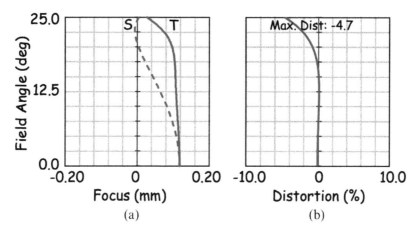

Figure 12.34 (a) Field curvature and (b) distortion plots for the OSsecureCam4 after optimization.

Finally, as is evident from Fig. 12.35, the MTF curves for all three fields are close to the diffraction limit and meet our design specification (remember that our specification for the design was an MTF greater than 50% at 50 cycles/mm for all fields). The lowest curve is the tangential curve for field 2 with a modulation of 54.3% at 50 cycles/mm (you can find the values on the text tab).

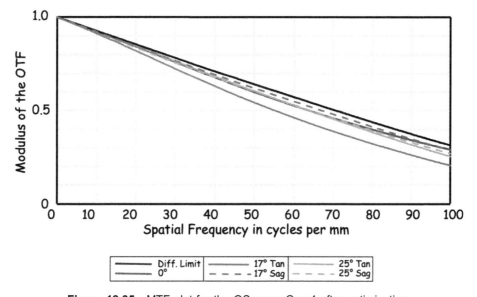

Figure 12.35 MTF plot for the OSsecureCam4 after optimization.

12.8 Step 5: Open Up the Lens

At this point, the lens performance is nearly diffraction limited. One way of looking at this is that the aperture stop is so small that the limitation on the lens performance is primarily diffraction (not aberrations). From a geometrical standpoint, the RMS spot sizes are small—much smaller than even the 7.5-μm sensor pixel in the 1/3" VGA format sensor.

To further increase the MTF, we need to open up the aperture to increase the diffraction limit. Opening this $f/10$ lens to $f/8$ (System Explorer > Aperture > Aperture Value) will result in an increase in the area of the stop of $(10/8)^2 = 1.25^2 = 1.56$, providing additional light to the sensor. This also opens the stop semi-aperture from 0.229 mm to 0.286 mm, as you can see when comparing the stop Clear Semi-Dia for the $f/10$ lens (Fig. 12.31) with that of the current design (Fig. 12.36). However, once you open up the lens, the design must be optimized again to try to correct the performance loss from the additional aberrations introduced by the larger aperture.

	Surface Type	Comment	Radius	Thickness	Material	Clear Semi-Dia
0	OBJECT Standard ▼		Infinity	Infinity		Infinity
1	Standard ▼		Infinity	1.000		1.714
2	Standard ▼		1.569 V	0.536 V	N-BK7	1.039
3	Standard ▼		5.515 V	0.644 V	N-SF4	0.869
4	Standard ▼		2.318 V	0.298 V		0.525
5	STOP Standard ▼		Infinity	1.955 V		0.286
6	Standard ▼		4.012 V	0.644 V	N-SK16	2.169
7	Standard ▼		5.969 V	2.402 V		2.205
8	IMAGE Standard ▼		Infinity	-		2.805

Title (OSsecureCam5); f/# (8); Field (0°,10°, 17°, 22°, 25°); Wavelength (F d C)

Figure 12.36 LDE for the OSsecureCam5 before optimization.

Before optimizing, one thing to notice about the design is that there are only two off-axis field points to cover a fairly large field. For example, Fig. 12.32 shows that the rays from the three field points (axis, 0.7, and full field) don't even come close to sampling the entire surface on each side of the field lens. Therefore, it makes sense to add at least two more intermediate field points to better sample and check the performance across the field. Open the field data editor (System Explorer > Fields > Field Data Editor) and insert two field points at 10° and 22°. The $f/8$ design with five fields is shown in Fig. 12.37. Title this lens OSsecureCam5.

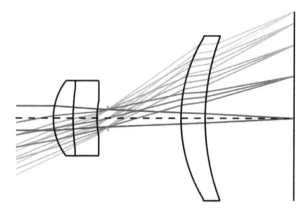

Figure 12.37 Lens cross-section for the OSsecureCam5 before optimization.

Because we increased the number of fields, we also need to update the default merit function before optimizing. Open the optimization wizard (Optimize > Optimization Wizard). Compare the current settings with Fig. 12.28. If nothing has changed (closing and reopening OpticStudio resets the settings for the Wizard back to their default values), then hit apply; otherwise, update the Wizard settings to match Fig. 12.28 and then hit apply. This generates two more sets of operands at the end of the merit function, one for field 4 and another for field 5.

This is also a good time in the design process to start thinking about the tolerance sensitivity (Chapter 13) of the design by checking the maximum ray angles at the surfaces. (In general, the tolerance sensitivity increases as the ray angles increase.) For example, the rays at the edge of the field in Fig. 12.37 appear to have very large angles of incidence on the field lens. To get an exact value, trace a real ray (Analyze > Rays & Spots > Single Ray Trace) for the full field ($Hy = 1$) and upper edge of the pupil ($Py = 1$). The real ray trace data output show an angle of incidence (Angle in) at surface 6 of 63.48°. This is a very large angle of incidence. Furthermore, most broadband AR coating designs start to lose performance at angles greater than 45°.

The good news is that we can control our ray angles by putting a limit on the angle of incidence during optimization with a new operand, MXAI, (MaXimum Angle of Incidence). In the MFE, insert a new operand below EFFL. Change the operand type to MXAI and give it a target of 50 and a weight of 1. This new operand holds the maximum ray angle of incidence for all surfaces (setting of 0), all wavelengths (setting of 0), and all fields (setting of 0) to be less than 50°. When you update the merit function, the current value for all rays is shown at 63.71°. The merit function at this point should have 288 total lines with the first 15 shown in Fig. 12.38.

	Type	Surf	Wave	Field	Symmetry	Data		Target	Weight	Value	% Contrib
								Merit Function:		1.76580404362062	
1	EFFL ▼		2					6.430	10.000	6.430	3.569E-15
2	MXAI ▼	0	0	0	0	0		50.000	1.000	63.710	99.999
3	BLNK ▼										
4	DMFS ▼										
5	BLNK ▼	Sequential merit function: RMS spot x+y centroid X Wgt = 1.0000 Y Wgt = 1.0000 GQ 3 rings 6 arms									
6	BLNK ▼	Default individual air and glass thickness boundary constraints.									
7	DIMX ▼	0		2	0			5.000	1.000	5.000	0.000
8	MNCA ▼	1		1				0.000	1.000	0.000	0.000
9	MXCA ▼	1		1				1000.000	1.000	1000.000	0.000
10	MNEA ▼	1		1	0.000	0		0.000	1.000	0.000	0.000
11	MNCG ▼	1		1				0.100	1.000	0.100	0.000
12	MXCG ▼	1		1				0.700	1.000	0.700	0.000
13	MNEG ▼	1		1	0.000	0		0.250	1.000	0.250	0.000
14	MNCA ▼	2		2				0.000	1.000	0.000	0.000
15	MXCA ▼	2		2				1000.000	1.000	1000.000	0.000

Figure 12.38 Top 15 lines of the merit function editor for the OSsecureCam5 before optimization with operand MXAI added.

The starting merit function value is 1.7658. After local optimization (Optimize > Optimize!), the merit function drops to 0.000806. The biggest change in the design is that the field lens has moved closer to the doublet. The distance between the stop (S5) and the front of the field flattener (S6) goes from ~2 mm to 1.7 mm, reducing the angle of incidence on its front surface. Save this lens as **OSsecureCam5**. The LDE for the lens is shown in Fig. 12.39, and the lens cross-section is shown in Fig. 12.40. The MTF for this faster design shows that opening up the lens raises the diffraction-limit curve (Fig. 12.41), but this also increases the aberrations; and, along with the imposition of the maximum angle of incidence constraint during optimization, the MTF is no longer nearly diffraction limited.

	Surface Type		Comment	Radius		Thickness		Material	Clear Semi-Dia
0	OBJECT	Standard ▼		Infinity		Infinity			Infinity
1		Standard ▼		Infinity		1.000			1.635
2		Standard ▼		1.692	V	0.460	V	N-BK7	0.989
3		Standard ▼		3.163	V	0.579	V	N-SF4	0.823
4		Standard ▼		2.124	V	0.319	V		0.537
5	STOP	Standard ▼		Infinity		1.693	V		0.303
6		Standard ▼		5.177	V	0.700	V	N-SK16	1.694
7		Standard ▼		30.779	V	3.182	V		1.801
8	IMAGE	Standard ▼		Infinity		-			2.774

Figure 12.39 LDE for the OSsecureCam5 after optimization.

Designing a Lens 285

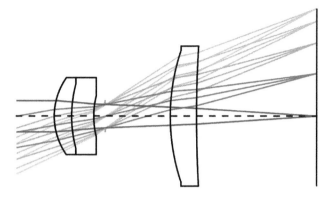

Figure 12.40 Lens cross-section for the OSsecureCam5 after optimization.

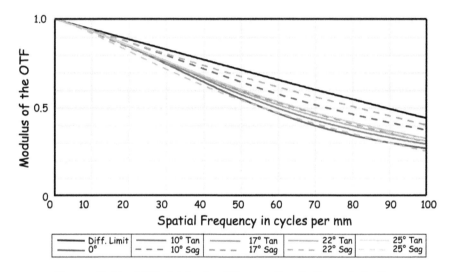

Figure 12.41 MTF plot for the OSsecureCam5 after optimization.

12.9 Step 6: Glass Substitution

At this point in the design process, if we wanted to further improve performance, we could try adding another lens; but before that we need to ask, "Have we really used *all* of our available variables?" What about the materials? We picked appropriate choices of glasses for our starting design, but are they really the best material choices for our final design? To answer these question, the glasses in the design should be allowed to vary just like their curvature and thickness, and then new combinations of glasses can be assessed as part of the optimization process. But, as you have seen in Fig. 9.2, the number of available catalog glasses plotted on the glass map is relatively small. All of the other points on the map represent glasses with possible combinations of refractive index and dispersion that are not in the commercial glass catalogs. So, it would seem impossible to optimize the material of a lens because the available glasses in the catalogs have discrete indices of refraction and dispersion.

OpticStudio has a nice feature called glass substitution that permits such glass variations. To enable glass substitution, click on the small cell next to N-BK7 in the material column of the LDE for the OSsecureCam5 and change the solve type from Fixed to Substitute in the drop-down menu. Leave the "Catalog" line entry blank. This then defaults to the existing catalog(s) defined in the system explorer (System Explorer > Material Catalogs). For our example, this is the entire Schott catalog. Do the same for the other two lens materials in the design. An "S" should now appear next to all of the glass names in the LDE (Fig. 12.42). Title this lens OSsecureCam6.

	Surface Type	Comment	Radius	Thickness	Material	Clear Semi-Dia
0	OBJECT Standard ▼		Infinity	Infinity		Infinity
1	Standard ▼		Infinity	1.000		1.635
2	Standard ▼		1.692 V	0.460 V	N-BK7 S	0.989
3	Standard ▼		3.163 V	0.579 V	N-SF4 S	0.823
4	Standard ▼		2.124 V	0.319 V		0.537
5	STOP Standard ▼		Infinity	1.693 V		0.303
6	Standard ▼		5.177 V	0.700 V	N-SK16 S	1.694
7	Standard ▼		30.779 V	3.182 V		1.801
8	IMAGE Standard ▼		Infinity	-		2.774

Figure 12.42 LDE for the OSsecureCam6 before glass substitution optimization.

To optimize with glass substitution, we do not need to create a new merit function (the ending merit function from Section 12.8 was 0.000806). However, we do have to switch from this local optimizer to one of their global optimizers (Hammer). The differences between local optimization and global optimization are discussed more in the text box "Global Optimization." The hammer optimization interface window is shown in Fig. 12.43. "Automatic" starts a local optimization (similar to Optimize!) and should be used before running hammer if the starting design has not been locally optimized. The Start button starts the global hammer optimization, while the Stop button pauses the optimization. As the optimization proceeds, the number of systems that the optimizer evaluates is constantly updated. If a new system is found with a lower merit function, that system is saved as the current system and the Current Merit Function value is updated accordingly. The execution time is also shown.

Start hammer (Optimize > Hammer > Start) and let it run through ~1 million systems (see Fig. 12.43) and then click the Exit button—this should take less than 2 minutes of execution time. Note: If you check the Auto Update box when you start Hammer, you can monitor the change in the lens shape in the lens layout during the hammer optimization, but this will slow it down. In most cases, running a short hammer will not exhaust the number of possible systems, but the time between updates to the merit function will become longer and longer, giving a good indication of when to exit the optimization.

Global Optimization

It is not uncommon for a lens to get stuck in a local minimum when optimizing with a local optimizer. So how do we get the optimizer unstuck? This is part of the art of the design process, and every experienced designer has their own set of techniques for getting out of local minima. For example, small changes in a radius or an air space before running the optimization will often send the design into a different design space. As a new designer, how do you gain that experience to know what to change and by how much? Try it and see what happens. You can always undo the changes and start over. Sometimes all it takes to get out of a local minimum is to add more field points (and therefore trace and evaluate more rays). For a broader search for better solutions, a more global approach is needed.

OpticStudio has two different global optimization algorithms, global search and hammer. While global search is beyond the scope of this text, hammer (lens designers often talk about "hammering" on a design to improve it) is an essential tool for exploring the design space and is particularly useful for varying materials with glass substitution. For simple systems, it can take just a few minutes to explore more than 1 million systems (a good rule of thumb for new users); however, for more complex lenses (with many more variables), it is not uncommon to need to let the hammer optimization "hammer" on the lens for several hours (usually overnight).

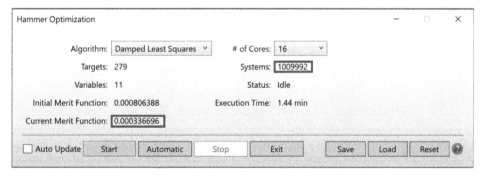

Figure 12.43 Hammer optimization window for the OSsecureCam6 after ~2 minutes of optimization.

In the example above, the merit function is about half of its original value when we exit Hammer. The final design (Fig. 12.44) looks very similar to its starting point (Fig. 12.40), but the lens materials (Fig. 12.45) have moved to higher-index (lanthanum) crown and flint glasses, and the lens performance has substantially improved. It's important to note that with this final hammer optimization, your results might significantly differ (e.g, different materials) from our results in this section because of the semi-random nature of how hammer explores the solution space. You can even see differences between two hammer optimization runs with identical starting points—try it for yourself.

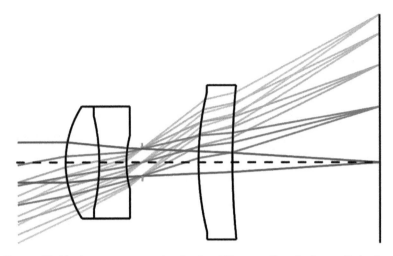

Figure 12.44 Lens cross-section for the OSsecureCam6 after optimization.

	Surface Type	Comment	Radius	Thickness	Material	Clear Semi-Dia
0	OBJEC Standard ▼		Infinity	Infinity		Infinity
1	Standard ▼		Infinity	1.000		1.777
2	Standard ▼		2.030 V	0.700 V	LASFN31 S	1.123
3	Standard ▼		-3.963 V	0.565 V	N-LASF46B S	0.933
4	Standard ▼		1.981 V	0.323 V		0.533
5	STOP Standard ▼		Infinity	1.173 V		0.272
6	Standard ▼		5.654 V	0.700 V	N-LAF34 S	1.393
7	Standard ▼		15.036 V	3.028 V		1.545
8	IMAGE Standard ▼		Infinity	-		2.971

Figure 12.45 LDE for the OSsecureCam6 after optimization.

The ray aberration curves and spot diagrams for the final system are shown in Fig. 12.46. All fields are now diffraction limited; the Airy disk circle has been added to the spot diagrams (Settings > Show Airy Disk) for confirmation. The RMS spot diameters are quite a bit smaller than the square pixel dimension (7.5 µm, or 0.0075 mm), leaving us room manufacturing errors.

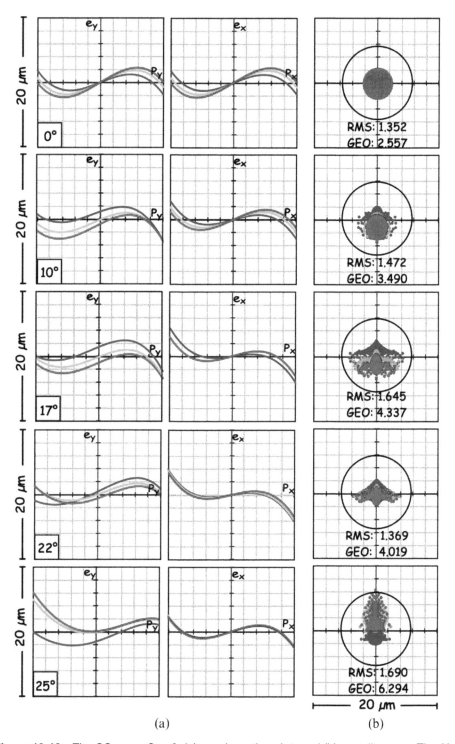

Figure 12.46 The OSsecureCam6: (a) ray aberration plots and (b) spot diagrams. The Airy disc is displayed as a black circle.

The MTF for the final design is shown in Fig. 12.47. All fields are now well above our 50% at 50 cycles/mm specification. Save this optimized design as **OSsecureCam6** so that we can use it to demonstrate how a complex optimized lens can be evaluated for errors due to manufacturing in the next chapter on tolerancing.

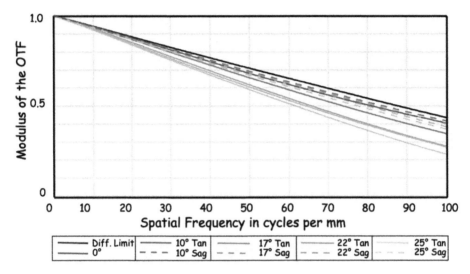

Figure 12.47 The OSsecureCam6: final MTF plot.

12.10 Wrap-up

In this chapter we built a lens "from scratch" so that we could thoroughly describe the design process. It is often easier to start from a classic design form with the right number of elements, a good negative/positive power balance for Petzval curvature correction, and a good selection of glasses for color correction. OpticStudio contains a large library of lenses that can be used as suitable starting points. Other lenses can be found in the patent literature. But now that we've designed a lens, is it possible to make one? That is the subject of our final chapter: Tolerancing.

Explorations

Exploration 12.1 Changes in the lens and the MTF with sensor size
Redo the design exercise in this chapter with a 2/3" VGA-format sensor (640×480). How do the specifications change? How does your final lens result compare?

Third Hiatus
Building a Lens

To properly tolerance a lens system (Chapter 13), you need to have a good understanding of how the lens will be built. This includes knowledge of how the lens elements will be fabricated, aligned into a mechanical housing, and tested. There are many ways to build a lens, ranging from traditional processes (completed primarily by hand) to modern, machine-driven [e.g., lens molding or computer numerical control (CNC) grinding and polishing] fabrication methods. While there are entire texts on this subject matter, we will briefly cover the more classical "by hand" methods here and leave the newer fabrication and testing techniques for you to explore on your own.

H3.1 Fabricating a Lens Element

Traditionally, a lens was made (often by hand) by grinding and polishing the lens on a rotating spindle. During the fabrication process, the lens parameters (e.g., radius and thickness) are tested to determine if they meet the design values. Ideally, each copy will be a replica of the optimized design.

Figure H3.1 illustrates the first steps in the fabrication of a singlet (OSsecureCam2). A disk of glass, appropriately called a **blank**, is attached to a mounting surface using a melted wax that hardens. A rotating metal tool with a convex surface is brought into contact with the blank [Fig. H3.1(a)]. This tool has been cut and polished to a high degree of accuracy to obtain a radius of curvature equal (but opposite in sign) to the radius of curvature of the second surface R_2 of the lens.

A slurry of abrasive metal oxides is applied to the surface, and the rotation of the tool against the blank grinds a rough surface. This process is called **generation**. Slurries with increasingly smaller grit sizes are applied to bring the surface from a "rough grind" to a "fine grind." The smaller the particle size in the slurry the less material is removed and the smoother the surface will be. Then the lens is demounted, flipped over, and reattached to the mount. A second set of grinding tools is used to generate R_1 [Fig. H3.1(b)]. In addition to generating R_1, this grinding stage must also obtain the correct thickness of the lens.

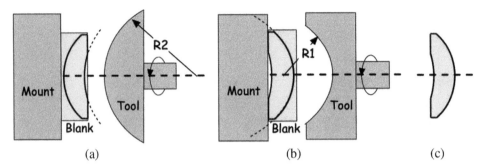

Figure H3.1 Initial steps in the fabrication of a singlet lens: (a) The blank is attached to a mount to grind a surface with a radius of curvature R_2. (b) The blank is reversed and reattached to be ground for radius R_1. (c) The lens is detached and sent to polishing.

Then the process is repeated with tools of the same radii of curvature but coated with a hard wax (also called pitch) and charged with a very fine abrasive to polish the lens to an optical finish. This process is called **pitch polishing**. During the grinding and polishing steps, the lens thickness and surface curvature are constantly tested to ensure that progress is being made on the lens and the dimensions are approaching the design values.

The progress of the polishing process is measured using a test plate, which is a polished piece of glass having a radius of curvature equal (but with opposite sign) to the design value (Fig. H3.2). The test plate is placed on the surface and illuminated with monochromatic light. The interference of light between the lens surface and the test plate surface produces fringes. The difference between the two radii can be calculated from the wavelength of light and the number of fringes that are visible over the diameter of the test plate surface. Because each fringe represents one half wavelength, the departure from the required surface radius at the edge of the lens shown in Fig. H3.2 is 1.5λ (three fringes are visible over the aperture). Note that the irregularity (difference from a sphere) of the surface can also be monitored/measured with the test plate by examining the shape of the fringes.

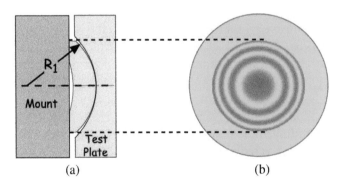

Figure H3.2 Testing a lens: (a) A test plate is placed against the R_1 surface and illuminated with monochromatic light. (b) The number of fringes in the resulting interference pattern is a measure of the error in the surface curvature of the part.

As the lens is being ground and polished to the correct shape, the lens thickness is also changing. A highly sensitive micrometer can be used to measure the center thickness of the lens. If the thickness value is not acceptable, the grinding (or polishing) continues while maintaining the radius of curvature until both the thickness and radius values are within specification. A final set of actions grinds the edge of the lens to produce the correct diameter (and mounting bevels) and locate the optical axis in the center of the lens to prevent wedge error (discussed further in Chapter 13). This process is known as **centering**. Once the lens has been centered, it is usually sent to the coating department to have an anti-reflection (AR) coating applied to each polished surface before final assembly.

H3.2 Mounting the Lens

Once the elements have been coated, they must be mounted into a mechanical housing to set the proper spacing between the lens elements and align their optical axes to each other. Low-cost assemblies are typically mounted in a cylindrical barrel with metal spacers between elements and a threaded retainer ring to hold the assembly together. For example, a possible mount for our OSsecureCam2 singlet with an external stop is shown in Fig. H3.3. If there is no wedge error in the element, the optical axis (OA) of the lens surfaces is the same as the lens axis (LA) through the center of the lens. There is also a mechanical axis (MA) of the mount, which is typically referenced to the outside diameter of the barrel (or an external flange surface). It is important to align the optical axis of the lens elements to this mechanical reference surface to ensure the best possible performance of the lens assembly. In a perfectly assembled lens, the optical axis of the lens surfaces, the lens axes of the elements, and the mount axis all coincide, as depicted in Fig. H3.3. In reality, assembly errors such as element tilt and element decenter can occur, introducing aberrations and degrading performance. These will be discussed further in Chapter 13.

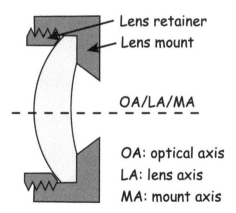

Figure H3.3 Example lens mount for OSsecureCam2.

H3.3 Testing the Lens

Many lens assemblies are performance tested after the lens has been completely assembled. If the lens passes, it is packaged for transport to the end user. If it fails, it is either thrown out or sent back for adjustment. Modern metrology systems use lasers and interferometry to measure lens performance and provide aberration feedback (e.g., Zernike polynomials) for active alignment.

In-process measurement technology has also improved quite a bit beyond the use of test plates and micrometers. For example, interferometric non-contact measurements of both radius and center thickness are common and reduce the chance of damaging the surface. In some cases, these methods have been incorporated directly into the CNC grinding and polishing machines, allowing for nearly continuous monitoring of the fabrication processes. Mechanical and laser-based probes can be used to measure the alignment of the lens surfaces during mounting. However, these types of measurements take time and effort (driving up the cost of the assembly).

Any deviation from the design prescription is expected to reduce the performance of the lens. After all, the whole point of optimization is to find the optimal solution. So here is the dilemma: it would be prohibitive in both time and cost to try to make all of the lens parameters perfect, but how close to perfect do we need to make them? Do we need to fabricate 1000 lenses and then correlate their measured lens parameter deviations with their measured final performance? Luckily, we don't need to do this because OpticStudio can simulate the complex fabrication process and its errors. Users can then determine the precision needed to make a batch of lenses that will produce a large fraction of acceptable lenses. This process will be addressed in Chapter 13.

Chapter 13
Tolerancing

So, an optimized security camera has been designed. But can it be built? For the performance of a manufactured lens system to be as good as the nominal design performance, it must be fabricated and assembled as close as possible to the nominal optical design prescription. The question is how precisely must the various lens parameters be made? If you copy (and paste) any cell value in the LDE, you will see that each parameter is expressed to 15 places beyond the decimal point. Although this level of precision is not needed to specify the lens parameter, the lens parameter must be made (see the Third Hiatus) to an accuracy or **tolerance** that will be good enough to meet the performance specifications. This provokes a third question: How do you determine what tolerances are "good enough?" For example, how carefully must we make the radius of curvature of a lens element to meet a specified RMS spot diameter across all fields?

The answers to these questions can be provided through a general process called **tolerancing**. While it is relatively easy to model one specific lens error, it is much more difficult to think about how multiple lens errors, all occurring at the same time (with random values within their tolerance ranges), will affect the final lens performance. This is where we turn to software and statistical tolerancing algorithms to predict the probability of producing a set of lenses with the desired performance (at a desired cost). In this chapter we will use our two security camera design examples from Chapter 12 to demonstrate how to use OpticStudio® to find appropriate tolerance values for an optical system, predict as-built performance, and then communicate our tolerances to a manufacturer using an optical drawing.

13.1 Statistical Tolerancing

The first step in any tolerancing process is to define a set of starting tolerance values for the various manufacturing errors. These are the ranges, usually symmetrical about the design value, by which the lens parameters can vary during manufacturing. For example, a tolerance value of ± 25 μm on the thickness of a 0.5-mm-thick lens would require a lens to be fabricated with a thickness between 0.475 and 0.525 mm.

Once an initial set of tolerance values has been selected, we need to understand how sensitive each of the lens parameters is to change by evaluating how much that change would degrade the lens performance. Both positive and negative changes in lens parameters are evaluated (one tolerance at a time). After determining the effect that each individual tolerance has on lens performance, the tolerances can be sorted according to their impact on performance. Sorting the tolerances in this manner helps determine which of the tolerances might need to be tightened later. This type of investigation is called a **sensitivity analysis**. A sensitivity analysis also helps identify potential **compensators**. Compensators are lens parameters that can be used to compensate for the performance loss caused by manufacturing errors. For example, adjusting the image plane distance (or sensor focus position) is a compensator that is almost always used to improve the as-built performance of a lens system.

However, this analysis does not tell the whole story as it ignores the interaction between multiple lens errors. For example, consider a lens system with both a radius error and a thickness error. The performance loss from the radius error might simply add to the loss from the thickness error, or the two errors could even cancel each other out (resulting in a lens with the nominal design performance). Usually, the impact of two tolerance errors is somewhere between these extremes. How then can we have any confidence that we can make any lens assembly (with a large number of tolerances) with acceptable performance? This is where statistical tolerancing comes in.

The most common statistical tolerancing algorithm is a **Monte Carlo** procedure. This procedure generates a large number of lens configurations with lens parameters (e.g., radius, thickness, and airspace) whose values are distributed randomly within their assigned tolerances ranges. Each lens configuration is called a **run** or a **trial**, and the use of a random distribution of lens parameters is the "Monte Carlo" part of the analysis. Then, the as-built performance (after compensation) of each perturbed configuration is analyzed. The data for all of the trials are collected and used to predict a statistical probability for the lens to achieve a particular performance with the assigned tolerances and compensators. While this method is very robust, it can take a long time (several hours, or even days if the system is very complex) to generate and evaluate enough trials (e.g., 5000) for statistical significance.

The primary goal of any statistical tolerance process is to simulate the build of a batch of lens systems, evaluate the performance of each system, and then compute the fraction of acceptable lenses in the batch, called the **yield.** For example, for a Monte Carlo analysis that tracks RMS spot radius, you can sort the trials by performance, starting with the smallest value and ending with the largest value. If you then plot the number of trials N with a specific RMS spot radius, the result will be a set of bell-shaped curves (one for each field), similar to the one shown in Fig. 13.1(a). You can also convert this data directly into a % yield curve [Fig. 13.1(b)] that plots the percentage of the total cases that can meet some maximum RMS spot radius. The data points shown on the graphs represent values derived from standard statistical modeling. These include the mean (i.e., 50% of systems will have this performance change or less), mean $+ 1\sigma$ (84%), mean $+ 2\sigma$ (98%), and mean $+ 3\sigma$ (99.9%).

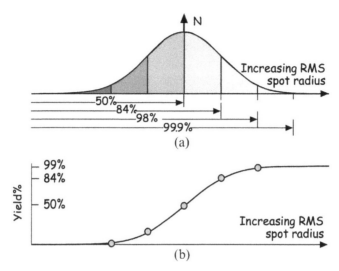

Figure 13.1 Construction of a yield curve: (a) distribution of toleranced lenses as a function of increasing RMS spot radius and (b) % yield curve.

13.2 The Tolerance Data Editor

In Chapter 10, we used the three main icons on the Optimize tab (the Merit Function Editor, the Optimization Wizard, and Optimize!) to establish a set of optimization operands, calculate a merit function, and then optimize the lens. The Tolerance tab in OpticStudio (see Fig. 13.2) has an analogous set of icons for tolerancing, called the Tolerance Data Editor, The Tolerance Wizard, and Tolerancing. These icons are used to define an appropriate set of tolerances (called **tolerance operands**), explore tolerance sensitivity, and predict yield with a Monte Carlo analysis. Like optimization, tolerancing is a complex process so we'll break it down by first introducing the **Tolerance Data Editor (TDE)** and discussing how to enter the various tolerance operands into the TDE.

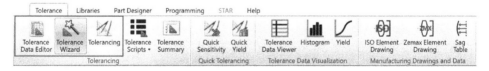

Figure 13.2 The Tolerancing Ribbon.

Let's start by looking at the security camera from Section 12.5, OSsecureCam2, a singlet with an external stop (Fig. 13.3). With just one lens and its stop, the number of tolerances for this lens is small. This permits us to demonstrate the various tolerancing features without being overwhelmed by the large amounts of data, lengthy tables, and long run-times required for more-complex designs. Restore your saved **OSsecureCam2** from Chapter 12 or key in the values given in the LDE in Fig. 13.4. If you restored the lens from a saved file, delete the 1-mm offset surface because it is not needed for tolerancing. In fact, it would only add extra, unwanted tolerances.

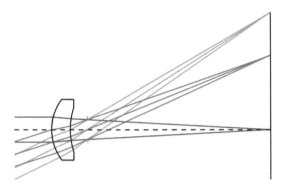

Figure 13.3 The OSsecureCam2 cross-section.

	Surface Type		Comment	Radius	Thickness	Material	Clear Semi-Dia
0	OBJECT	Standard ▼		Infinity	Infinity		Infinity
1		Standard ▼		1.265 V	0.500	N-BK7	0.782
2		Standard ▼		1.768 V	0.467 V		0.571
3	STOP	Standard ▼		Infinity	4.851 V		0.255
4	IMAGE	Standard ▼		Infinity	-		3.031

Title (OSsecureCam2); f/# (10); Fields (0°, 17°, 25°); Wavelength (F,d,C)

Figure 13.4 LDE for the OSsecureCam2 with the extra visualization surface deleted.

The Tolerance Data Editor displays all of the active tolerances to be used in tolerancing. When you open a new TDE (Tolerance > Tolerance Data Editor), there are no tolerance operands displayed, just a place keeper, TOFF (Fig. 13.5). Like the entry of optimization operands in the MFE, four-letter tolerance operands can be entered in the TDE by selecting the Type of operand from the drop-down menu (or typing the operand name directly).

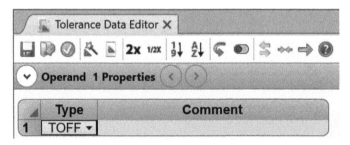

Figure 13.5 An empty Tolerance Data Editor with no tolerance operands.

For example, enter a ±50-μm tolerance on the radius of the first surface of the lens (Surf 1) using the TRAD operand by entering "TRAD" in the Type column and 1 in the Surf column. OpticStudio automatically fills in the radius of curvature in the "Nominal" column. The "Min" and "Max" tolerance values can then be entered by the user. Depending on the operand, it may also have a special instruction (Code) associated with it. For this example, enter "−0.05" in the Min column and "0.05" in the Max column and set the Code to "0." The resulting TDE is shown in Fig. 13.6.

	Type	Surf	Code	Nominal	Min	Max	Comment
1	TRAD ▼	1	0	1.265	-0.050	0.050	

Figure 13.6 Radius-of-curvature tolerance operand, TRAD, given in millimeters.

Other tolerances can be entered into the TDE in a similar manner (one by one). For example, a simple singlet like the OSsecureCam2 has around 15 tolerances that need to be analyzed. But what are they and how are they entered into the TDE? To help answer these questions, the most common tolerance operands (and how they apply to the OSsecureCam2) will be described in the next two sections on element fabrication errors and lens assembly errors. After that, the Tolerance Wizard is introduced in Section 13.6 to help automate the entry of tolerances (saving a significant amount of time for more complex systems).

13.3 Element Fabrication Errors

This section defines a set of tolerances associated with the fabrication process of a single lens element (described earlier in the third Hiatus, H3.1). These include the radii of curvature, the thickness of the element, and the material properties of glass. Even if the surface curvatures and the thickness of an element were made perfectly, its performance can still be compromised by two other lens shape errors. One error breaks the axial symmetry of the lens (wedge), and the other breaks the spherical symmetry of the lens surfaces (irregularity).

Tolerances that don't break the rotational symmetry of the system (like radius or element thickness) primarily cause a focus shift that is compensated by moving the sensor. However, if the fabrication error is large enough, spherical aberration may be introduced that will begin to degrade the image quality. Tolerances that break rotational symmetry (like wedge) primarily cause a pointing error (or boresight error), and if the fabrication error is large enough, coma can be observed. If needed, more complex compensation like space changes (to correct the spherical) and element push-arounds (to correct the coma) can be implemented during lens testing to improve lens performance.

13.3.1 Radius of curvature: TRAD and TFRN

The radius of curvature of a lens can be toleranced in several different ways. If the tolerance is measured in lens units (e.g., millimeters) the tolerance operand is TRAD (as shown above in Fig. 13.6). The radius tolerance can also be specified as a percentage of the nominal radius with TRAD by changing the operand Code (in the second column of the TDE) from 0 to 1. Figure 13.7 shows an example where the radius tolerance is ±5% of the nominal radius.

	Type	Surf	Code	Nominal	Min	Max	Comment
1	TRAD ▼	1	1	1.265	-5.000	5.000	

Figure 13.7 Radius-of-curvature tolerance operand, TRAD, given in % radius.

If the radius of curvature is measured using interferometry (e.g., with a test plate during fabrication), an alternative way to specify the radius tolerance is in units of fringes with the TFRN operand. The number of fringes is measured at the maximum semi-diameter of the surface at the test wavelength specified by the operand TWAV. In the example below (Fig. 13.8), the wavelength of a HeNe laser (0.633 μm) is specified by the first operand with a ±7 fringe tolerance on the radius of the first surface (operand 2).

	Type		Nominal	Wave		Comment
1	TWAV ▼			0.633		
2	TFRN ▼	1	0.000	-7.000	7.000	

Figure 13.8 Radius-of-curvature tolerance operand, TFRN.

You might ask, why are there so many ways to tolerance the radius of curvature? If a lens element is made using a classic fabrication process (see the Third Hiatus), radius-of-curvature errors can arise from two sources, the fit to the test plate (known as the power error and given in fringes) and the error in making the test plate radius of curvature. This would require the use of both TFRN and TRAD. However, more modern measurement techniques do not require a test plate to measure radius of curvature, and in this case, only one radius tolerance is needed. So, which one do you use? Tolerancing with fringes (or % radius) is typically preferred as it provides a more uniform sensitivity response for both small and large values of radius rather than tolerancing directly on the radius of curvature in lens units.

13.3.2 Thickness: TTHI

The tolerance operand for element thickness is TTHI. This operand represents the range of center thickness values that an element can have after fabrication. Like TRAD, the TTHI operand has a surface number, nominal value, and min and max tolerance values. Figure 13.9 shows an example of a thickness tolerance for surface 1 where the thickness tolerance is set at ±0.1 mm.

	Type	Surf	Adjust	Nominal	Min	Max	Comment
1	TTHI ▼	1	1	0.500	-0.100	0.100	

Figure 13.9 Element thickness tolerance operand, TTHI.

It is important to note that the TTHI operand can be used to tolerance both the center thickness of a lens *and* the absolute position of a lens element in an opto-mechanical assembly. For some lens systems, such as those assembled on fixed lens seats [Fig. 13.10(a)], a change in element thickness will cause an equal (but opposite sign) change in the airspace following the first element

[Fig. 13.10(b)]. If, instead, the lenses are assembled by stacking elements with a spacer in a tube [Fig. 13.10(c)], a change in element thickness has no impact on the airspace following the element because everything after the element is shifted by an amount equal to the tolerance [Fig. 13.10(d)].

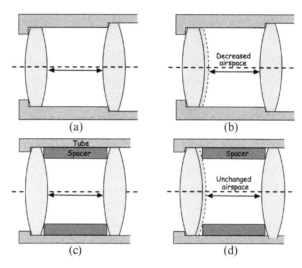

Figure 13.10 Opto-mechanical assembly methods and the impact of an element thickness change on the airspace following the element with (a) two lenses mounted on fixed seats, (b) decreased airspace, (c) two lenses separated by a spacer, and (d) an unchanged airspace.

Both types of opto-mechanical assembly methods can be modeled during tolerancing using the Adjust code in the TTHI tolerance operand. For example, in Figure 13.11, the TTHI operand for surface 1 adjusts the airspace on surface 2 so that the total length of the assembly stays constant. If the Adjust value is set to the same surface number as the tolerance operand (as shown in Fig. 13.9), then the only thing changed by the thickness tolerance is the element thickness itself.

	Type	Surf	Adjust	Nominal	Min	Max	Comment
1	TTHI	1	2	0.500	-0.100	0.100	

Figure 13.11 Element thickness tolerance operand, TTHI (with an airspace adjustment surface).

13.3.3 Material: TIND and TABB

The two common tolerances applied to a material are the index of refraction (TIND) and V-number (TABB). Material tolerances are typically set by the manufacturer and can be found in the glass company's catalog. For example, standard glass tolerances for N-BK7 are shown in Fig. 13.12 with a tolerance of ± 0.0005 on the index of refraction and ± 0.642 (or 1% of the nominal value) on the V-number. Depending on the annealing processes used to make the glass, the refractive-index and *V*-number tolerances can be tightened but at considerable cost. *V*-number (TABB) tolerances are also only valid for visible wavelength optical systems and depending on the size of the tolerance can impact the chromatic aberrations in the lens.

	Type	Surf	Nominal	Min	Max	Comment
1	TIND	1	1.517	-5.000E-04	5.000E-04	
2	TABB	1	64.167	-0.642	0.642	

Figure 13.12 Index-of-refraction (TIND) and *V*-number (TABB) tolerances.

13.3.4 Wedge: TSTX, TSTY (or TIRY, TIRX)

When spherical surfaces are ground on each side of the lens blank, an optical axis is established and defined by the line between the centers of curvatures of the two surfaces. However, if the edge of the lens is ground while it rotates on an axis that is not the optical axis, the optical axis (OA) of the lens element does not align with the axis through the center of the lens apertures, its lens axis (LA). One way to describe this type of lens error is that the two spherical surfaces are tilted with respect to each other, and the space between them, instead of being defined by two parallel surfaces, is a **wedge** (see Fig. 13.13).

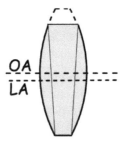

Figure 13.13 Example of wedge error in a lens element.

The amount of wedge in a lens element can be quantified by measuring the angular deflection of a laser beam sent through the center of the lens or by measuring the edge thickness difference (ETD) between the top of the lens and the bottom of the lens. However, both methods can be difficult to implement in practice, especially when trying to accurately measure small changes in angles or small changes in edge thickness.

A third way to measure wedge starts by mounting the lens in a rotating fixture (Fig. 13.14). The lens axis is first aligned to the rotating fixture by sliding the lens (about the first surface of the lens) on the lens mount until the outside diameter of the lens is centered on the axis of rotation. Then a stylus is placed at the second surface of the lens (typically at the edge of the clear aperture). The amount of change in the stylus position (as the lens is rotated) is referred to as total indicator runout (TIR). For small wedge angles, the measured TIR can be quickly converted to a surface tilt (in radians) by dividing the TIR by the clear aperture over which the TIR was measured.

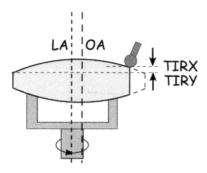

Figure 13.14 Measurement of wedge using TIR.

OpticStudio has two different wedge tolerance operands. For a TIR measurement, use TIRX and TIRY, where the values are given in lens units. For a wedge error given by a surface tilt (measured in degrees), use TSTX and TSTY. Figure 13.15 shows the entry of both types of tolerance operands in the TDE with a ±0.1° surface tilt (TSTX and TSTY) and a ±25-μm TIR (TIRX and TIRY) on surface 2. In practice, only one type of tolerance is needed (depending on how the lens wedge is specified and measured during fabrication).

	Type	Surf	Nominal	Min	Max	Comment
1	TSTX	2	0.000	-0.100	0.100	
2	TSTY	2	0.000	-0.100	0.100	
3	TOFF					
4	TIRX	2	0.000	-0.025	0.025	
5	TIRY	2	0.000	-0.025	0.025	

Figure 13.15 Two different types of wedge tolerance operands.

13.3.5 Irregularity: TIRR

Because of the way spherical surfaces of a lens are fabricated, it may be difficult to see how anything other than a spherical surface could result from the process. However, no process is perfect, and complex variations (known as irregularity) from a spherical shape can occur. Irregularity tolerancing is more problematic than other types of tolerancing because the irregularity is not deterministic (like a change in radius) but, by nature, is random. Therefore, some basic assumptions about the type of irregularity are made to model irregularity and then perform the tolerance analysis.

Figure 13.16 shows the shape of two common irregularity errors, astigmatism and spherical. The first error, astigmatism, is one of the simplest departures from a spherical surface, where the surface is a section of a toroid wherein the surface curvatures are different in the x and y directions. The second error, spherical, maintains rotational symmetry, but the surface shape is no longer a simple sphere, having a single "ripple" in it (higher-order spherical errors have more ripples). Figure 13.16(c) shows what the surface irregularity would look like with a combination of the two errors in equal amounts.

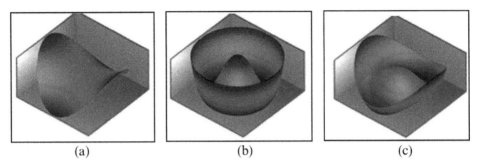

Figure 13.16 Surface irregularity errors: (a) astigmatism, (b) spherical, and (c) both errors combined in equal amounts.

OpticStudio's irregularity tolerance operand is TIRR (Fig. 13.17) and is composed of half spherical and half astigmatism. With effort, more-complex surface irregularity shapes can be modeled and toleranced (e.g., TEZI is a powerful Zernike surface irregularity tolerance operand), but, in general, this is a good place to start. The Min and Max values are the irregularity in units of fringes measured at the maximum semi-diameter of the surface. The TWAV operand is used to define the test wavelength. During a Monte Carlo analysis, the angle of the astigmatism is chosen randomly between 0° and 90°. This enables the simulation of a randomly oriented astigmatic error, which is less severe and more realistic than placing all of the astigmatism along the y axis of each element.

	Type	Surf	Nominal	Min	Max	Comment
1	TWAV			0.633		
2	TIRR	1	0.000	-2.000	2.000	

Figure 13.17 Irregularity tolerance operand, TIRR.

13.4 Lens Assembly Errors

At this point, we've discussed tolerances associated with an individual lens element. This lens element (and any other components associated with the full lens assembly) must be enclosed in a mount that can be handled as a single unit and be located within a completed device using the mount's reference surfaces. Therefore, the assembly of the lens element (or multiple lens elements) within its mounting structure must also be toleranced. This is to account for any tilt or displacement errors that a component suffers during assembly in relation to other components.

To isolate the mounting errors from lens fabrication errors, let's assume that the lens element is made properly and there are no shape errors (e.g., wedge or irregularity). So, the optical axis of the lens element is the same as the lens axis. For a perfectly assembled lens, the lens axis (LA) and the mount axis (MA) all coincide, as depicted in Fig. 13.18(a). The opening in the mount for the lens must be slightly larger than the lens. Otherwise, it is physically difficult to slide the lens into the mount. However, if the mount opening is too large or the assembly is not done carefully, assembly errors such as an element decenter [Fig 13.18(b)] or element tilt [Fig. 13.18(c)] can occur. Since these types of tolerance errors "break" the optical axis of the system, they are modeled in OpticStudio with coordinate breaks during tolerancing (see the text box "Understanding Coordinate Breaks" for more information on their implementation).

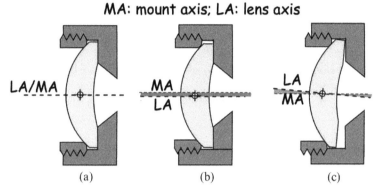

Figure 13.18 Lens assembly: (a) perfect alignment, (b) element decenter, and (c) element tilt.

When modeling an element tilt or an element decenter during tolerancing, OpticStudio automatically inserts coordinate breaks into the lens system as required to model the perturbation. For example, element decenter is achieved by inserting a pair of coordinate breaks around the element (one to decenter the lens and another to bring the coordinate system back to its original location). Note: It is usually a good idea to check the coordinate breaks in any tolerance model for accuracy to make sure they model the perturbation(s) you intended.

13.4.1 Element decenter: TEDX, TEDY, and TEDR

In the example shown in Fig. 13.18(b), the LA was displaced vertically from the MA. The tolerance operand TEDY can be used to model this type of assembly error. This operand represents a displacement of a lens element perpendicular to the optical axis in the Y direction. To enter this tolerance in the Tolerance Data Editor, Surf1 and Surf2 indicate the first and last surfaces of a lens element (or group of elements) to decenter. Min and Max specify the minimum and maximum decentration in lens units. Figure 13.19 shows an element decenter tolerance (in y) of ± 100 μm.

	Type	Surf	Surf2	Nominal	Min	Max	Comment
1	TEDY	1	2	0.000	-0.100	0.100	

Figure 13.19 Element y decenter tolerance operand (TEDY).

Understanding Coordinate Breaks

Coordinate breaks are special surfaces that are used to define a new coordinate system in terms of the current one. Coordinate break surfaces are not drawn and are always considered to be "dummy" surfaces for ray tracing. A coordinate break can be entered manually in the LDE by inserting a new surface and changing the surface type from Standard to Coordinate Break in the Surface Properties dialog box. However, OpticStudio also has a handy Tilt/Decenter Element tool on the LDE menu bar (look for the icon that looks like a "+" sign made of double-sided arrows) that inserts coordinate breaks to tilt and/or decenter a range of surfaces once the required values are entered into a dialog box. This tool is especially useful for modeling complex coordinate breaks where the number of coordinate break surfaces and other dummy surfaces needed depends on the specific tilt and decenter that is requested.

Coordinate breaks have seven inputs to describe the new coordinate system: x decenter, y decenter, tilt about x, tilt about y, tilt about z, a thickness (decenter in z), and a flag (Order) to indicate the order of tilting and decentration.

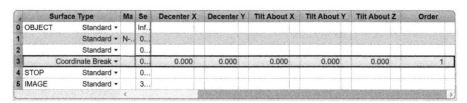

Coordinate break decenters are always specified in lens units, while coordinate tilts are specified in degrees. When applying the coordinate break, OpticStudio first decenters in x and y if the order flag is set to zero. Then OpticStudio tilts about the local x axis, then tilts about the new y axis, then finally tilts about the new z axis. If the "order" flag is any other value (such as unity), then the tilts are done first, in the reverse order (z, then y, then x), and then finally the decenters are done. This "order" flag is extremely useful because a single coordinate break can undo an earlier coordinate break, even for compound tilts and decenters (like those used in tolerancing). The thickness of a coordinate break surface defines the position of the following surface in the new coordinate system and is applied last regardless of the order flag.

The other two decenter tolerance operands TEDX and TEDR are entered in a similar manner to analyze the decentration of a lens group in X or a random radial direction, respectively. TEDR is recommended for systems with rotationally symmetric optics and mounts such as traditional camera lenses. TEDX and TEDY operands are recommended either for anamorphic systems or for systems with rectangular optics and rectangular mounts.

13.4.2 Element tilt: TETX, TETY

If the element's lens axis is tilted with respect to its mechanical axis (or the lens axes of the other components), alignment errors will occur. For example, the image will no longer be centered on the sensor. If the element tilt is large enough, aberrations may occur that degrade the lens performance. This type of tilt can arise from either poor assembly of the lens in the mount, as shown in Fig. 13.18(c), or due to a manufacturing error in the mount itself.

The tolerance operands TETX and TETY are used to analyze tilts of either a surface or a lens group about the X or Y axes, respectively. In the tolerance data editor Surf1 and Surf2 indicate the first and last surfaces of an element (or group of lens elements) to be tilted. The default pivot point of the tilt is about the front vertex of the lens group, but this can be changed if needed by the appropriate addition of a dummy surface. Min and Max specify the minimum and maximum tipping angles in degrees. The example in Fig. 13.20 shows an element tilt tolerance (in y) of $\pm 0.5°$.

	Type	Surf	Surf2	Nominal	Min	Max	Comment
1	TETY	1	2	0.000	-0.500	0.500	

Figure 13.20 Element y tilt tolerance operand (TETY).

13.4.3 Airspace: TTHI

Lens displacement errors measured along z, the optical axis, (referred to as air space errors) are toleranced with the same tolerancing operand as the element thicknesses, TTHI. Again, it is important to check (and change) the Adjust surface to match the expectations from the opto-mechanical assembly. For example, if the lens elements are separated using mechanical spacers, the Adjust surface should be set to the same value as the airspace being toleranced because this will shift subsequent elements just as a spacer would. If lenses are mounted on machined flanges within a barrel, a different Adjust surface may be more appropriate. Just like all tolerances described here, it is important to understand the opto-mechanical design and the fabrication and assembly processes so that the most appropriate tolerances can be applied.

13.5 The Tolerance Wizard

In the previous two sections, we used the Tolerance Data Editor to introduce some common tolerance operands for potential fabrication and assembly errors. In effect, these operands answer the question, "What could possibly go wrong?" A basic singlet like the OSsecureCam2 has around 15 total tolerances. However, as elements are added to a lens assembly, more errors are introduced and the number of things that can go wrong skyrockets. A three-element lens system, like OSsecureCam6, has more than 50 tolerances. Entering in these tolerances one at a time would be tedious, and it would be easy to miss a few. Luckily, OpticStudio has a **Tolerance Wizard** that can quickly generate a set of starting tolerances for the current lens system.

If you don't have it open, restore your saved **OSsecureCam2** (deleting the offset surface) or key in the values given in the LDE in Fig. 13.4. To open the tolerance wizard, click on its icon in the tolerance menu bar (Tolerance > Tolerance Wizard). You can also get to the wizard from the TDE by selecting the Operand Properties drop-down menu (just above the Type column of the TDE) and clicking on the tolerance wizard tab. The tolerancing operation in OpticStudio resembles optimization in that the tolerance operands and Tolerance Wizard share a general window (but different tabs), just as the optimization operands and the Optimization Wizard did. Fig. 13.21 shows the Tolerance Wizard tab with its default settings.

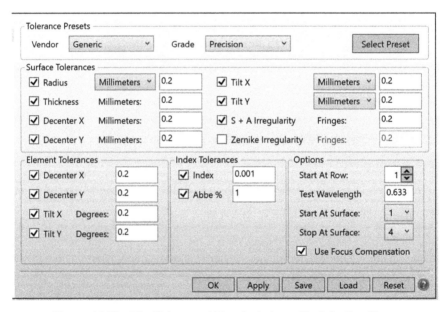

Figure 13.21 The Tolerance Wizard window with default settings.

From here, users can select the tolerances they want to use (or not use) for their lens system by checking (or unchecking) the box next to a tolerance and then entering an appropriate value for each tolerance. The tolerances are grouped into surface, element, and index tolerances and will be applied uniformly to every surface in the range between the "Start at Surface" and "Stop at Surface" (under Options at the bottom right of the window). The default/starting value for all tolerances is somewhat arbitrarily set at "0.2" with every tolerance active (checked) except Zernike Irregularity. It is also assumed that focus compensation is allowed during tolerancing as the "Use focus compensation" box is checked.

To help in selecting appropriate tolerance values, the tolerance wizard has several **tolerance presets** to choose from. To use these presets, a vendor (e.g., Asphericon, Edmund Optics, Generic, LaCroix, or Optimax) and a grade (e.g., Commercial, Precision, or High Precision) must be chosen from their respective drop-down menus. The tolerance values for the "Generic" vendor are reasonable starting values for each grade. However, they are not associated with a specific manufacturing company or process and, therefore, should be verified with your actual vendor. Changing from commercial to precision to high precision decreases the tolerance values but can significantly increase cost. One handy feature is that you can also create your own tolerance set, Save it, and Load it any time you want to use it. For our example, choose Generic for Vendor and Commercial for Grade from the drop-down menus of Tolerance Presets. Once the "Select Preset" button is pushed, the tolerance values in the window will be updated accordingly (Fig. 13.22).

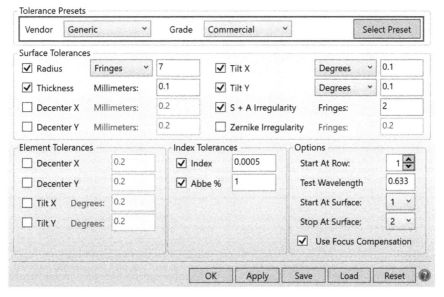

Figure 13.22 The Tolerance Wizard window after the Generic/Commercial Preset is selected.

Note that the element tolerances (e.g., decenter and tilt) after applying the preset are now all unchecked. This is because the majority of the preset vendors primarily make lens elements and not lens assemblies. This means that assembly tolerances like element decenter and tilt need to be manually added based on your opto-mechanical design. If the mechanical design is not complete yet (which is usually the case during the initial stage of tolerancing), it is up to you to make some reasonable starting guesses for tolerance values. For our lens assembly, we are going to assume that the element decenter tolerance is ± 100 μm and the element tilt tolerance is $\pm 0.5°$. Update your element tolerance wizard settings to match Fig. 13.23. Also, change the surface range to go from 1 to 3 to include the stop surface in the tolerancing.

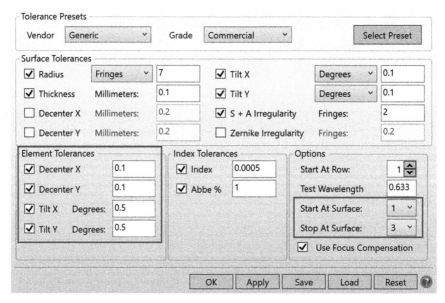

Figure 13.23 The Tolerance Wizard window after element tolerances have been added and the surface range has been updated.

Execute the tolerance wizard by clicking OK. The Tolerance Data Editor (Fig. 13.24) automatically pops open filled with a focus compensator (COMP) and a test wavelength (TWAV) in lines 1 and 2, and 15 different tolerance operands in lines 3 through 18. Because the "Start At Row" was 1 in the tolerance wizard, any previous tolerance operands in the TDE were overwritten. Note: The V-number tolerance (Abbe %) is entered as 1% in the tolerance wizard. This is converted to a min/max value in the TDE based on the nominal V-number for the material.

	Type	Surf	Adjust	Nominal	Min	Max	Comment
1	COMP	3	0	4.851	-0.200	0.200	Default compensator on back focus.
2	TWAV				0.633		Default test wavelength.
3	TFRN	1		0.000	-7.000	7.000	Default radius tolerances.
4	TFRN	2		0.000	-7.000	7.000	
5	TTHI	1	2	0.500	-0.100	0.100	Default thickness tolerances.
6	TTHI	2	3	0.467	-0.100	0.100	
7	TEDX	1	2	0.000	-0.100	0.100	Default element dec/tilt tolerances 1-2.
8	TEDY	1	2	0.000	-0.100	0.100	
9	TETX	1	2	0.000	-0.500	0.500	
10	TETY	1	2	0.000	-0.500	0.500	
11	TSTX	1		0.000	-0.100	0.100	Default surface dec/tilt tolerances 1.
12	TSTY	1		0.000	-0.100	0.100	
13	TSTX	2		0.000	-0.100	0.100	Default surface dec/tilt tolerances 2.
14	TSTY	2		0.000	-0.100	0.100	
15	TIRR	1		0.000	-2.000	2.000	Default irregularity tolerances.
16	TIRR	2		0.000	-2.000	2.000	
17	TIND	1		1.517	-5.000E-04	5.000E-04	Default index tolerances.
18	TABB	1		64.167	-0.642	0.642	Default Abbe tolerances.

Figure 13.24 The Tolerance Data Editor for the OSsecureCam2 after applying the Tolerance Wizard.

At this point in the tolerancing process, the tolerance operands should be checked for appropriateness against an opto-mechanical design (or planned mounting strategy) and, if needed, updated by the user using the TDE. For example, the wedge tolerance (TSTX/TSTY) was applied to *both* surfaces of a lens element by the wizard. In practice, a wedge tolerance is usually only needed on one side of the element. It is up to the user to select that surface based on the opto-mechanical mounting scheme and delete the extra tolerances on the other side of the element created by the wizard. For the OSsecureCam2's opto-mechanical design [shown in Fig. 13.17(a)], it makes sense to put the wedge tolerances on side 2 and delete the TSTX and TSTY tolerances on side 1. Then for the given mounting strategy, we also need to change Adjust surfaces for the thickness (TTHI) tolerances (operands #5 and #6) so that the Surf2 cell for the TTHI matches its Surf cell. That way, the airspaces after the tolerance surfaces won't get modified. Finally, because our lens is rotationally symmetric and we are mounting the lens in a barrel, it makes sense to change the element decenter tolerances from X and Y (TEDX and TEDY) to a radial value (TEDR). Open the Tolerance Data Editor (Tolerance > Tolerance Data Editor) and make these changes. The resulting TDE should look like Fig. 13.25 with a total of 15 tolerance operands.

	Type	Surf	Surf2	Nominal	Min	Max	Comment
1	COMP	3	0	4.851	-0.200	0.200	Default compensator on back focus.
2	TWAV				0.633		Default test wavelength.
3	TFRN	1		0.000	-7.000	7.000	Default radius tolerances.
4	TFRN	2		0.000	-7.000	7.000	
5	TTHI	1	1	0.500	-0.100	0.100	Default thickness tolerances.
6	TTHI	2	2	0.467	-0.100	0.100	
7	TEDR	1	2	0.000	0.000	0.100	Default element dec/tilt tolerances 1-2.
8	TETX	1	2	0.000	-0.500	0.500	
9	TETY	1	2	0.000	-0.500	0.500	
10	TSTX	2		0.000	-0.100	0.100	Default surface dec/tilt tolerances 2.
11	TSTY	2		0.000	-0.100	0.100	
12	TIRR	1		0.000	-2.000	2.000	Default irregularity tolerances.
13	TIRR	2		0.000	-2.000	2.000	
14	TIND	1		1.517	-5.000E-04	5.000E-04	Default index tolerances.
15	TABB	1		64.167	-0.642	0.642	Default Abbe tolerances.

Figure 13.25 The tolerance data editor for OSsecureCam2Tol.

Several of the tolerances (e.g., TFRN and TIRR) depend on the clear aperture of the part so it is also important to change any automatically calculated clear apertures to fixed apertures before starting any tolerancing. For our lens, its LDE must be modified by clicking on the solve box next to values in the Clear Semi-Dia column and change the apertures on surfaces 1 through 3 from Automatic to Fixed in the drop-down menu. A "U" should appear next to each clear semi-dia, as shown in Fig. 13.26. Any solves should also be removed before tolerancing or else OpticStudio will return an error message. Save this lens (and its tolerances) as **OSsecureCam2Tol**. It will be used for both a sensitivity analysis and a Monte Carlo analysis in the next two sections.

	Surface Type	Comment	Radius	Thickness	Material	Clear Semi-Dia
0	OBJECT Standard ▼		Infinity	Infinity		Infinity
1	(aper) Standard ▼		1.265 V	0.500	N-BK7	0.782 U
2	(aper) Standard ▼		1.768 V	0.467 V		0.571 U
3	STOP Standard ▼		Infinity	4.851 V		0.255 U
4	IMAGE Standard ▼		Infinity	-		3.031

Figure 13.26 LDE for OSsecureCam2Tol with fixed apertures.

13.6 Sensitivity Analysis

Now that the initial tolerances for the OSsecureCam2 have been defined, we can investigate the various software options for tolerancing. We'll start with a sensitivity analysis where the impact of each tolerance is evaluated independently. For example, if the radius of curvature of a surface has a nominal value of 100 mm with a minimum tolerance of −0.1 mm, the radius is set to 99.9 (the extreme minimum value). Then, if compensation is active, any compensators are optimized, and the performance criterion is evaluated and tracked. The procedure is repeated for the extreme maximum tolerance (+0.1 mm). This procedure is then repeated for each tolerance operand in the TDE.

After all of the individual tolerance sensitivities are computed, OpticStudio computes a variety of statistics, the most important of which is the RSS Estimated Change in performance, where RSS is root sum of squares. For each tolerance, the change in performance from the nominal is squared and then averaged between the Min and Max tolerance values. Then the average of the minimum and maximum squared change is found because the minimum and maximum tolerances cannot occur simultaneously, and summing the squares would result in an overly pessimistic prediction. The total change can then be computed from a sum over all tolerances of their averaged squared values. From these, the final as-built performance prediction is computed.

Note: In a sensitivity analysis, the tolerances are unchanged by the analysis. It is also possible to run an **inverse sensitivity** where each tolerance is iteratively changed until some predefined performance criterion is met. The as-built performance is still computed in the same way as a sensitivity analysis. The goal of an inverse sensitivity analysis is to eliminate critically sensitive tolerances in the design and have all tolerances contribute equally to the overall performance loss. In practice, this is often difficult to achieve due to tolerance limits imposed by manufacturing, and it is typically easier to select a manufacturer/manufacturing process, run a sensitivity analysis, and tighten any sensitive tolerance by hand (if allowed).

13.6.1 Tolerancing Dialog Box Settings

If you don't have it open, restore your saved **OSsecureCam2Tol** from Section 13.5. To start a sensitivity analysis, the main tolerancing feature: Tolerance > Tolerancing is launched. This opens a complex dialog box (Fig. 13.27) with several different tabs for choosing the type of tolerancing that is to be performed. This might seem a bit overwhelming at first. We're not going to cover everything this powerful tool can accomplish in this text! Instead, we will focus on two of its key analyses, sensitivity (covered in this section) and Monte Carlo (covered in the next section). For each analysis, we'll demonstrate how to use OpticStudio's tolerancing in the most efficient manner (tab by tab).

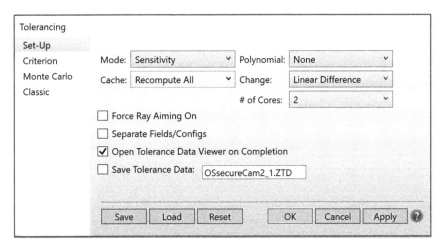

Figure 13.27 Tolerancing dialog box showing the Set-Up tab.

The first tab in the Tolerancing dialog box is Set-Up (Fig. 13.27). We will leave most of the settings on this tab (e.g., Polynomial or Cache) at their defaults. The two items that should be attended to are the Mode selection and the check box for Separate Fields/Configs. The Mode drop-down options are Sensitivity, Inverse Limit, Inverse Increment, and Skip Sensitivity. Any initial tolerance run starts in the default Mode, Sensitivity. This computes the change in a performance criterion for each of the extreme values of the tolerance (as defined in the TDE). The last mode, Skip Sensitivity, bypasses the sensitivity analysis and proceeds directly to a Monte Carlo analysis. (Once the Sensitivity analysis is finished, there is no need to run Sensitivity again before running the Monte Carlo analysis.) For now, keep the default mode selection, Sensitivity.

The checkbox for Separate Fields/Configs is useful for switching between a quick sensitivity analysis and a more detailed one. If the box is unchecked, the performance criterion is averaged over all fields in all configurations (the way a merit function is constructed). It provides a quick look at the ranking of tolerances using a single performance number. When the box is checked, a more detailed sensitivity analysis is done. The performance is computed and displayed for all field positions so that one can see how each tolerance affects the performance at each field point. For now, leave it unchecked.

The second tolerancing tab, Criterion, shown in Fig. 13.28, specifies the performance criteria to be used for tolerancing (for both Sensitivity and Monte Carlo). Many of the performance metrics we've discussed in past chapters are listed under the drop-down menu for Criterion. The default criterion is RMS Spot Radius, which is the fastest computation choice as it does not require complex diffraction-based calculations (like MTFs). Because of the statistical nature of tolerancing, the speed of the calculation becomes very important when choosing a performance criterion. RMS Spot Radius is a good choice for systems that are not diffraction limited (like ours). For diffraction-limited systems, RMS Wavefront is a good choice and is nearly as fast as RMS Spot Radius, while a Criterion like MTF is significantly slower.

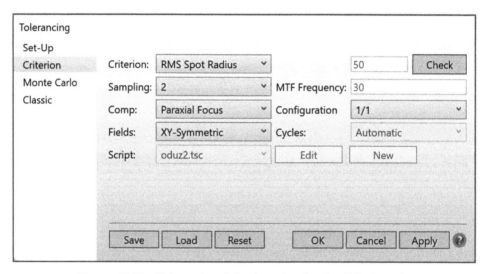

Figure 13.28 Tolerancing dialog box showing the Criterion tab.

The Sampling number sets the number of rays that are traced when computing the selected performance criterion. A larger number equals higher sampling, providing more accurate results, but increases the execution time. For example, for RMS Spot Radius, the sampling value is an integer that refers to the number of rays traced along a radial arm of the pupil in Gaussian quadrature. Usually, a sampling of 3 or 4 is sufficient.

To check to see if the sampling is sufficient, select a sampling value and then press the "Check" button. The value of the current performance criterion is returned in the box next to the Check button for the current sampling choice. Figure 13.29 shows the results for the OSsecureCam2Tol when the sampling is changed from 2 to 3 to 4. There is a change in the fourth decimal place (0.00568 versus 0.00576) when switching from 2 to 3. However, the change is in the ninth decimal place when increasing from 3 to 4. For our security camera, a sampling of 3 is definitely enough to evaluate RMS Spot Radius!

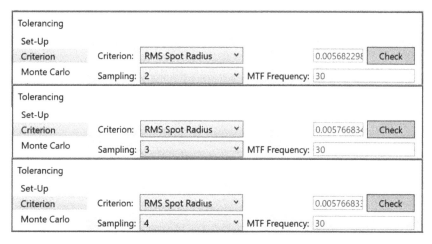

Figure 13.29 Impact of Sampling on the calculation of RMS Spot Radius.

The next drop-down menu in the Criterion tab (just below Sampling) is "Comp" (see Fig. 13.28). This determines how the compensators are handled during tolerancing and which optimization algorithm is used for compensation. If "None" is selected, no compensation will be performed, and any defined compensators in the TDE will be ignored. The default setting is "Paraxial Focus," where only the change in paraxial back focus is considered as a compensator. This is much faster than either of the optimization algorithms [e.g., damped least squares (DLS) or orthogonal decent (OD)] but should only be used for rough tolerancing as it also ignores any defined compensators in the TDE.

The final active drop-down menu on the Criterion tab is "Fields." This setting can have a significant impact on execution time as extra fields are added to the lens system (before tolerancing) to increase the accuracy of the results when analyzing tolerances like tilt or decenter. The default setting is "XY-Symmetric," which adds fields in both X and Y. "Y-Symmetric" finds the maximum field setting and then defines new field points at $+1.0$, $+0.7$, 0.0, -0.7, and -1.0 times the maximum field coordinate, in the Y direction only. All X field values are set to zero. These field values override the field assignments that were specified in the FDE so adding extra fields is a good option for rotationally symmetric lenses. "User-defined" uses whatever fields are currently defined in the lens.

For now, we will look at a sensitivity analysis that is **uncompensated**. So, change the "Comp" setting to "None." Then change the sampling to 3 and change the field setting to "Y-Symmetric." Because our system is rotationally symmetric, the "Y-Symmetric" field setting will help improve the speed of the calculation. Check (by hitting "Check") that the RMS Spot Radius hasn't changed significantly with the change in fields. The nominal RMS Spot Radius (average over all fields) is 0.005766. This is the value from which we want to track changes during our sensitivity analysis. Before moving on to the next tab, the Criterion tab should look like Fig. 13.30.

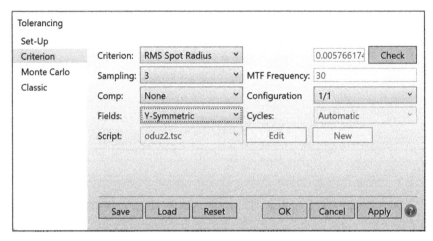

Figure 13.30 Tolerancing dialog box showing the Criterion tab with updated settings for an uncompensated sensitivity analysis.

The third tab is for a "Monte Carlo" analysis, and its settings will be discussed in detail in the next section. For now, if the number of Monte Carlo Runs is nonzero (the default number of Monte Carlo Runs is 20), then set it to 0 (Fig. 13.31). This will omit the Monte Carlo analysis from the tolerancing summary report so we can explore only the sensitivity results while saving execution time. The fourth tab "Classic" is inactive by default and will not be addressed in this text.

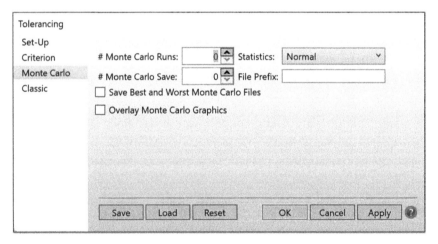

Figure 13.31 Tolerancing dialog box showing the Monte Carlo tab with the number of Monte Carlo Runs set to 0.

13.6.2 Uncompensated Sensitivity Analysis

Now that the tolerance settings have been defined for an uncompensated sensitivity analysis, click the "OK" button. This runs the analysis and automatically opens a "Tolerancing Results: 1" window. The window has two tabs, a Sensitivity table and a data Summary. The sensitivity table is shown in

Fig. 13.32. For each tolerance, it lists the nominal value, the tolerance value, an evaluation of the performance criterion for that tolerance value, and the value of any compensators. These tables can be extremely useful when identifying the most sensitive tolerance parameters that potentially cause the largest loss in performance and that may need to be tightened.

Type	Surf		Nominal	Tol Val	Comment	RMS Spot Radius	COMP
TFRN	1		0.000	-7.000	Default radius tole	0.008	4.851
TFRN	1		0.000	7.000	Default radius tole	0.008	4.851
TFRN	2		0.000	-7.000		0.010	4.851
TFRN	2		0.000	7.000		0.010	4.851
TTHI	1	1	0.500	0.400	Default thickness	0.017	4.851
TTHI	1	1	0.500	0.600	Default thickness	0.016	4.851
TTHI	2	2	0.467	0.367		0.007	4.851
TTHI	2	2	0.467	0.567		0.007	4.851
TEDR	1	2	0.000	0.000	Default element de	0.006	4.851
TEDR	1	2	0.000	0.100	Default element de	0.007	4.851
TETX	1	2	0.000	-0.500		0.006	4.851
TETX	1	2	0.000	0.500		0.006	4.851
TETY	1	2	0.000	-0.500		0.006	4.851
TETY	1	2	0.000	0.500		0.006	4.851
TSTX	2		0.000	-0.100	Default surface de	0.006	4.851
TSTX	2		0.000	0.100	Default surface de	0.006	4.851
TSTY	2		0.000	-0.100		0.006	4.851
TSTY	2		0.000	0.100		0.006	4.851
TIRR	1		0.000	-2.000	Default irregularity	0.006	4.851
TIRR	1		0.000	2.000	Default irregularity	0.006	4.851
TIRR	2		0.000	-2.000		0.006	4.851
TIRR	2		0.000	2.000		0.006	4.851
TIND	1		1.517	1.516	Default index toler	0.006	4.851
TIND	1		1.517	1.517	Default index toler	0.006	4.851
TABB	1		64.167	63.526	Default Abbe toler	0.006	4.851
TABB	1		64.167	64.809	Default Abbe toler	0.006	4.851

Figure 13.32 Uncompensated sensitivity analysis for OSsecureCam2Tol (unsorted).

The RMS Spot Radius column in Fig. 13.32 contains a range of different values, given in millimeters. But which of these tolerance operands are the worst offenders? In its menu bar, OpticStudio provides some easy-to-use options for sorting data in this table. First, the data can be sorted from worst to best by clicking on the performance criterion column (in this case, RMS Spot Radius) and then clicking on the menu icon. In this case, "worst" means largest, i.e., those operands that cause the largest RMS spot radius and, therefore, are the worst offending operands. Note: If needed, the data can also be sorted from best to worst using the icon.

It is often easier to look at the change in performance from the nominal design value by clicking on the RMS Spot Radius column and then clicking on the △ icon. Figure 13.33 shows the resulting tolerances when the data are sorted from the largest to the smallest change in performance. (Note: The △ icon is now checked (☒.) The main cause of performance loss is clearly the thickness of the lens element (TTHI Surf 1), with approximately the same change in RMS spot radius for a thinner (−0.1 mm) lens as for a thicker (0.1 mm) lens. The second most sensitive tolerance is the radius of the second surface (TFRN), again, with a similar change in performance for both the +7 fringe and −7 fringe tolerance change.

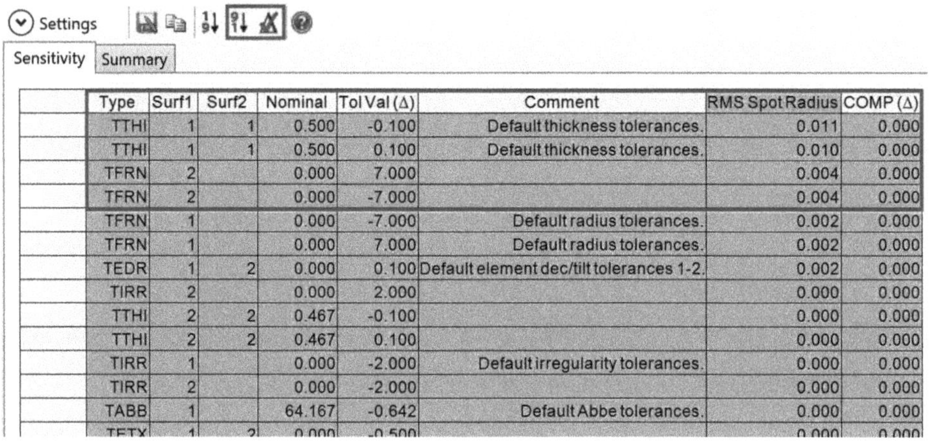

Type	Surf1	Surf2	Nominal	Tol Val (Δ)	Comment	RMS Spot Radius	COMP (Δ)
TTHI	1	1	0.500	-0.100	Default thickness tolerances.	0.011	0.000
TTHI	1	1	0.500	0.100	Default thickness tolerances.	0.010	0.000
TFRN	2		0.000	7.000		0.004	0.000
TFRN	2		0.000	-7.000		0.004	0.000
TFRN	1		0.000	-7.000	Default radius tolerances.	0.002	0.000
TFRN	1		0.000	7.000	Default radius tolerances.	0.002	0.000
TEDR	1	2	0.000	0.100	Default element dec/tilt tolerances 1-2.	0.002	0.000
TIRR	2		0.000	2.000		0.000	0.000
TTHI	2	2	0.467	-0.100		0.000	0.000
TTHI	2	2	0.467	0.100		0.000	0.000
TIRR	1		0.000	-2.000	Default irregularity tolerances.	0.000	0.000
TIRR	2		0.000	-2.000		0.000	0.000
TABB	1		64.167	-0.642	Default Abbe tolerances.	0.000	0.000
TETX	1	2	0.000	-0.500		0.000	0.000

Figure 13.33 Uncompensated sensitivity analysis for OSsecureCam2Tol sorted from the most sensitive to the least sensitive, with the performance Δ from nominal listed.

The data summary tab also contains some very useful information. The tolerance sensitivity for each operand along with the "Worst offenders" is listed again (with more decimal places than the table on the sensitivity tab). But more importantly, at the end of the summary is the estimated RMS performance change based on an RSS method of combining the individual sensitivities that can be used to quickly predict as-built performance (before running the more time-consuming Monte Carlo analysis). Table 13.1 shows the result for this uncompensated run. It is predicted that the RMS Spot Radius (averaged across the field) will be more than three times greater than the nominal RMS Spot Radius with these tolerances (and no focus compensation).

Table 13.1 Tolerance sensitivity summary for OSsecureCam2Tol (uncompensated).

```
Estimated Performance Changes based on the
Root-Sum-Square method:
Nominal RMS Spot Radius      :   0.00576617
Estimated change             :   0.01178843
Estimated RMS Spot Radius    :   0.01755460
```

13.6.3 Compensated Sensitivity Analysis

Two possible methods can be used for improving the as-built performance at this point. One would be to tighten the tolerances on the sensitive lens parameters, e.g., the lens thickness (T1) and the rear surface curvature (R2). However, in most cases, doing this would increase the cost of fabricating the element. It would be less costly to change the distance between the stop and the image plane because this is only a matter of focusing an image on a sensor (remember that this distance was a key variable for improving performance during optimization). Because these types of changes can compensate for fabrication errors, they are referred to as **compensators**.

The first sensitivity analysis was done uncompensated—note the zeros in the last (COMP) column of the sensitivity table (Fig. 13.33). To implement the focus compensator and run the sensitivity analysis compensated, open the Criterion dialog box (Tolerance > Tolerancing > Criterion) and change the Comp drop-down menu to "Optimize All (DLS)" (see Fig. 13.34). Once this is activated, it is very important to check the (now active) Cycles drop-down menu. This determines how rigorously OpticStudio will attempt to optimize the compensator values. The default setting is "Automatic," which can run for many cycles, sometimes taking hours (or days) to complete. For a simple focus compensation, the analysis will run much quicker if Cycles is changed from Automatic to 1 (or even 2) cycles. For this example, change the number of cycles to 1. Update the Criterion tab to match Fig. 13.34 and click "OK" to run the new analysis.

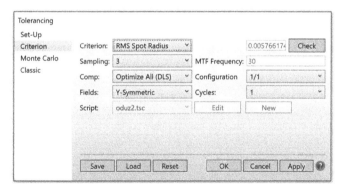

Figure 13.34 Tolerancing dialog box showing the Criterion tab with updated settings for a compensated sensitivity analysis.

The result of the compensated sensitivity analysis is displayed in a second tolerancing window labeled "Tolerance Results: 2." Figure 13.35 shows the results, where the changes in the RMS Spot Radius (Δ's) from nominal are sorted from largest to smallest. The amount of focus compensation needed for each tolerance is given in the last COMP (Δ) column. While the element thickness is still the most sensitive tolerance, its effect on RMS spot radius has been cut by more than half (0.004 mm versus 0.011 mm) using a focus compensator (Fig. 13.35). The focus compensator has also significantly improved the estimated performance change from all tolerances (Table 13.2). However, at this point it's also important to look at the amount of compensation required by each tolerance. For example, the two most sensitive tolerances from the uncompensated run require a compensator movement of 0.201 mm. That is the edge of the allowed compensation—recall that the Min/Max values for the focus compensator were set at ±0.2 mm in the TDE (Fig. 13.25).

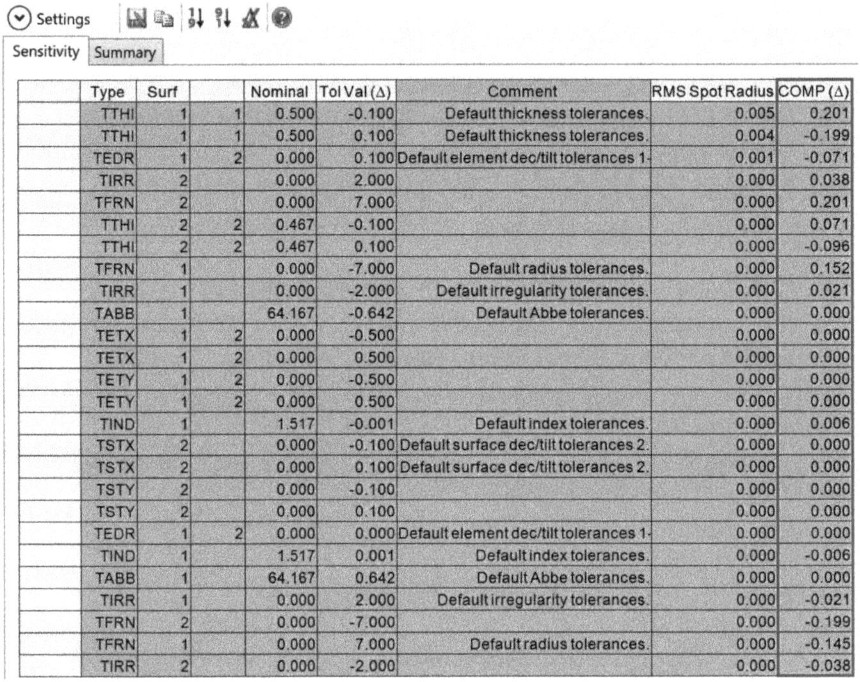

Type	Surf		Nominal	Tol Val (Δ)	Comment	RMS Spot Radius	COMP (Δ)
TTHI	1	1	0.500	-0.100	Default thickness tolerances.	0.005	0.201
TTHI	1	1	0.500	0.100	Default thickness tolerances.	0.004	-0.199
TEDR	1	2	0.000	0.100	Default element dec/tilt tolerances 1-	0.001	-0.071
TIRR	2		0.000	2.000		0.000	0.038
TFRN	2		0.000	7.000		0.000	0.201
TTHI	2	2	0.467	-0.100		0.000	0.071
TTHI	2	2	0.467	0.100		0.000	-0.096
TFRN	1		0.000	-7.000	Default radius tolerances.	0.000	0.152
TIRR	1		0.000	-2.000	Default irregularity tolerances.	0.000	0.021
TABB	1		64.167	-0.642	Default Abbe tolerances.	0.000	0.000
TETX	1	2	0.000	-0.500		0.000	0.000
TETX	1	2	0.000	0.500		0.000	0.000
TETY	1	2	0.000	-0.500		0.000	0.000
TETY	1	2	0.000	0.500		0.000	0.000
TIND	1		1.517	-0.001	Default index tolerances.	0.000	0.006
TSTX	2		0.000	-0.100	Default surface dec/tilt tolerances 2.	0.000	0.000
TSTX	2		0.000	0.100	Default surface dec/tilt tolerances 2.	0.000	0.000
TSTY	2		0.000	-0.100		0.000	0.000
TSTY	2		0.000	0.100		0.000	0.000
TEDR	1	2	0.000	0.000	Default element dec/tilt tolerances 1-	0.000	0.000
TIND	1		1.517	0.001	Default index tolerances.	0.000	-0.006
TABB	1		64.167	0.642	Default Abbe tolerances.	0.000	0.000
TIRR	1		0.000	2.000	Default irregularity tolerances.	0.000	-0.021
TFRN	2		0.000	-7.000		0.000	-0.199
TFRN	1		0.000	7.000	Default radius tolerances.	0.000	-0.145
TIRR	2		0.000	-2.000		0.000	-0.038

Figure 13.35 Compensated sensitivity analysis for OSsecureCam2Tol sorted from the most sensitive data to the least sensitive data with the change (Δ) from nominal listed.

Table 13.2 Sensitivity Summary for OSsecureCam2Sens (compensated).

```
Estimated Performance Changes based on the
Root-Sum-Square method:
Nominal RMS Spot Radius      :   0.00576595
Estimated change             :   0.00428367
Estimated RMS Spot Radius    :   0.01004962
```

The limits on the focus compensator were somewhat arbitrarily set to ±0.2 mm by the tolerance wizard. What happens if we allow a larger focus compensation? Open the TDE and increase the limits on the focus compensator (Fig. 13.36) to ±0.5 mm. Rerun the compensated sensitivity analysis with the same settings as were used in the last run. The results are shown in Fig. 13.37 and Table 13.3. With the increased focus compensation, the predicted as-built RMS spot radius (averaged across the field) is only 20% larger than nominal (a significant improvement). We are now ready to move on to a Monte Carlo analysis. Save this lens as **OSsecureCam2Monte**.

Tolerancing

	Type	Surf	Code	Nominal	Min	Max	Comment
1	COMP	3	0	4.851	-0.500	0.500	Default compensator on back focus.
2	TWAV					0.633	Default test wavelength.
3	TFRN	1		0.000	-7.000	7.000	Default radius tolerances.
4	TFRN	2		0.000	-7.000	7.000	
5	TTHI	1	1	0.500	-0.100	0.100	Default thickness tolerances.
6	TTHI	2	2	0.467	-0.100	0.100	
7	TEDR	1	2	0.000	0.000	0.100	Default element dec/tilt tolerances 1-2.
8	TETX	1	2	0.000	-0.500	0.500	
9	TETY	1	2	0.000	-0.500	0.500	
10	TSTX	2		0.000	-0.100	0.100	Default surface dec/tilt tolerances 2.
11	TSTY	2		0.000	-0.100	0.100	
12	TIRR	1		0.000	-2.000	2.000	Default irregularity tolerances.
13	TIRR	2		0.000	-2.000	2.000	
14	TIND	1		1.517	-5.000E-04	5.000E-04	Default index tolerances.
15	TABB	1		64.167	-0.642	0.642	Default Abbe tolerances.

Figure 13.36 Tolerance data editor for OSsecureCam2Tol with increased focus range compensation.

Type	Surf		Nominal	Tol Val (Δ)	Comment	RMS Spot Radius	COMP (Δ)
TEDR	1	2	0.000	0.100	Default element dec/tilt to	0.001	-0.071
TTHI	1	1	0.500	-0.100	Default thickness toleran	0.000	0.419
TIRR	2		0.000	2.000		0.000	0.038
TTHI	2	2	0.467	-0.100		0.000	0.071
TFRN	2		0.000	7.000		0.000	0.222
TTHI	2	2	0.467	0.100		0.000	-0.096
TFRN	1		0.000	-7.000	Default radius tolerances	0.000	0.152
TIRR	1		0.000	-2.000	Default irregularity tolera	0.000	0.021
TTHI	1	1	0.500	0.100	Default thickness toleran	0.000	-0.402
TABB	1		64.167	-0.642	Default Abbe tolerances	0.000	0.000
TETX	1	2	0.000	-0.500		0.000	0.000
TETX	1	2	0.000	0.500		0.000	0.000
TETY	1	2	0.000	-0.500		0.000	0.000
TETY	1	2	0.000	0.500		0.000	0.000
TIND	1		1.517	-0.001	Default index tolerances	0.000	0.006
TSTX	2		0.000	-0.100	Default surface dec/tilt to	0.000	0.000
TSTX	2		0.000	0.100	Default surface dec/tilt to	0.000	0.000
TSTY	2		0.000	-0.100		0.000	0.000
TSTY	2		0.000	0.100		0.000	0.000
TEDR	1	2	0.000	0.000	Default element dec/tilt to	0.000	0.000
TIND	1		1.517	0.001	Default index tolerances	0.000	0.006
TABB	1		64.167	0.642	Default Abbe tolerances	0.000	0.000
TIRR	1		0.000	2.000	Default irregularity tolera	0.000	-0.021
TFRN	1		0.000	7.000	Default radius tolerances	0.000	-0.145
TFRN	2		0.000	-7.000		0.000	-0.207
TIRR	2		0.000	-2.000		0.000	-0.038

Figure 13.37 Compensated sensitivity analysis for OSsecureCam2Tol with increased focus range, sorted from the most sensitive data to the least sensitive with the performance change (Δ) from nominal listed.

Table 13.3 Tolerance Sensitivity Summary for OSsecureCam2Tol (compensated with increased focus range).

```
Estimated Performance Changes based on the
Root-Sum-Square method:
Nominal RMS Spot Radius      :     0.00576595
Estimated change             :     0.00103064
Estimated RMS Spot Radius    :     0.00679659
```

13.7 Monte Carlo Analysis

Now that we have determined the sensitivity of our tolerance operands and increased the allowed range of the focus compensator, we're ready to run a Monte Carlo analysis and statistically calculate the expected as-built yield. Unlike sensitivity, a Monte Carlo analysis simulates the effect of all perturbations simultaneously. For each Monte Carlo trial, lens parameters with specified tolerances are randomly perturbed using the defined tolerance range of the parameter and a statistical model of the distribution of that parameter (see text box "Tolerance Probability Distributions") over the specified range. For example, a radius of 100.00 mm with a tolerance of +4.0 / −0.0 mm (and the default probability distribution) will be assigned a random radius between 100.00 and 104.00 mm with a normal distribution centered on 102.00 mm and a standard deviation of 1.0 mm. Once all lens parameters are assigned values, the compensators are adjusted, and the performance criterion is evaluated and stored. Then the design is returned to its nominal state and the process is repeated for a given number of trials or runs. After all of the Monte Carlo trials are completed, a statistical summary is provided.

13.7.1 Simulation #1 (10 trials): RMS spot radius

Restore the **OSsecureCam2Monte** from Section 13.6.3. On the Set-up tab (Tolerance > Tolerancing > Set-up), change the Mode from Sensitivity to "Skip Sensitivity." On the Monte Carlo tab (Tolerance > Tolerancing > Monte Carlo), set the number of Runs to 10. While this number of runs is not statistically significant, it demonstrates some important features of a Monte Carlo analysis before running a much larger number of trials. Click "OK" to start the analysis. The result is displayed in a new tolerancing window (Tolerance Results: 4), as shown in Fig. 13.38.

	RMS Spot Radius Statistics	COMP Statistics	TFRN Statistics	TFRN Statistics	TTHI Statistics	TTHI Statistics	TEDR Statistics	TETX Statistics	TETY Statistics	TSTX Statistics
	Field: 0	Surf: 3	Surf: 1	Surf: 2	Surf: 1	Surf: 2	Surf1: 1	Surf1: 1	Surf1: 1	Surf: 2
	Config: 0	Code: 0	Nominal: 0	Nominal: 0	Adjust: 1	Adjust: 2	Surf2: 2	Surf2: 2	Surf2: 2	Nominal: 0
	Nominal: 0.00576	Nominal: 4.84921	Min: -7	Min: -7	Nominal: 0.5	Nominal: 0.467	Nominal: 0	Nominal: 0	Nominal: 0	Min: -0.1
	Sampling: 3	Min: -0.5	Max: 7	Max: 7	Min: -0.1	Min: -0.1	Min: 0	Min: -0.5	Min: -0.5	Max: 0.1
	Comp: Optimize A	Max: 0.5	Comment: Def	Comment:	Max: 0.1	Max: 0.1	Max: 0.1	Max: 0.5	Max: 0.5	Comment: De
	Fields: Y Symmet	Comment: Default			Comment: De	Comment:	Comment: Def	Comment:	Comment:	
MC_1	0.006	4.598	1.512	-6.295	0.500	0.502	-0.041	0.054	-0.428	0.027
MC_2	0.006	4.883	-0.006	-4.357	0.450	0.464	0.058	0.244	-0.087	0.060
MC_3	0.006	4.818	1.936	5.415	0.545	0.416	-0.032	-0.071	-0.033	-0.021
MC_4	0.006	4.718	-0.648	-2.151	0.492	0.560	0.021	0.286	0.132	0.098
MC_5	0.006	5.117	-1.325	4.131	0.461	0.498	0.051	-0.055	0.023	-0.054
MC_6	0.006	4.905	-3.177	2.156	0.505	0.535	0.010	-0.072	0.080	0.057
MC_7	0.007	4.969	0.830	-2.419	0.443	0.445	-0.071	0.155	-0.142	0.050
MC_8	0.006	5.242	-3.972	3.215	0.441	0.519	0.027	-0.363	-0.265	-0.015
MC_9	0.006	4.759	-1.432	-5.480	0.498	0.413	0.007	-0.346	0.357	-0.039
MC_10	0.006	4.944	-2.145	2.090	0.496	0.472	-0.020	0.063	0.241	-0.067

Figure 13.38 Monte Carlo RMS Spot Radius results for 10 trials for OSsecureCam2Monte.

Tolerance Probability Distributions

By default, all tolerance parameter distributions are assumed to follow a modified Gaussian "normal" distribution (a bell-shaped curve) with a total width of four standard deviations (σ) between the extreme minimum and maximum allowed values. The default operation can be changed using a STAT operand in the Tolerance Data Editor. All tolerance operands below the STAT operand will use the statistical probability distribution defined by STAT. The default distribution is STAT 0. For "uniform," the randomly selected value (STAT 1) will lie somewhere between the specified extreme tolerances with uniform probability. A parabolic distribution (STAT 2) yields selected values that are more likely to be at the extreme ends of the tolerance range. Note that going from normal to uniform to parabolic statistics yields a successively more pessimistic analysis and thus more conservative tolerances.

Because the grinding and polishing of lenses are processes of removal, the thickness of a lens gets smaller with increased time. Once the thickness is within the desired tolerance range, and provided that the surface curvatures are correct, the work is finished. This means that the lens thickness will usually be on the high side of the design value. In a more complex tolerance analysis, a probability distribution can be applied to the tolerance range to take into account this bias using a user-defined distribution by using STAT 3.

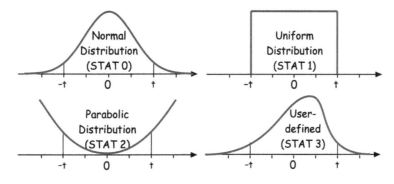

Examine Fig. 13.38 carefully to understand what is listed there. Each row starting at "MC_1" represents an individual Monte Carlo trial. The first column evaluates the performance criteria (RMS Spot Radius) for that trial. The second column gives the focus compensator value for that trial, and then each column after that lists the values for individual tolerance operands for a given trial. Similar to sensitivity, this data can also be displayed as Δ's from nominal (see Fig. 13.39). Clicking on "Statistics" for an individual column brings up a new window with detailed statistics (and more significant digits) for that column (see Fig. 13.40). Because these Monte Carlo runs are based on randomly determined lens parameter values, the results of your individual trials will differ from those shown in Figs. 13.38 and 13.39. But the Nominal Mean and Std Dev in the Summary should be comparable (if enough trials are run).

	RMS Spot Radius	COMP	TFRN	TFRN	TTHI	TTHI	TEDR	TETX	TETY	TSTX
	Statistics	Statistics	Statistics	Statistics	Statistics	Statistics	Statistics	Statistics	Statistics	Statistics
	Field: 0	Surf: 3	Surf: 1	Surf: 2	Surf: 1	Surf: 2	Surf1: 1	Surf1: 1	Surf1: 1	Surf: 2
	Config: 0	Code: 0	Nominal: 0	Nominal: 0	Adjust: 1	Adjust: 2	Surf2: 2	Surf2: 2	Surf2: 2	Nominal: 0
	Nominal: 0.00576	Nominal: 4.84921	Min: -7	Min: -7	Nominal: 0.5	Nominal: 0.467	Nominal: 0	Nominal: 0	Nominal: 0	Min: -0.1
	Sampling: 3	Min: -0.5	Max: 7	Max: 7	Min: -0.1	Min: -0.1	Min: 0	Min: -0.5	Min: -0.5	Max: 0.1
	Comp: Optimize A	Max: 0.5	Comment: Def	Comment:	Max: 0.1	Max: 0.1	Max: 0.1	Max: 0.5	Max: 0.5	Comment: D
	Fields: Y Symmet	Comment: Default			Comment: De	Comment:	Comment: Def	Comment:	Comment:	
MC_1	0.000	-0.251	1.512	-6.295	0.000	0.034	-0.041	0.054	-0.428	0.027
MC_2	0.001	0.034	-0.006	-4.357	-0.050	-0.003	0.058	0.244	-0.087	0.060
MC_3	0.000	-0.031	1.936	5.415	0.045	-0.051	-0.032	-0.071	-0.033	-0.021
MC_4	0.000	-0.131	-0.648	-2.151	-0.008	0.092	0.021	0.286	0.132	0.098
MC_5	0.001	0.268	-1.325	4.131	-0.039	0.030	0.051	-0.055	0.023	-0.054
MC_6	0.000	0.056	-3.177	2.156	0.005	0.068	0.010	-0.072	0.080	0.057
MC_7	0.001	0.120	0.830	-2.419	-0.057	-0.023	-0.071	0.155	-0.142	0.050
MC_8	0.000	0.393	-3.972	3.215	-0.059	0.052	0.027	-0.363	-0.265	-0.015
MC_9	0.000	-0.090	-1.432	-5.480	-0.002	-0.054	0.007	-0.346	0.357	-0.039
MC_10	0.000	0.095	-2.145	2.090	-0.004	0.004	-0.020	0.063	0.241	-0.067

Figure 13.39 Monte Carlo RMS Spot Radius results for ten trials for OSsecureCam2Monte (values displayed as Δ's from nominal).

Summary statistics for RMS Spot Radius (Nominal Subtracted)

Sample Size	10
Min	6.05831149016241E-06
Max	0.00126118423493355
Mean	0.000378224746664184
Standard Deviation	0.000377181708120443
Standard Deviation (population)	0.000357825986824029
Variance	1.42266040940655E-07
Sample Error	0.000119275328941343
Histogram Underflow	0

Figure 13.40 Ten trial Monte Carlo results for OSsecureCam2Monte, showing Summary statistics for the RMS spot radius column.

As with Sensitivity, there is also much more statistical data on the Summary tab of the tolerance results. For example, Table 13.4 shows the statistics on both performance and compensation for the 10-trial Monte Carlo analysis. While these tables of statistics provide detailed information about the effects of manufacturing errors on the yield of a fabrication run, OpticStudio also provides some graphical assessments of the trial results. A Yield plot (Tolerance > Yield), discussed in Section 13.1, for the 10-trial Monte Carlo analysis is shown in Fig. 13.41. This plot shows that the (field-averaged) RMS spot radius for the lenses generated by the 10-trial run ranged from about 5.68 to 6.15 μm. If the as-built performance target was <6 μm, then this 10-lens simulation would indicate that only 50% of the lenses would be usable.

Table 13.4 Monte Carlo Summary for OSsecureCam2Monte.

```
Monte Carlo Trials: 10

Nominal              0.00576595
Best                 0.00568477     Trial 2
Worst                0.00614731     Trial 8
Mean                 0.00592860
Std Dev              0.00014995

Compensator Statistics:
Thickness Surf 3:

Nominal:                   4.849213
Minimum:                   4.593692
Maximum:                   4.961371
Mean:                      4.742230
Standard Deviation:        0.109499

90% >                0.00611472
80% >                0.00605706
50% >                0.00596970
20% >                0.00576836
10% >                0.00569268
```

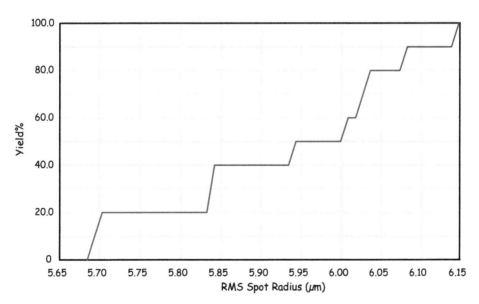

Figure 13.41 Yield% plot for the ten-trial Monte Carlo analysis. Note the discontinuities in the plot, indicating that not enough trials have been run.

13.7.2 Simulation #2 (500 trials): RMS spot radius

The discontinuties in the Yield plot (Fig. 13.41) clearly indicate that 10 trials are nowhere near enough, even for a simple singlet. In general, the number of runs should be increased until the yield plots are "smooth." The analysis will just take longer to run. For example, let's change the number of Runs on the Monte Carlo tab (Tolerance > Tolerancing > Monte Carlo) from 10 to 500. Now it should take a few minutes to run, depending on the speed of your computer's microprocessor.

The summary results are shown in Table 13.5, and a (much smoother) yield plot (Tolerance > Yield) is shown in Fig. 13.42. It is interesting to note that the 90% yield here (0.00677666) approximately agrees with the estimated RMS spot radius (0.00679480) from the compensated sensitivity RSS in Table 13.3. Note: The default number of Monte Carlo runs is 20, and this number should be increased for any lens once you have a better understanding of how many trials you need and how long it will take to run! If you don't run enough trials the first time, the tolerance data can also be saved and copied to an Excel spreadsheet, making it easy to combine tolerance data from multiple Monte Carlo runs.

Table 13.5 Monte Carlo Summary for OSsecureCam2Monte: 500 trials

```
Monte Carlo Trials: 500
Nominal           0.00576595
Best              0.00554750    Trial 436
Worst             0.00820010    Trial  85
Mean              0.00620914
Std Dev           0.00044139

Compensator Statistics:
Thickness Surf 3:
Nominal:                       4.849213
Minimum:                       4.350594
Maximum:                       5.350594
Mean:                          4.820465
Standard Deviation:            0.210632

98% >                          0.00757718
90% >                          0.00677666
80% >                          0.00650640
50% >                          0.00608976
20% >                          0.00587536
10% >                          0.00577957
 2% >                          0.00567811
```

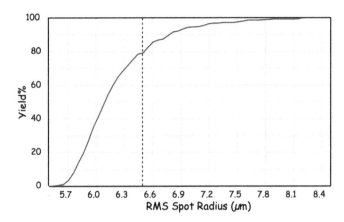

Figure 13.42 Yield% plot for the 500-trial Monte Carlo analysis.

Monte Carlo results like these can be used to predict the yield for high-volume production. To be more concrete—suppose that you need a batch of lenses with a field-averaged, as-built RMS spot radius of less than 0.0065 mm. The yield for this specification (see Fig. 13.42) is predicted to be around 80%. So, you can expect to throw away 20% of your product. If you are only building one lens, this plot will still tell you the probability of success for that one lens.

If we want to improve yield, we will need to tighten the tolerances and/or add compensators. Moreover, the slope of the line shows how much the performance can be improved when the changes in RMS spot radius become smaller as the slope of the plot becomes steeper. The ideal yield plot would be one where the RMS spot radius does not deteriorate at all, and the plot would be a vertical line at nominal performance.

Another way to graphically examine the Monte Carlo data is to plot a histogram of the performance data (Tolerance > Histogram). The result (Fig. 13.43) shows that most of the runs are clustered around the 6.0-μm RMS spot radius with only a few outliers that would need to be rejected.

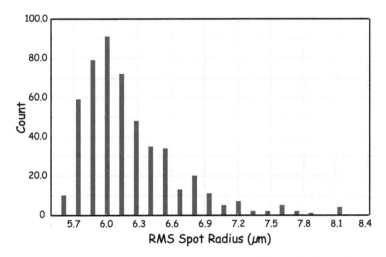

Figure 13.43 Histogram plot for the 500-trial Monte Carlo analysis.

13.7.3 Simulation #3 (500 trials): MTF

Our original performance metric to gauge the quality of the OSsecureCam2 was the MTF. We can get a feel for what the as-built MTF is with Monte Carlo tolerancing, it just takes a bit longer to run. First, we need to select a target frequency and minimum contrast for the performance evaluation. From Chapter 12, the Nyquist frequency of our security camera sensor was 66 cycles/mm. For an as-built performance target, we can use 70% of the Nyquist frequency (66 cycles/mm × 0.7), or 45 cycles/mm, and aim for a contrast of greater than 30% with >98% yield. The choice of a 30% modulation is a recognition that modulation in an image is still distinguishable at 30%, but it is difficult to see at 10% (or below), as shown in Fig. 13.44.

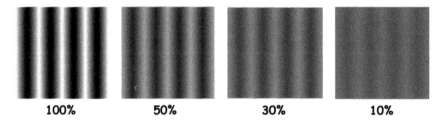

Figure 13.44 Sine wave pattern at differing contrasts.

A plot of the nominal MTF for the OSsecureCam2 (Fig. 13.45) shows that the 25° tangential field limits the performance of the lens, and the contrast drops off quickly above 30 cycles/mm for this field. At 45 cycles/mm, the nominal design has a contrast of 53% on axis and 3% (T) and 50% (S) at full field. A common metric to track is the average of the S and T curves, which yields a value of 26% for full field. So, nominally, this design doesn't meet our as-built specification (this is why we continued to add elements to our design in Chapter 12), but we can still use this simple example to see how much worse the MTF gets after tolerancing.

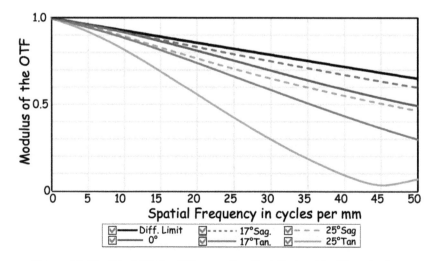

Figure 13.45 The MTF for OSsecureCam2 plotted out to 50 cycles/mm.

Open the Criterion dialog box (Tolerance > Tolerancing > Criterion) and change the performance criterion to "Diff. MTF Avg," as shown in Fig. 13.46. Enter "45" in the MTF Frequency box. This will calculate a (field-averaged) "Average" MTF at 45 cycles/mm, where "Average" is the average of the tangential and sagittal values for a given field. Change the sampling from 3 to 2 to speed up the analysis. If you click on the "Check" button, it returns a nominal value of 0.427 (42.7%).

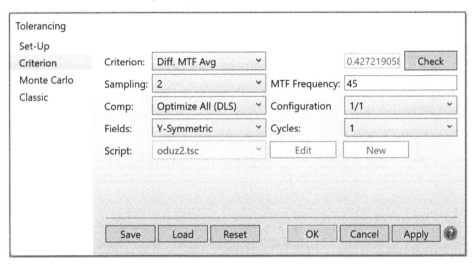

Figure 13.46 Tolerancing dialog box showing the Criterion tab with updated settings for an MTF Monte Carlo run.

However, Fig. 13.45 shows that the MTF performance values for the nominal design at individual fields are significantly different. Therefore, it makes sense to also track the as-built performance for individual fields. Go to the Set-up dialog box (Tolerance > Tolerancing > Set-up) and check the "Separate Fields/Configs" box. On the Monte Carlo tab, enter "500" for the number of Monte Carlo Runs (if it is not already set at 500) and click "OK." This type of diffraction calculation will take longer; e.g., it can take almost 10 minutes to complete 500 trials for this simple singlet. Once the trials are completed, this lens should be saved as **OSsecureCam2Monte500**. In addition to saving the lens, OpticStudio will also save the 500-trial tolerancing results so that if you close the lens or quit the program, you can do the additional analyses without needing to run the long simulation again.

The Monte Carlo summary results are shown in Table 13.6. There are now five fields shown in the output as the Monte Carlo analysis simulates the performance at both the + and – fields. The 98% yield performance on-axis (Field 1) is 0.53 MTF and meets specification. However, at full field (Fields 4 and 5), the 98% yield is only 0.22 MTF (as expected, they don't meet the specifications). A design with more elements and a field flattener (like OSsecureCam6) is needed.

Table 13.6 Monte Carlo results for 500 trials for Diff. MTF Avg at 45 cycles/mm.

```
    Number of traceable Monte Carlo files generated: 500
              Average Field 1 Field 2 Field 3 Field 4 Field 5
Nominal       0.4272  0.5906  0.4968  0.4968  0.2494  0.2494
Best   #298   0.4443  0.6555  0.5217  0.5218  0.3261  0.3180
Worst  #84    0.0821  0.0891  0.0704  0.0704  0.0675  0.1067
Mean          0.4222  0.5724  0.4784  0.4792  0.2618  0.2609
Std Dev       0.0213  0.0379  0.0312  0.0324  0.0203  0.0193

Compensator Statistics:
 Thickness Surf 3:
Nominal:                 4.881742
Minimum:                 4.350509
Maximum:                 5.350875
Mean:                    4.872139
Standard Deviation:      0.224286

%Yield        Average Field 1 Field 2 Field 3 Field 4 Field 5
98% >         0.3856  0.5250  0.4144  0.4163  0.2257  0.2241
90% >         0.4078  0.5481  0.4442  0.4438  0.2419  0.2413
80% >         0.4171  0.5606  0.4614  0.4607  0.2482  0.2471
50% >         0.4263  0.5787  0.4848  0.4855  0.2596  0.2596
20% >         0.4316  0.5899  0.5002  0.5006  0.2759  0.2750
10% >         0.4338  0.5939  0.5058  0.5065  0.2877  0.2861
 2% >         0.4383  0.6012  0.5136  0.5144  0.3054  0.2998
```

Yield plots (Tolerance > Yield) for the axis (field 1), 0.7 field (field 3), and full field (field 5) are shown in Fig. 13.47. To create this plot, you need to first generate a yield plot for each field and then overlay them by right clicking on one of the plots and using the "Add/remove/overlay series" option from the resulting plot menu (Fig. 13.48). Then select the series you want to overlay and check "primary" to add it to the primary axis. The x axis of the plot has also been scaled from 1 to 0 using the "Edit axis options" from the plot menu (see Fig. 13.48) by unchecking the Auto boxes for the X axis and replacing the Min and Max values with "0" and "1." All three yield curves show about a 10% contrast loss from nominal for high yield (>98%).

Tolerancing 331

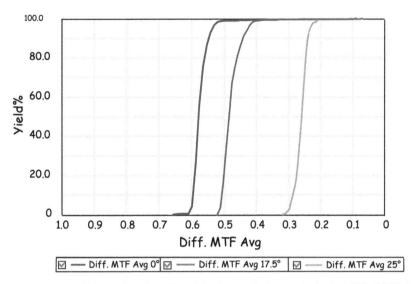

Figure 13.47 Yield% plot for the 500-trial Monte Carlo analysis for Diff. MTF Avg at 45 cycles/mm for 0°, 17.5°, and 25° fields.

Figure 13.48 Example plot menu (right click on the Yield% plot to open this menu).

13.7.4 Advanced Monte Carlo Settings

On the Monte Carlo tab (Fig. 13.49), there are a few other useful tools that can be turned on (usually at the cost of additional processing time). For example, one nice feature that can be activated is the "Save Best and Worst Monte Carlo Trials." This saves the best and the worst trials so that they can be examined in detail later to investigate the underlying cause of performance loss in the worst trial. If needed, "# Monte Carlo Save" can be used to save some (or all) of the trials for further analysis. Useful tip: If you set the number of Monte Carlo Runs to 1 and the number of Monte Carlo Saves to 1 and run the tolerancing routine, OpticStudio will save one Monte Carlo file (MC_T0001.ZMX) that is the same as the nominal design with all of the coordinate breaks in it. This can be very useful for debugging complex systems to see exactly how OpticStudio interprets the different tolerancing operands!

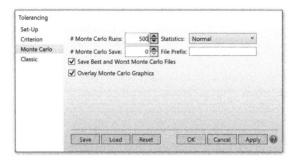

Figure 13.49 Advanced Monte Carlo settings.

Another useful advanced feature of the Monte Carlo analysis is the ability to overlay different trials on performance graphics (see the last check box on the Monte Carlo tab in Fig. 13.49). If this option is activated, it is best to close any graphic windows that you don't want to overlay to improve the execution time and significantly reduce the number of runs. For example, Fig. 13.50 shows an MTF plot for the OSsecureCam2 with 25 Monte Carlo trials overlayed on the nominal MTF. While not quantitative, these overlay plots can give the user a visual cue as to how tightly clustered the performance values will be. The plots are also particularly useful for MTF evaluation where the performance metric tracked for the Monte Carlo statistics is the contrast at only one specified lp/mm at a time; and these plots simultaneously show the as-built performance (visually) over the full range of lp/mm.

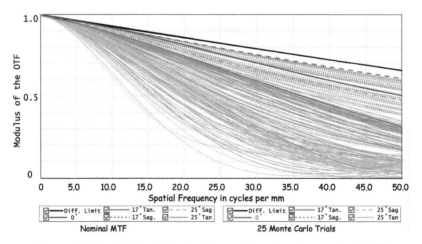

Figure 13.50 25 Monte Carlo trials overlayed on the nominal MTF.

13.8 Design Example: The OSsecureCam6

Now that we've introduced the basics of tolerancing using the OSsecureCam2, we will tolerance a more complex design: the OSsecureCam6 (Fig. 12.45). Restore the $f/8$ **OSsecureCam6**. Since the OSsecureCam6 was optimized with Hammer optimization, your final result may not exactly match the result we showed in Chapter 12. To follow along with the text, it may make sense to type in a new lens from the LDE data given in Fig. 13.51.

	Surface Type	Comment	Radius	Thickness	Material	Clear Semi-Dia
0	OBJECT Standard ▼		Infinity	Infinity		Infinity
1	Standard ▼		Infinity	1.000		1.777
2	Standard ▼		2.030 V	0.700 V	LASFN31 S	1.123
3	Standard ▼		-3.963 V	0.565 V	N-LASF46B S	0.933
4	Standard ▼		1.981 V	0.323 V		0.533
5	STOP Standard ▼		Infinity	1.173 V		0.272
6	Standard ▼		5.654 V	0.700 V	N-LAF34 S	1.393
7	Standard ▼		15.036 V	3.028 V		1.545
8	IMAGE Standard ▼		Infinity	-		2.971

Title: OSsecureCam6; f/# (8); Fields (0°, 10°, 17°, 22°, 25°); Wavelengths (F d C)

Figure 13.51 The LDE for OSsecureCam6.

With the addition of two lenses and achromatic correction, the nominal performance of the OSsecureCam6 (Fig 12.47) is much better than that of the OSsecureCam2, and we can set a higher as-built performance target for this design example. From Chapter 12, the Nyquist frequency of our security camera sensor was 66 cycles/mm. For an as-built performance target, we will use 75% of the Nyquist frequency (66 cycles/mm $\times$ 0.75), or 50 cycles/mm, and aim for a contrast of greater than 30% with a >98% yield.

Following a similar set of steps to what we did for the OSsecureCam2, we need to prepare the lens for tolerancing. These steps are outlined in Actions 13.1. For this more complex design example, we are using tighter tolerances for both element fabrication (Precision vs Commercial Grade) and assembly. Once you've followed the steps, save this lens as **OSsecureCam6Tol**.

Actions 13.1 Prepare OSsecureCam6 for tolerancing.

Delete the offset surface (surface #1).
Open the tolerance wizard (Tolerance > Tolerance Wizard).
 Select the Generic (Vendor) Precision (Grade) Tolerance Preset.
 Enter element decenter tolerances of ±25 μm.
 Enter element tilt tolerances of ±0.1°.
 Check that the surface range starts at surface 1 and ends at surface 7.
 Check that focus compensation is enabled.
 Click OK.
Open the Tolerance Data Editor (Tolerance > Tolerance Data Editor).
 Adjust the 5 TTHI operands in the TDE as you did with Cam2, setting the Surf2 cell value for the TTHI to match that of the Surf cell before it.
 Delete the extra wedge tolerances (TSTX and TSTY) on S2 and S6.
You should now have a total of 37 tolerance operands.
Change all clear semi-diameters in the LDE from Automatic to Fixed.
Delete all solves (including changing the materials from "S" to Fixed).

Once the lens has been prepared and the tolerances have been defined, we can now proceed to tolerancing. Actions 13.2 summarizes the steps needed to set up a 500-trial Monte Carlo analysis that tracks the average MTF performance at 50 cycles/mm for each field. These are the same steps as we used for the OSsecureCam2; however, it will likely take more than 30 minutes to run this more complex Monte Carlo example. The results of the run are given in Table 13.7, and yield plots are shown in Fig. 13.52. With the current tolerances, we can easily achieve 30% contrast for all fields with a yield of 90% but not our original goal of 98%. To increase the yield to 98%, we need to tighten some of the individual tolerances. These are best identified through a sensitivity analysis (see Section 13.6), and this exercise is left to the reader to explore.

Actions 13.2 Setup of a 500-trial Monte Carlo for the MTF.

Open the tolerancing dialog box (Tolerance > Tolerancing).
    Set-Up > Mode > Skip sensitivity
    Set-Up > Separate Fields/Configs > Checked
    Criterion > Criterion > Diff. MTF Avg
    Criterion > Sampling > 2
    Criterion > MTF Frequency > 50
    Criterion > Comp > Optimize All (DLS)
    Criterion > Fields > Y-Symmetric
    Criterion > Cycles > 1
    Monte Carlo > # Monte Carlo Runs > 500
Click OK to start the run.

Table 13.7 Monte Carlo results after 500 trials for Diff. Avg. MTF at 50 cycles/mm for the OSsecureCam6.

```
   Number of traceable Monte Carlo files generated: 500
            Average Field 1 Field 2 Field 3 Field 4 Field 5
Nominal     0.6629  0.7004  0.6544  0.6544  0.6371  0.6371
Best   #76  0.6646  0.7066  0.6724  0.6724  0.6480  0.6478
Worst #222  0.0429  0.0440  0.0448  0.0414  0.0300  0.0354
Mean        0.6255  0.6587  0.6216  0.6215  0.6002  0.6004
Std Dev     0.1145  0.1265  0.1143  0.1146  0.1076  0.1079

Compensator Statistics:
 Thickness Surf 3:
Nominal:                   3.043380
Minimum:                   2.826424
Maximum:                   3.228759
Mean:                      3.036960
Standard Deviation:        0.116546
```

%Yield	Average	Field 1	Field 2	Field 3	Field 4	Field 5
98% >	0.0492	0.0560	0.0508	0.0504	0.0415	0.0408
90% >	0.6225	0.6370	0.6123	0.6161	0.5879	0.5914
80% >	0.6455	0.6763	0.6365	0.6356	0.6079	0.6091
50% >	0.6554	0.6954	0.6502	0.6505	0.6247	0.6252
20% >	0.6596	0.7031	0.6588	0.6586	0.6360	0.6356
10% >	0.6610	0.7045	0.6620	0.6622	0.6397	0.6394
2% >	0.6625	0.7057	0.6659	0.6659	0.6443	0.6447

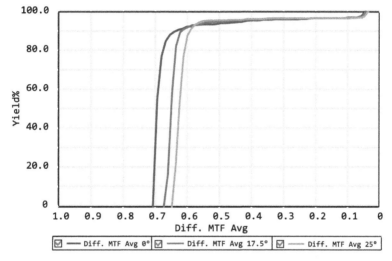

Figure 13.52 Yield% plot for the 500-trial Monte Carlo analysis for Diff. Avg. MTF at 50 cycles/mm for the OSsecureCam6 at 0°, 17.5°, and 25° fields.

13.9 Some Final Comments

Now what? You've designed a lens and toleranced it—toleranced in the sense of establishing a set of limits for the fabrication and assembly of the lens by the vendors. Someone, maybe you, perhaps a senior engineer, then sends your masterpiece out to vendors for bids. The higher the yields for your lens the fewer will need to be fabricated to fulfill your order of, say, 1500 lenses.

At this point, this product of your work on OpticStudio is transformed into a real-world object that refracts light and projects images onto a sensor—something beyond the computer and the pages of this text. It's a start! This may be the end of this text, but our intention was not to produce a compendium describing all of the concepts and procedures needed to master optical design. That cannot be done—the field is too large and contains so many subfields that it takes enterprises such as SPIE Press books and Field Guides to address them.

You can gain some idea of the concepts and analyses that we didn't address by examining the submenus in OpticStudio. For instance, the analytical tools available are quite large and varied. Finally, in none of our examples did we avail ourselves of any non-spherical surfaces. In all of our LDEs, the Surface Type was always "Standard." Go ahead and check out the other types of surfaces available under "Surface Properties." See, there's still a lot to learn. Have a good time!

Appendix: The Lens Drawing

All of the applied tolerances, whether they are part of the default set or assigned by a designer, are meaningful only if an optical manufacturer can meet them. Unrealistic tolerances serve no purpose. Therefore, it is very important to get an estimate of the time and costs of fabricating and mounting a set of lenses for a given set of tolerances from an optical shop so that a designer can decide if the design is complete or if the design process needs to continue to find a less-sensitive/less-costly design.

Once acceptable tolerances have been found, this information must be communicated to the optical shop. Within OpticStudio, there are two features for creating element drawings. Tolerance > Zemax Element Drawing returns the drawing shown in Fig. 13.53. The lens parameters and tolerances are given at the top of the figure. Other instructions below the prescription boxes may include the dimensional units, polishing details, military specifications for material and surface quality, and beveling directions for the edges to avoid chipping the lens during handling. The drawing contains only the lens element. An assembly drawing is also needed to incorporate the lens, aperture stop, and sensor into a mechanical housing.

An alternative format for specifying a lens uses the International Standard Organization's optical drawing standard, ISO 10110. OpticStudio can also generate a plot (Fig. 13.54) in this format with Tolerance > ISO Element Drawing. Because some of the quantities are given using codes, the specifications are not as obvious as they are in the Zemax element plot. A complete discussion of this standard can be found in *Modern Optics Drawings: The ISO 10110 Companion* by Eric Herman, David M. Aikens, and Richard N. Youngworth (SPIE Press, 2022).

Tolerancing 337

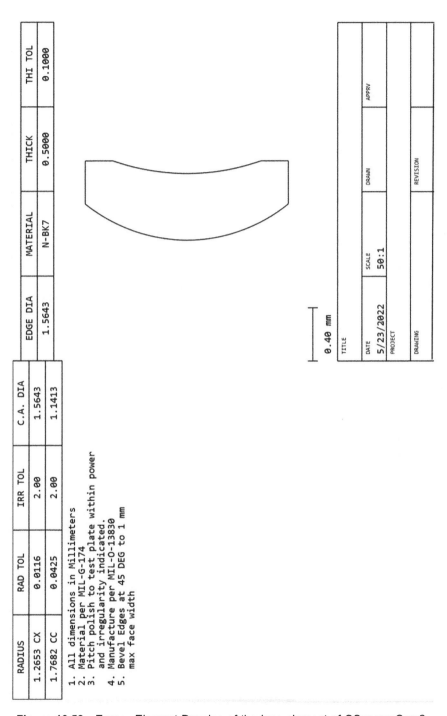

Figure 13.53 Zemax Element Drawing of the lens element of OSsecureCam2.

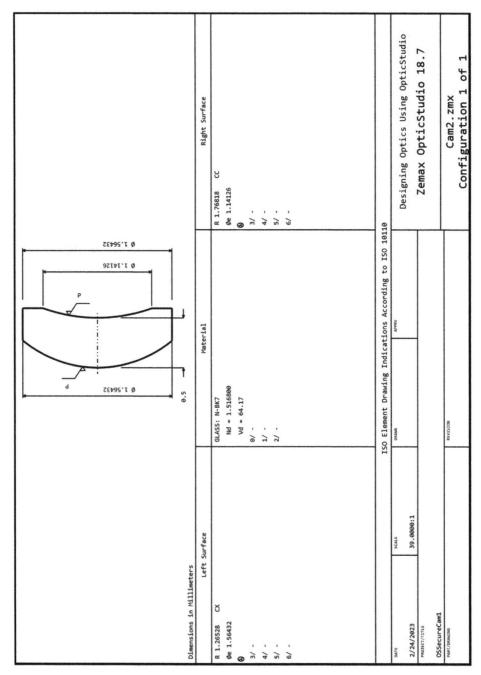

Figure 13.54 ISO 10110 plot of the lens element of OSsecureCam2.

Appendix
Macros: FIRST, THIRD

First, let's find out if the FIRST and THIRD macros described in Sections 3.4 and 6.6 are already installed in your copy of OpticStudio. Click on the Programming tab and then select the Macros List drop-down menu at the far left of the Programming ribbon. This will display a list of all preinstalled macros. Check to see if FIRST.ZPL and THIRD.ZPL are present, and if they are, click on them to use them.

If you are running a version of OpticStudio that does not come with the FIRST and THIRD macros preinstalled, you can create your own. First, open the Programming tab and click on the New Macro icon (sunburst) to open a New ZPL Macro window. Then copy the entire text of one of the macros below and paste it into the text window. Note that the text in OpticStudio is now displayed in a color-coded format. Click the Save As icon (second from left). This opens the Zemax > MACROS folder. Type the file name (FIRST or THIRD) and Save. You can then check that the file is now included in the Macros List in the Programming tab.

FIRST.zpl LISTING

```
! FIRST.zpl
! v0.5 (2022-03-22)
! Written by Don O'Shea 12-24-18
! Differences from prescription data output include:
!    F/# is calculated by EFL/EPD (standard textbook definition)
!    Exit pupil is measured from last surface vertex (not image plane)

GETSYSTEMDATA 1

! get other data
bfl = OPEV(OCOD("CARD"), 1, NSUR(), PWAV(), 0, 3, 0) + THIC(NSUR()-1)
ffl = OPEV(OCOD("CARD"), 1, NSUR(), PWAV(), 0, 2, 0)
m = OPEV(OCOD("PMAG"), 0, PWAV(), 0, 0, 0, 0)
LL = OPEV(OCOD("TTHI"), 1, NSUR() - 2, 0, 0, 0, 0)
TL = OPEV(OCOD("TTHI"), 0, NSUR(), 0, 0, 0, 0)
```

```
img_dist = THIC(NSUR()-1)
ang = ACOS(OPEV(OCOD("PARC"), 0, PWAV(), 0, 1, 0, 0)) * 90 / ACOS(0)
! image space paraxial direction cosine
b = OPEV(OCOD("PARB"), NSUR(), PWAV(), 0, 0, 0, 1)
c = OPEV(OCOD("PARC"), NSUR(), PWAV(), 0, 0, 0, 1)
pwfn = 1 / (2 * INDX(NSUR()) * ABSO(SINE(b / c)))

isInfinite = 0
IF THIC(0) > 1e9 THEN isInfinite = 1

! Header information
PRINT "First Order Report"
PRINT
PRINT "File : ", $FILENAME()
temp = SYPR(16)
PRINT "Title: ", $BUFFER()
PRINT "Date : " $DATE()
PRINT

FORMAT 12.4

PRINT "Infinite Conjugates"
PRINT "   Effective Focal Length    ", VEC1(7)
PRINT "   Back Focal Length          ", bfl
PRINT "   Front Focal Length         ", ffl
PRINT "   F/#                        ", VEC1(7) / VEC1(11)
IF isInfinite
    PRINT "   Image Distance          ", img_dist
    PRINT "   Lens Length             ", LL
    PRINT "   Paraxial Image"
    PRINT "      Height               ", VEC1(15)
    PRINT "      Angle                ", ang
    PRINT "   Entrance Pupil"
    PRINT "      Diameter             ", VEC1(11)
    PRINT "      Location             ", VEC1(12)
    PRINT "   Exit Pupil"
    PRINT "      Diameter             ", VEC1(13)
    PRINT "      Thickness            ", VEC1(14)+THIC(NSUR()-1)
ELSE
    PRINT "   At Used Conjugates"
    PRINT "      Magnification              ", m
    PRINT "      Paraxial Image Height      ", VEC1(15)
    PRINT "      Paraxial Working F/#       ", pwfn
    PRINT "      Object Distance            ", THIC(0)
    PRINT "      Lens Length                ", LL
    PRINT "      Image Distance             ", img_dist
    PRINT "      Total Length               ", TL
```

```
    PRINT "     Entrance Pupil"
    PRINT "        Diameter              ", VEC1(11)
    PRINT "        Thickness             ", VEC1(12)
    PRINT "     Exit Pupil"
    PRINT "        Diameter              ", VEC1(13)
    PRINT "        Thickness             ",
VEC1(14)+THIC(NSUR()-1)
ENDIF

PRINT
PRINT "Object space positions are measured with respect to the
first surface vertex"
PRINT "Image space positions are measured with respect to the last
surface vertex"
PRINT
```

THIRD.zpl LISTING

```
! this macro calculates the transverse ray aberrations
! v0.1 MRH: 2020-08-13
! 2020-09-17: updated the non-color terms to have the correct sign
! JLB 2021-01-18 Reordered the coefficients and names to match ZOS
! JLB 2022-07-23 updated formatting to fit SPIE text page

REWIND
PRINT "THIRD"
dummy = SYPR(16)
str$ = $BUFFER()
IF SLEN(str$) > 0 THEN PRINT "Title: ", $BUFFER()
!PRINT
PRINT "Primary Wavelength:   ", WAVL(PWAV()) , " um"
!PRINT

w = WAVL(PWAV()) / 1000
DECLARE t, DOUBLE, 1, 11

s$ = "    "
PRINT "Surf ", " ", "TSPH", s$, "TTCO", s$, "TAST" s$, "TPFC", s$,
"TSFC", s$, "TTFC", s$, "TDIS", s$, "TAXC", s$, "TLAC"

! get paraxial values for image space
m = OPEV(OCOD("PARB"), NSUR(), PWAV(), 0, 0, 0, 1)
n = OPEV(OCOD("PARC"), NSUR(), PWAV(), 0, 0, 0, 1)
nu = INDX(NSUR()) * m / n
```

```
! get the stop surface
GETSYSTEMDATA 1
ss = VEC1(23)

! get the min/max index values for axial and lateral color
mm = 1
xx = 1
FOR i = 1, SYPR(201), 1
    IF WAVL(i) < WAVL(mm) THEN mm = i
    IF WAVL(i) > WAVL(xx) THEN xx = i
NEXT
DECLARE nm, DOUBLE, 1, NSUR() + 1
DECLARE nx, DOUBLE, 1, NSUR() + 1
twav = PWAV()
PWAV mm
FOR i = 0, NSUR(), 1
    nm(i + 1) = INDX(i)
NEXT
PWAV xx
FOR i = 0, NSUR(), 1
    nx(i + 1) = INDX(i)
NEXT
PWAV twav

nus = 0
FOR surf = 1, NSUR(), 1
    ! get wavefront coefficients
    w040 = OPEV(OCOD("SPHA"), surf, PWAV(), 0, 0, 0, 0)
    w131 = OPEV(OCOD("COMA"), surf, PWAV(), 0, 0, 0, 0)
    w222 = OPEV(OCOD("ASTI"), surf, PWAV(), 0, 0, 0, 0)
    w220 = OPEV(OCOD("FCUR"), surf, PWAV(), 0, 0, 0, 0)
    w311 = OPEV(OCOD("DIST"), surf, PWAV(), 0, 0, 0, 0)

    ! calculate seidel coefficients
    s1 = (8 * w) * w040
    s2 = (2 * w) * w131
    s3 = (2 * w) * w222
    ! there is a bug between the reported Seidel and TRA...TRA is correct
    s4 = (4 * w) * w220      # as reported in 'Seidel Aberration Coefficients' section
    !s4 = (4 * w) * w220 - (2 * w) * w222
    s5 = (2 * w) * w311

    ! calculate transverse ray aberrations
    tsph = s1 / (2 * nu)
```

```
    tsco = s2 / (2 * nu)
    ttco = 3 * s2 / (2 * nu)
    tast = s3 / (nu)
    tpfc = s4 / (2 * nu)
    tsfc = (s3 + s4) / (2 * nu)
    tstf = (3 * s3 + s4) / (2 * nu)
    tdis = s5 / (2 * nu)

    !    cl = -ni_m * ys_m * dn
    !        ni_m = nu_m + INDX * y_m * CURV
    !        nu_m = u_m * INDX = m_m / n_m * INDX
    !        dn = (nm2 - nx2) / n2 - (nm1 - nx1) / n1
    !    ct = -ni_c * y_c * dn
    m_m = OPEV(OCOD("PARB"), surf, PWAV(), 0, 0, 0, 1)
    n_m = OPEV(OCOD("PARC"), surf, PWAV(), 0, 0, 0, 1)
    y_m = OPEV(OCOD("PARY"), surf, PWAV(), 0, 0, 0, 1)
    m_c = OPEV(OCOD("PARB"), surf, PWAV(), 0, 1, 0, 0)
    n_c = OPEV(OCOD("PARC"), surf, PWAV(), 0, 1, 0, 0)
    y_c = OPEV(OCOD("PARY"), surf, PWAV(), 0, 1, 0, 0)

    nu_m = 0
    IF ABSO(n_m) > 0 THEN nu_m = m_m / n_m * INDX(surf)
    ni_m = nu_m + INDX(surf) * y_m * CURV(surf)

    nu_c = 0
    IF ABSO(n_c) > 0 THEN nu_c = m_c / n_c * INDX(surf)
    ni_c = nu_c + INDX(surf) * y_c * CURV(surf)

    dn = ((nm(surf + 1) - nx(surf + 1)) / INDX(surf)) - ((nm(surf) - nx(surf)) / INDX(surf - 1))

    cl = -ni_m * y_m * dn
    ct = -ni_c * y_m * dn

    taxc = cl / nu
    tlac = ct / nu

    ! keep a running sum of all transverse ray aberrations
    t(1)  = t(1) + tsph
    !t(2) = t(2) + tsco
    t(3)  = t(3) + ttco
    t(4)  = t(4) + tast
    t(5)  = t(5) + tpfc
    t(6)  = t(6) + tsfc
    t(7)  = t(7) + tstf
    t(8)  = t(8) + tdis
    t(9)  = t(9) + taxc
    t(10) = t(10) + tlac

s$ = "   "
```

```
        FORMAT 3.0
        s$ = $STR(surf) + " "
        s$ = $LEFTSTRING(s$, 4)

        IF surf == ss THEN s$ = "STO "
        IF surf == NSUR() THEN s$ = "IMA "
        PRINT s$, "",

        FORMAT 9.4
        PRINT tsph,
        !PRINT tsco,
        PRINT ttco,
        PRINT tast,
        PRINT tpfc,
        PRINT tsfc,
        PRINT tstf,
        PRINT tdis,
        PRINT taxc,
        PRINT tlac
NEXT
PRINT "TOT ", t(1), t(3), t(4), t(5), t(6), t(7), t(8), t(9), t(10)
```

Index

100% diameter, 123, 124
3D Layout Settings Window, 87
3D viewer, 86

Abbe diagram, 178
Abbe number, 176
aberration, 119
  image surface, 157
aberration coefficients,
  third-order, 129
absolute indices, 3
achromatic doublet, 188
afocal systems, 31
air–water interface, 74
Airy disk, 243
Airy pattern, 116
  perfect lens, 246
  OStripletMod, 246
anastigmatic lens, 164
angle solves, 53
angular magnification, 32
aperture stop, 90
Aperture Type, 40
  Image Space $f/\#$, 106
  Paraxial Working $f/\#$, 107
  Entrance Pupil Diameter (EPD), 40
Aperture Value, 40
aplanat, 147
Appendix, 336
artificially flattening,
  tangential field, 225
aspect ratio, 240

astigmatism, 149
  correction, 224
axial color, 179
axial ray, 23, 89
Axis (F1), 85

back focal length (BFL), 49, 66
back focal plane, 66
back focal point (BFP), 20, 65
back principal plane (BBP), 66
back principal surface (BPS), 66
bending, lens, 131
bending factor, 12, 133
blank, 291

cardinal point, 75
center of curvature, 5, 10
center of curvature, 14, 22
center ray, 14
centering, 293
chief ray, 93, 97
chromatic aberration, 175
  longitudinal, 179
  longitudinal, plot, 183
chromatic focal shift, 186
Clear Semi-Diameter (CSD), 92
Clear Semi-Diameter Margin %, 92
CMOS, 240, 257
collimation, 19
coma, 141, 146
  reduction, 212
common digital sensors,
  table of, 241

compensator, 296
concave mirror, 26
conjugate pairs, 29
convex mirror, 26
coordinate breaks, 306
coordinate origin, 5
coordinate system, 2, 120
   rectangular, 120
   polar, 120
Cross-Section, 46
crown, 177

decenter, 305
default merit function (DMF), 205
defocus, 124, 206
   optimization, 206
design forms, 57
diffraction, 115–117
diffraction-limited lens, 244
diopter, 11
dispersion, 175
distances, 5
distortion, 167
  barrel, 167
  examples, 173
  grid, 170
  pincushion, 167
  plot, 218
  reduction, 217
distortion-free lens, 223
Double Gauss, 60, 70
dummy surface, 213

effective focal length (EFL), 28
element fabrication errors, 299
Emsley model, human eye, 77–78
entrance pupil diameter (EPD), 40
exit pupil (EnP), 99
exit window, 101
eye, 78
eyepiece, 198

f-number ($f/\#$), 105
F2 glass, 185
fabrication,
  grinding, 291
  initial steps, 291
  lens element, 291
  pitch polishing, 292
$f/\#$, 105
Fast Fourier Transform, 253
FFT MTF settings, 253
field curvature, 161
  reduction, 224
  sagittal, 158
  tangential, 158
field curves, 162, 166
Field Data Editor (FDE), 85
field flattener, 227
field lens, 228
field of view, 103, 258
field point,
  Axis (F1), 85
  Midfield (f2), 85
  Full Field (F3), 85
field specification, 103
field specification options,
  Angle, 104
  Object Height, 104
fields, 83
field stop, 101
field type, 85, 104
  Object Height, 85, 221
  Angle, 89
finite object distance, 23
first-order properties, 50
flint glass, 179
flipping the lens, 128, 215
FIRST.ZPL, 50
  Macro, 339
focal length, 9
  mirror, 27
  back (BFL), 49

front (FFL), 67
focal point, 9
  back, 20
  front, 20
focal ray, 22
focus compensator, 310
footprint diagram, 110
front principal plane (FPP), 68
Fraunhofer, Joseph von, 176
front focal length (FFL), 67

Galilean telescope, 31
glass map, 178
glass number, 179
graphic user interface (GUI), 39
graphics overlays, 186

hiatus, 68
histogram plot,
  Monte Carlo 500 trial, 329
human eye, 77

image quality, 206
  wavefront, 206
  spot, 206
images,
  real, 15
  virtual, 25, 52
image height, 14, 25
image space, 19
imaging, 13
immersed system, 27, 73
  cardinal point, 76
index of refraction, 3
infinity, object at, 20
infrared, 177
interface, 4
inverse sensitivity, 312
ISO 10110 plot, 338

Keplerian telescope, 31

lateral color, 179, 194, 200
law of magnification, 14

law of reflection, 6
law of refraction, 4
lens,
  positive, 9
  negative, 9
  plano-concave, 12
lens assembly errors, 299
lens bending, 131, 146
Lens Data Editor (LDE), 39, 42
lens file formats, 56
lens mount, 293
lens plot, 47
lens prescription, 42
lensmaker's formula, 11
light, collimated, 19

macro,
  FIRST.ZPL, 339
  THIRD.ZPL 341
magnification, 14
magnifying power (MP), 32
Make Surface Stop, 90
marginal ray, 44, 90
meridional ray, 89
merit function, 203
  adding boundary value operands, 276
Merit Function Editor (MFE), 204
minimum spatial frequency, 259
mirror, 26, 52
  concave, 26
  convex, 26
  spherical, 26
Modulation Transfer Function (MTF),
  curve, 250
  chart, 250
  OStripletMod, 251
  Monte Carlo 500 trials, 326
Monte Carlo analysis, 296, 322
  OSsecureCam2Monte, 308
  OSsecureCam6Monte, 332

mounting,
   lens element, 293
multiple elements, 31

N-BK7 glass, 176
nodal point, 71
nodal slide, 72
numerical aperture, 103
   object space, 107
Nyquist frequency, 252
Nyquist theorem, 253

object at infinity, 20, 24
Object Height, 84, 104
object space, 19
off-axis ray, 24
offset, 47
on-axis ray errors, 122
operand, 204
   user-entered, 205
optical axis, 5
optical power, 8
   mirror, 26
OpticStudio, navigating, 40
optimization, 203
   Global, 286
   hammer, 286
Optimization Wizard, 205
Optimize!, 207
OSachromat, 189
OSachromatOpt, 191
OSaplanat, 157
OSasdoublet, 91–95
OSasdoubletAxis, 91–93
OSasdoubletSym, 96, 99
OSdoubleGauss, 60, 70–71
OSequiconvex, 84
OSeye, 79
OSfieldflattener, 230
OSlandscape, 214, 227
   optimized, 226
   reversed, 218
OSlandscapeRearStop, 216

OSlens, 44,
   ray trace, 48
   cardinal points, 72
OSprotar, 59
OSramsden, 197
OSrapidrect, 59
OSschwarzschild, 60, 159
OSsecureCam0, 260–263
OSsecureCam1, 264–266
OSsecureCam2, 267–269
OSsecureCam3, 270–273
OSsecureCam4, 274–281
OSsecureCam5, 282–285
OSsecureCam6, 286–290
OSsinglet, 55, 141, 166, 208
OSsingletRev, 131, 175
   third-order coefficients, 129
OSsingletRevBB, 180
OSsingletRevF10, 185
OSsymmetrical, 220, 228
OStriplet, 59, 164–166,
OStripletMod, 242–248

parallel ray, 20
paraxial focus, 49
paraxial image plane, 45
paraxial transfer equation, 8
Paraxial Working $f/\#$, 107
perfect lens, 157
   Airy pattern, 246
Petzval aberration coefficient, 164
Petzval curvature correction, 230
Petzval curvature, 162
Petzval surface, 162
pickup, 219
pincushion, 167
pinhole camera, 117
pitch polishing, 292
pixel, 240
planar optics, 29
Point Spread Function, 242
prescription data, 42, 181
principal planes, 69

Protar, 59
Point Spread Function (PSF), 242
   Huygens-PSF plot, 246
   OStripletMod, off-axis, 247
pupil, 97
   entrance, 98
   exit, 99
   specification, 103

Quick Focus, 124, 166

radius of curvature, 10
Ramsden eyepiece, 197
Rapid Rectilinear, 55
rays,
   axial, 89
   chief, 93
   object at infinity, 89
   tangential, 87
   meridional, 87
   sagittal, 88
   skew, 89
   special, 87
   skew, 89
ray aiming, 94
ray density, 110, 123, 132
ray sketching, 19
   pupils, 100
ray tracing, 2
Rayleigh criterion, 115
real image, 15
reflections, 5
refraction, 4
refraction, angles, 4
relative index, 3
resolution, 248
Reverse Elements, 131, 215
right-handed coordinate system, 3

sagittal fan, 88
sagittal ray, 88
Scale Lens, 220
Schwarzschild mirror, 57

secondary color, 192
security camera, 257
   field of view, 258
Seidel aberration coefficients, 121
semi-diameters, 10, 44
sensitivity analysis, 296, 312
   uncompensated, 316
   compensated, 319
sensitivity table, 316
sensors, 239
   CCD, 240
   CMOS, 240, 257
   full frame, 240
   dimensions, 241
shape factor, 133
sign conventions, 4, 5
sine wave pattern, 328
skew ray, 89
Slider, 134, 150
Solve Type, 93
solves, angle, 53
spatial frequency, 249
special rays, 22, 87
speed, 105
spherical aberration, 119
   transverse, 129
   longitudinal, 123
   reduction, 208
spherical mirror, 26
   ray trace, 57
spherochromatism, 192
spot diagram, 123
stop,
   aperture, 90
   field, 101
stop shifting, 213
stop down the lens, 130
Strehl ratio, 247
Surface Properties, 90
Surface Type, 43
symmetrical lens, 220
system data, 40
System Explorer, 40

tangential field,
    artificially flattening, 225
tangential rays, 87
telescope,
    Keplerian, 31
    Galilean, 31
test plate, 292
testing a lens, 294
thickness,
    center, 10
    edge, 10
thickness solve, 45
thin lens approximation, 11
thin lenses, 21
THIRD, 129
third-order aberrations, 120
tilt, 302
TIR measurement, 307
Tolerance Data Editor (TDE), 298
    OSsecureCam2, 297
    OSsecureCam2Tol, 311
    TABB, V-number, 301
    TEDX and Y, element decenter, 305
    TETX and Y, element tilt, 307
    TIND, index of refraction, 302
    TIRR, irregularity, 303
    TRAD, radius of curv., 300
    TSTX and Y, wedge, 302
    TTHI, thickness, 300, 307
    TWAV, wavelength, 300
tolerance presets, 309
Tolerance Wizard, 307
Tolerance Wizard, window, 308
Tolerancing, 295
    statistical, 295
    dialog box, 312
        Set-Up tab, 313
        Criterion tab, 314
        Sampling, 315
    Monte Carlo, 316
Tolerancing Ribbon, 299
total indicator runout (TIR), 302
Tracing Rays, 32

tracing single rays, 169
transfer equation, 7
transverse aberration coefficients, 129
transverse coma (TTCO), 146
transverse ray curve, 126
transverse ray errors, 121
transverse ray plot, 125
triplet, 62

ultraviolet, 177,
uncompensated, 318,
USAF 1951 bar chart, 250

V-number, 177
vertex, 10
vertex, ray, 29
vignetting, 109
vignetting Parameters, 112
virtual image, 15, 52
    Treating, 25
von Seidel, Phillip Ludwig, 120

wavelength bands,
    visible, 177
    ultraviolet, 177
    infrared, 177
Wavelength box, 41
wavelengths,
    preset, 180
window,
    entrance, 101
    exit, 101
working $f/\#$, 108

yield curve, 296
    construction, 297
Yield% plot
    Monte Carlo 500 trial, 331
    OSsecureCam6Monte, 335

Zemax Element Drawing, 337
Zemax Programming Language
    (ZPL), 50

OpticStudio Lens LDE

OSsinglet, 55
OSprotar, 62
OSrapidrect, 62
OStriplet, 62
OSdoubleGauss, 63
OSschwarzschild, 62
OSeye, 79
OSequiconvex, 84
OSasdoubletAxis, 91
OSasdoublet, 93
OSasdoubletSym, 96
OSsingletRev, 131
OSaplanat, 148
OSsingletRevBB, 180
OSachromat, 190

OSramsden, 197
OSlandscape, 214
OSlandscapeRearStop, 217
OSsymmetrical, 220
Ossymmetrical (50 mm), 221
OSfieldflattener, 231
OstripletMod, 242
OSsecureCam0, 260
OSsecureCam1, 264
OSsecureCam(Opt), 265
OSsecureCam2, 267
OSsecureCam3, 271
OSsecureCam4, 279
OSsecureCam5, 284
OSsecureCam6, 288

Donald C. O'Shea is Professor Emeritus for the School of Physics at the Georgia Institute of Technology. He is a Fellow of SPIE and of the Optical Society (OSA). He served as editor of SPIE's flagship journal, *Optical Engineering*, from 1998–1999 and 2001–2009. During 2000, he served as President of SPIE.

O'Shea received a B.S. in Physics from the University of Akron (1960), a M.S. in Physics from Ohio State (1963), and a Ph.D. in Physics from Johns Hopkins (1968). He did postdoctoral work on laser spectroscopy at the Gordon McKay Laboratory at Harvard from 1968–1970. In 1970, he joined the faculty at Georgia Tech, where he created an optics curriculum and published over 50 papers on optics and optics education. He was a Visiting Scholar at the Optical Sciences Center of the University of Arizona and at the University of Oulu, Finland.

He co-authored an undergraduate textbook on lasers, *An Introduction to Lasers and Their Applications*, with W. R. Callen and W. T. Rhodes (Addison-Wesley, 1977). He published a textbook on optical design, *Elements of Modern Optical Design* (Wiley, 1985), and a SPIE Tutorial Text on diffractive optics, *Diffractive Optics: Design, Fabrication, and Test*, with T. J. Suleski, A. D. Kathman, and D. W. Prather (SPIE Press, 2004). This current text is a companion to *Designing Optics Using CODE V®* (SPIE Press, 2018). He created the Optics Discovery Kit for OSA for use in precollege education. In 1996, he was awarded the Esther Hoffman Beller Award by OSA for "excellence in the field of optics education."

Julie L. Bentley is a Professor at the Institute of Optics, University of Rochester, and has taught courses in geometrical optics, optical design, and product design for over 25 years. She is a Fellow of SPIE and OSA and a former president of the Rochester Optical Society. She has served as an SPIE board member, an associate editor for *Optics Express*, and the chair for the International Optical Design Conference (IODC) and SPIE's optical fabrication conference (Optifab). In 2023, she was elected into the presidential chain for SPIE and will serve as its president in 2026.

Bentley received a B.S. in Optics (1990), a M.S. in Optics (1992), and a Ph.D. in Optics (1995) from the University of Rochester. She holds several U.S. patents and co-authored her first book, *Field Guide to Lens Design*, with S. Craig Olson (SPIE Press, 2012). Her second book, *Designing Optics Using CODE V®* (SPIE Press, 2018)), is a companion to this text, and was co-authored with Don O'Shea. In 2014, she received the University of Rochester's Goergen Excellence in Undergraduate Teaching award. In 2022, she received both the SPIE Director's Award and Optica's Esther Hoffman Beller Medal for Educational Excellence.

Her expertise is in the area of optical design and tolerancing of precision optical instruments. At the university, her research group is focused on three different aspects of optics: optical design with freeform GRIN (gradient index) materials, the design of next-generation adaptive optics (AO) ophthalmoscopes, and improved zoom lens design and optimization. Outside of teaching and research, she runs a successful optical design consulting business, Bentley Optical Design, where she designs lens systems for a wide variety of applications ranging from medical devices for cancer detection to visible and infrared military optics.